I0072214

Agronomy: Agricultural Practices in a Changing World

Agronomy: Agricultural Practices in a Changing World

Edited by Roy Tucker

SYRAWOOD
PUBLISHING HOUSE

New York

Published by Syrawood Publishing House,
750 Third Avenue, 9th Floor,
New York, NY 10017, USA
www.syrawoodpublishinghouse.com

Agronomy: Agricultural Practices in a Changing World
Edited by Roy Tucker

© 2018 Syrawood Publishing House

International Standard Book Number: 978-1-68286-578-1 (Hardback)

This book contains information obtained from authentic and highly regarded sources. Copyright for all individual chapters remain with the respective authors as indicated. All chapters are published with permission under the Creative Commons Attribution License or equivalent. A wide variety of references are listed. Permission and sources are indicated; for detailed attributions, please refer to the permissions page and list of contributors. Reasonable efforts have been made to publish reliable data and information, but the authors, editors and publisher cannot assume any responsibility for the validity of all materials or the consequences of their use.

Trademark Notice: Registered trademark of products or corporate names are used only for explanation and identification without intent to infringe.

Cataloging-in-Publication Data

Agronomy : agricultural practices in a changing world / edited by Roy Tucker.
    p. cm.
Includes bibliographical references and index.
ISBN 978-1-68286-578-1
1. Agronomy. 2. Agriculture. I. Tucker, Roy.
SB91 .A37 2018
631--dc23

TABLE OF CONTENTS

PREFACE

Agronomy refers to that sub-field of agriculture, which deals with the use of plants to get fiber, fuel, and food. It includes specialized areas like plant breeding, weed control, pest control, irrigation, crop rotation, soil classification, etc. Agronomy uses the elements of plant genetics, plant meteorology, ecology, biology, soil science, etc. This book provides significant information of this discipline to help develop a good understanding of the subject and its related fields. Some of the diverse topics covered in it address the varied branches that fall under this category. Scientists, researchers and students actively engaged in this field will find the book full of crucial and unexplored concepts.

The researches compiled throughout the book are authentic and of high quality, combining several disciplines and from very diverse regions from around the world. Drawing on the contributions of many researchers from diverse countries, the book's objective is to provide the readers with the latest achievements in the area of research. This book will surely be a source of knowledge to all interested and researching the field.

In the end, I would like to express my deep sense of gratitude to all the authors for meeting the set deadlines in completing and submitting their research chapters. I would also like to thank the publisher for the support offered to us throughout the course of the book. Finally, I extend my sincere thanks to my family for being a constant source of inspiration and encouragement.

Editor

Effect of Postsowing Compaction on Cold and Frost Tolerance of North China Plain Winter Wheat

Caiyun Lu,[1,2] **Chunjiang Zhao,**[1,2] **Xiu Wang,**[1,2] **Zhijun Meng,**[1,2] **Jian Song,**[1,2] **Milt McGiffen,**[3] **Guangwei Wu,**[1,2] **Weiqing Fu,**[1,2] **Jianjun Dong,**[1,2] **and Jiayang Yu**[1,2]

[1]*Beijing Research Center of Intelligent Equipment for Agriculture, Beijing 100097, China*
[2]*Beijing Research Center for Information Technology in Agriculture, Beijing 100097, China*
[3]*Department of Botany and Plant Sciences, University of California, Riverside, CA 92521-0124, USA*

Correspondence should be addressed to Chunjiang Zhao; zhaocj@nercita.org.cn

Academic Editor: Maria Serrano

Improper postsowing compaction negatively affects soil temperature and thereby cold and frost tolerance, particularly in extreme cold weather. In North China Plain, the temperature falls to 5 degrees below zero, even lower in winter, which is period for winter wheat growing. Thus improving temperature to promote wheat growth is important in this area. A field experiment from 2013 to 2016 was conducted to evaluate effects of postsowing compaction on soil temperature and plant population of wheat at different stages during wintering period. The effect of three postsowing compaction methods—(1) compacting wheel (CW), (2) crosskill roller (CR), and (3) V-shaped compacting roller after crosskill roller (VCRCR)—on winter soil temperatures and relation to wheat shoot growth parameters were measured. Results showed that the highest soil midwinter temperature was in the CW treatment. In the 20 cm and 40 cm soil layer, soil temperatures were ranked in the following order of CW > VCRCR > CR. Shoot numbers under CW, CR, and VCRCR treatments were statistically 12.40% and 8.18% higher under CW treatment compared to CR or VCRCR treatments at the end of wintering period. The higher soil temperature under CW treatment resulted in higher shoot number at the end of wintering period, apparently due to reduced shoot death by cold and frost damage.

1. Introduction

Autumn-sown field crops trend to higher yields than spring varieties due to their early spring development [1], an advantage over spring-germinating weeds when competing for limited moisture [2, 3]. In addition, overwintering crops confer some advantages, including reduced soil erosion and nutrient leaching. However, overwintering crops are vulnerable to unfavourable weather for a longer period, potentially reducing winter survival rate, crop vigour, and thereby ultimate yield [4–7].

In North China Plain, the temperature falls to 5 degrees below zero, even lower in winter with a frost-free period of around 190 days. Winter wheat may suffer low-temperature damage from both prolonged exposure to relatively mild but suboptimal temperatures and short exposure to extremely low temperature [8]. Cold and frost damage occur occasionally, and large declines in temperature reduce wheat growth

ultimately [9]. Thus improving temperature to resist damage is necessary in this area.

Soil properties are usually influenced by tillage, compaction, and other soil management factors [10–15]. Moderate seedbed compaction is required for optimal soil-seed contact and to reduce soil moisture loss following planting [16]. Several studies have demonstrated that compaction significantly affected soil temperature [17–20]; however, previous research has generally focused on wheel traffic compaction and its effect on soil temperature only a few days after compaction. There is little knowledge about effect of postsowing compaction on soil temperature during winter several months after postsowing compaction, which is more important for cold and frost tolerance. Therefore, the objective of this research is to explore the effects of three postsowing compaction methods (compacting wheel, crosskill roller, and V-shaped compacting roller after crosskill roller) on cold and frost tolerance capacity of North China

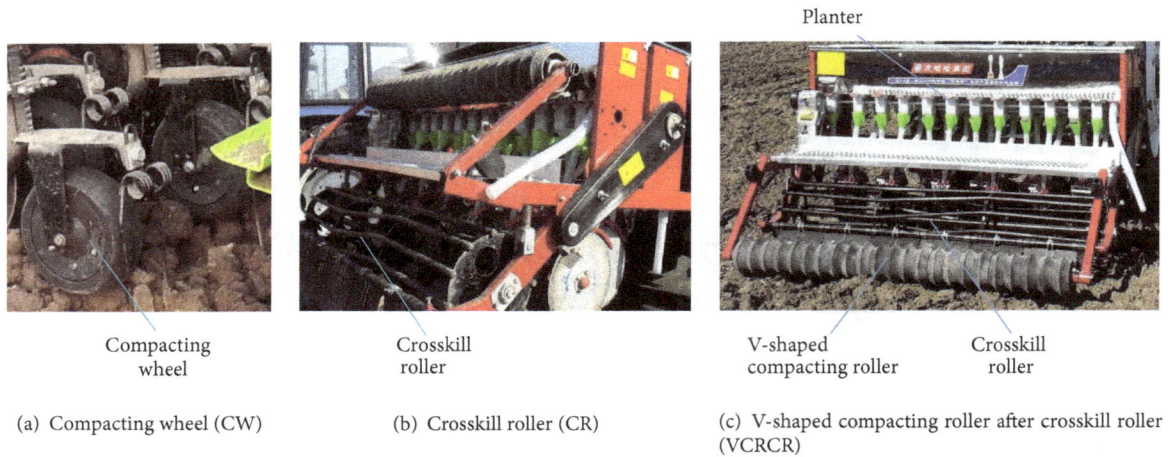

(a) Compacting wheel (CW)

(b) Crosskill roller (CR)

(c) V-shaped compacting roller after crosskill roller (VCRCR)

FIGURE 1: Postsowing compaction methods.

Plain winter wheat. The cold and frost tolerance capacity is measured by soil temperature variation in 0–60 mm soil depth and winter wheat growth parameters (shoot number, length, and fresh and dry weight) during wintering period. The research will provide a basis for improving cold and frost tolerance capacity of winter wheat.

2. Materials and Methods

2.1. Site Description. The study was conducted from 2013 to 2016 near Beijing, China, in an area with a temperate continental climate and four distinct seasons. Mean annual temperature in the region is 11°C, with a frost-free period of around 190 days. Average annual rainfall is 600 mm, 75% of which occurs during summer. To ensure the same conditions, only winter wheat is sown in October and harvested in June in each year. Soil at the site was a loam with an average soil organic matter content of 18.7 g/kg, total nitrogen of 0.115%, 16.7 mg/kg available phosphorus, and 96 mg/kg available potassium.

2.2. Experimental Design. For each wheat season, the site was uniformly rotary tilled and laser-levelled, and then fertilizer was applied at the rate of 150 kg/ha N, 140 kg/ha P, and 85 kg/ha K before sowing. Winter wheat (Jingdong 8) was then sown at the rate of 187.5 kg/ha. This study assessed three types of postsowing compaction: (1) compacting wheel (CW), (2) crosskill roller (CR), and (3) V-shaped compacting roller after crosskill roller (VCRCR). The compacting wheel and rollers were installed on the planter behind disc openers (Figure 1). CW only compressed the intrarow soil surface just behind the opener; both CR and VCRCR compressed both the intra- and interrow soil surface. The urea fertilizer was applied at rate of 225 kg/ha at jointing stage.

2.3. Measured Parameters. Soil temperature and shoot growth parameters were obtained from November, 2015, to March, 2016, to determine the effect of postsowing

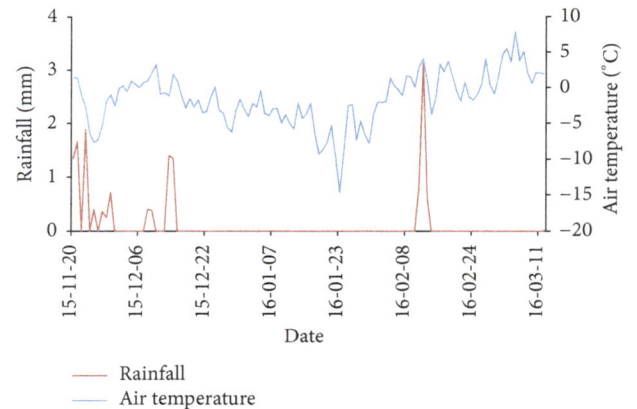

FIGURE 2: Distribution of rainfall and temperature at experiment site.

compaction on cold and frost tolerance. Winter rainfall and air temperature are shown in Figure 2.

2.3.1. Soil Temperature. Soil temperature was continuously recorded to a depth of 60 cm from sowing to green stage. In each plot, five soil temperature sensors were inserted into the intrarow and interrow soil surface and random position at depths of 20 cm, 40 cm, and 60 cm. The sensors were connected to a solar-powered automatic weather station that recorded data hourly. A 24-hour period at the beginning (November 27, 2015), middle (January 27, 2016), and end (March 4, 2016) of wintering period was selected for statistical comparison of the effect of postsowing compaction on soil temperature.

2.3.2. Shoot Growth Parameters. Shoot length and fresh and dry weight were measured at the end (March 4, 2016) of the wintering period. Shoot number was measured at the beginning (November 27, 2015) and the end (March 4, 2016) of wintering period. Shoot dry weights were determined after

(a) CW

(b) CR

(c) VCRCR

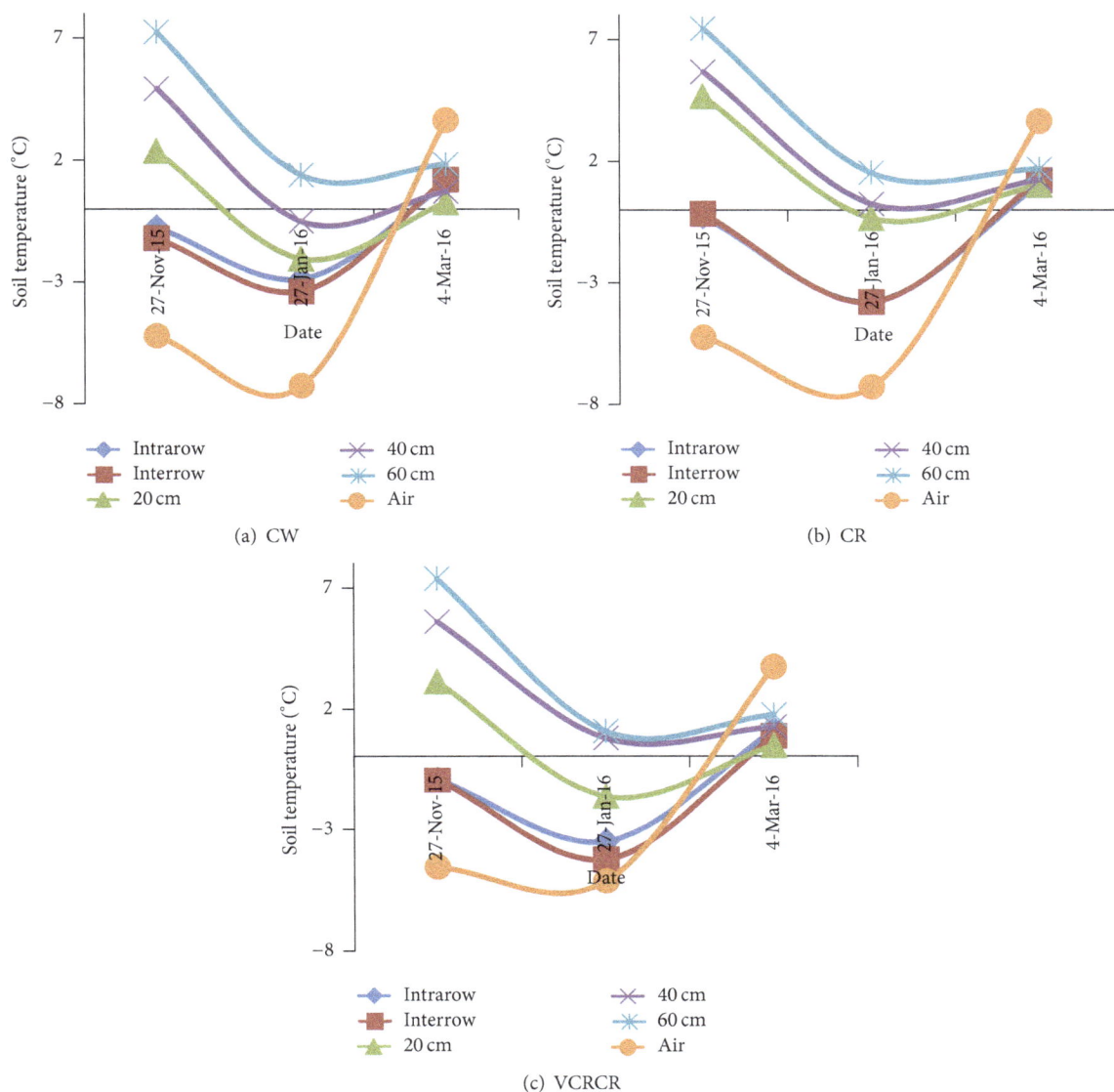

FIGURE 3: Mean soil and air temperature during wintering period.

heating at 105°C for 15 min, followed by incubation at 80°C for 24 h [21–24]. All measurements were performed in triplicate and the data thus obtained was averaged.

2.4. Statistical Analysis. Mean values were calculated for each of the measured variables, and ANOVA was used to assess the treatment effects. When ANOVA indicated a significant F-value, multiple comparisons of mean values were performed by the least significant difference test at $\alpha = 0.05$.

3. Results

3.1. The Relationship between Soil Temperature and Air Temperature. Midwinter was the coldest of the three sampling stages (Figure 3). All sampling depths followed a similar pattern of temperature change with air temperature. There was

less temperature differences between sampling dates at the deeper sampling depths, and temperature generally increased with sampling depth during entire winter period. At the beginning and middle of wintering period, soil temperature for each depth was significantly higher than air temperature; however, at the end of wintering period, air temperature was greater than soil temperature at any depth.

The compaction treatments affected soil temperature. Under CW treatment, there were significant differences in soil temperatures at the beginning and middle of wintering period, with 60 cm > 40 cm > 20 cm > intrarow > interrow. Soil temperatures in the 60 cm soil layer were 2.33°C, 4.88°C, 8.45°C, and 8.00°C higher at the beginning and 1.91°C, 3.45°C, 4.78°C, and 4.28°C higher at the middle of wintering period than 40 cm, 20 cm, interrow, and intrarow soil surface layer, respectively. Under CR treatment, soil temperature in 60 cm soil layer was significantly 1.78°C, 2.80°C, 7.60°C, and 7.70°C

(a) November 27, 2015

(b) January 27, 2016

(c) March 4, 2016

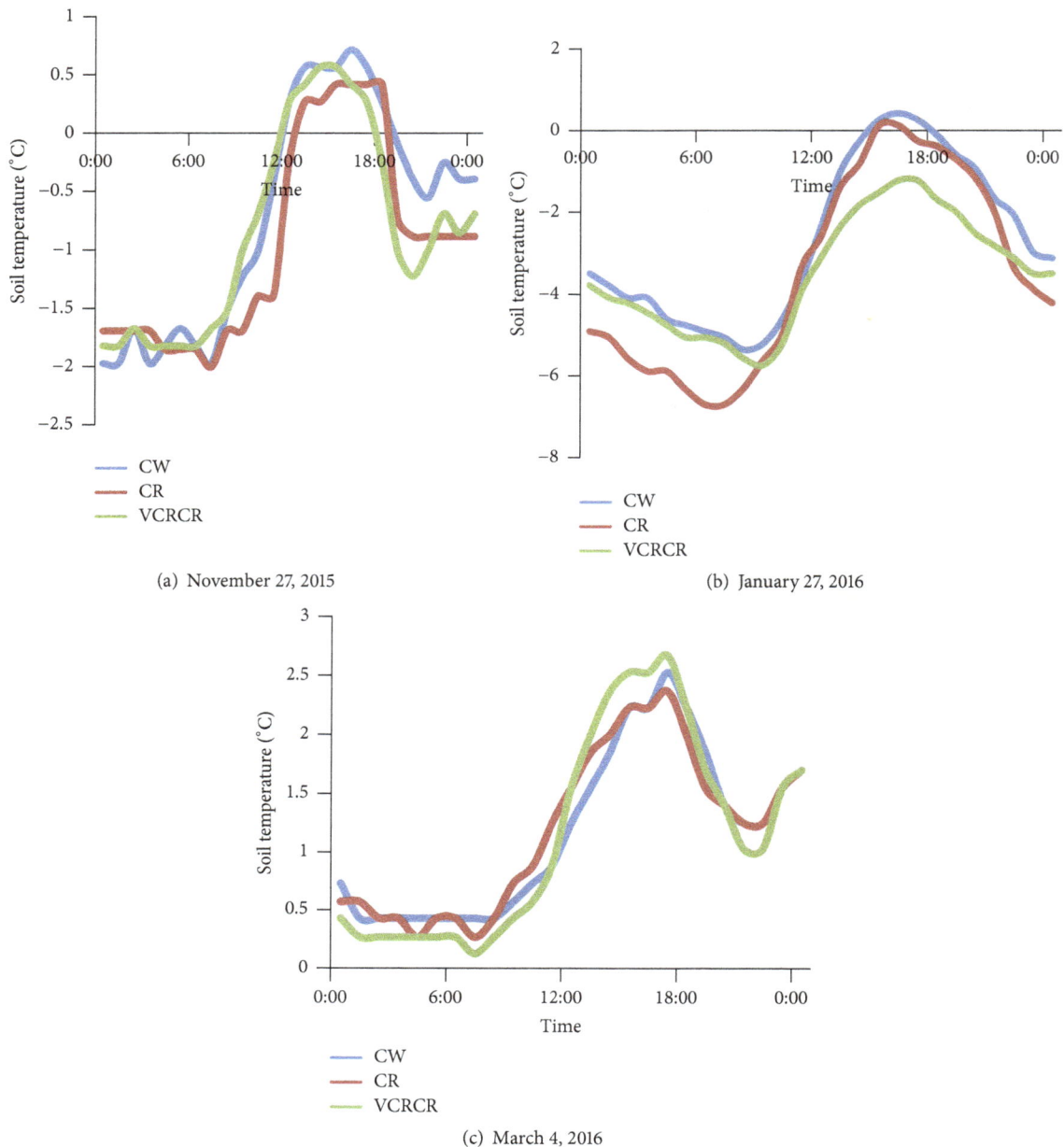

FIGURE 4: Temporal soil temperature changes on intrarow soil surface under different postsowing compaction treatments.

higher at the beginning and 1.28°C, 1.88°C, 5.33°C, and 5.31°C higher at the middle of wintering period than 40 cm, 20 cm, interrow, and intrarow soil surface layer, respectively. Soil temperature on the intrarow soil surface was equivalent to the interrow soil surface. However, surface soil temperatures were remarkably lower than deeper soil layers. Under VCRCR treatment, soil temperature in 60 cm soil layer was separately 1.75°C, 4.20°C, 8.24°C, and 8.21°C higher than 40 cm, 20 cm, interrow, and intrarow soil surface layer at the beginning of wintering period. And significant differences only occurred between 60 cm soil layer and 20 cm layer or on the soil surface. Soil temperature on the intrarow soil surface was 20.84% higher than that on interrow soil surface layer in the middle

of wintering period and remarkably lower on the surface soil than deeper soil layers. At the end of wintering period, soil temperature had no remarkable distinction between soil layers for all treatments.

3.2. Temporal Soil Temperature Changes on Intrarow Soil Surface. Soil surface temperatures on the intrarow followed the typical sinusoidal pattern (Figure 4), with minimum temperatures at 6:00–8:00 and maximum soil temperature at 15:00–17:00. At the beginning of wintering period, the maximum soil temperature under CW was 0.3°C and 0.15°C higher than CR and VCRCR treatments, and the minimum soil temperature under the CW treatment was 0.03°C higher than

CR treatment but 0.15°C lower than VCRCR treatment. In the middle of wintering period, the maximum soil temperature under CW treatment was 0.3°C and 1.65°C higher than CR and VCRCR treatments, and the minimum soil temperature under CW treatment was 1.35°C and 0.38°C higher than CR treatment and VCRCR treatment, respectively. At the end of wintering period, the maximum soil temperature under CW treatment was 0.16°C higher than CR treatment and 0.14°C lower than VCRCR treatment, and the minimum soil temperature under CW treatment was 0.16°C and 0.3°C higher than CR treatment and VCRCR treatment, respectively. At the beginning and end of the wintering period, soil temperature was unaffected by compaction. However, treatment differences were significant in midwinter. CR had the lowest soil temperatures throughout the midwinter sampling period and CW the highest.

3.3. Temporal Soil Temperature Changes on Interrow Soil Surface. Temporal soil temperature changes on interrow soil surface during wintering period were showed at Figure 5. The trend of temporal soil temperature changes on interrow soil surface was similar to those on intrarow surface. At the beginning of wintering period, CR and VCRCR treatments had the highest maximum soil temperatures, 0.15°C higher than CW treatment. The lowest minimum soil temperature occurred in the CW treatment, 0.26°C and 0.53°C lower than CR and VCRCR treatments, respectively. In the middle of wintering period, CR had the highest maximum soil temperature, 0.53°C higher than CW or VCRCR treatments. The highest minimum soil temperature occurred under the CW treatment, which was 0.97°C and 1.27°C higher than CR and VCRCR treatments, respectively. At the end of wintering period, the highest maximum soil temperature was under CR treatment, 0.46°C and 1.13°C higher than CW and VCRCR. CW treatment had the same minimum soil temperature with CR, which was 0.16°C higher than VCRCR treatment.

3.4. Temporal Soil Temperature Changes in Different Soil Depth. Temporal soil temperature in the 20–60 cm soil layer increased with increasing soil depth (Figure 6). Only the 20 cm soil layer in the middle of wintering period displayed significant diurnal temperature variation and also the greatest differences between treatments. In the 20 cm and 40 cm soil layer, soil temperatures were ranked in the following order of CW > VCRCR > CR during the entire wintering period. Soil temperature under CW treatment was significantly higher than that under CR treatment at the beginning and middle of wintering period and remarkably higher than that under VCRCR treatment in the middle of wintering period in the 20 cm soil layer. CW treatment had remarkably higher soil temperature than CR during the entire wintering period and higher than VCRCR in the middle of wintering period; soil temperature under VCRCR treatment was notably higher than that under CR treatment at all stages in the 40 cm soil layer. There were no significant differences in the 60 cm soil layer.

3.5. Shoot Growth Parameters at the Beginning and End of Wintering Period. Shoot number at the beginning of the wintering period (November 27, 2015) had no significant difference between every two treatments (Table 1). At the end of wintering period (March 4, 2016), shoot number was 12.40% and 8.18% higher under CW compared to CR and VCRCR treatments, and only CW and CR treatments were significantly different.

Shoot length was 8.68% and 4.78% taller under the CW treatment than CR and VCRCR treatments (Table 2). Fresh weight of 10 shoots was 9.57% and 7.66% lower under CR treatment than CW and VCRCR treatments; shoot dry weight was 14.30% and 13.79% lower under CR treatment than CW and VCRCR treatments, respectively. The difference between CW and VCRCR treatments was not significant both for shoot fresh and for dry weight.

4. Discussion

Consistent with previous studies [25–27], we found that the hourly change in soil temperature followed that of daily air temperature, although the amplitude was smaller than air temperature. Temporal soil temperature on the intra- and interrow surface soil layer had a similar diurnal pattern as with air temperature, but also with a smaller fluctuation; this same pattern has been reported by several previous authors [28–30]. However, temporal soil temperature in the 20–60 cm depths was stable in a 24-hour cycle; however, it had an analogical change rule with the surface soil temperature and air temperature in the whole wintering period, which manifested that temperature in deeper soil layers had a hysteresis compared to surface soil and air temperature. This may be because soil temperature change is the result of energy absorption from solar radiation and air temperature and the subsequent release, which results in smaller soil temperature fluctuations as soil depth increased. These results were in agreement with Liao et al. [31], who reported that variation in soil temperature was lower at deeper soil as compared to that in the surface soil, indicative of thermal insulation provided by vegetation, water, and surface soil layers.

There is limited information on the effects of postsowing compaction on soil temperature variation.

In this study, we observed that soil temperature on the intrarow soil surface under CW treatment was higher than CR and VCRCR treatments, especially at the coldest stage in the middle of wintering period. This indicated that CW had the ability to increase soil temperature, particularly at colder weather. Our study demonstrated that soil temperature on the intrarow soil surface was similar to that on the interrow soil surface under CR treatment; however, it was 20.84% higher on the intrarow soil surface than interrow soil surface layer in the middle of wintering period under VCRCR treatment. This indicated that different levels of compaction (CR treatment compacted planted soil once, and VCRCR treatment compacted planted soil twice) resulted in differences between intrarow and interrow soil surface, with heavier compaction causing higher soil temperature on the intrarow soil surface. In addition, soil temperature on the intrarow soil surface was significantly higher than that on the interrow soil surface during wintering period, especially in the middle of the period under the CW treatment, which

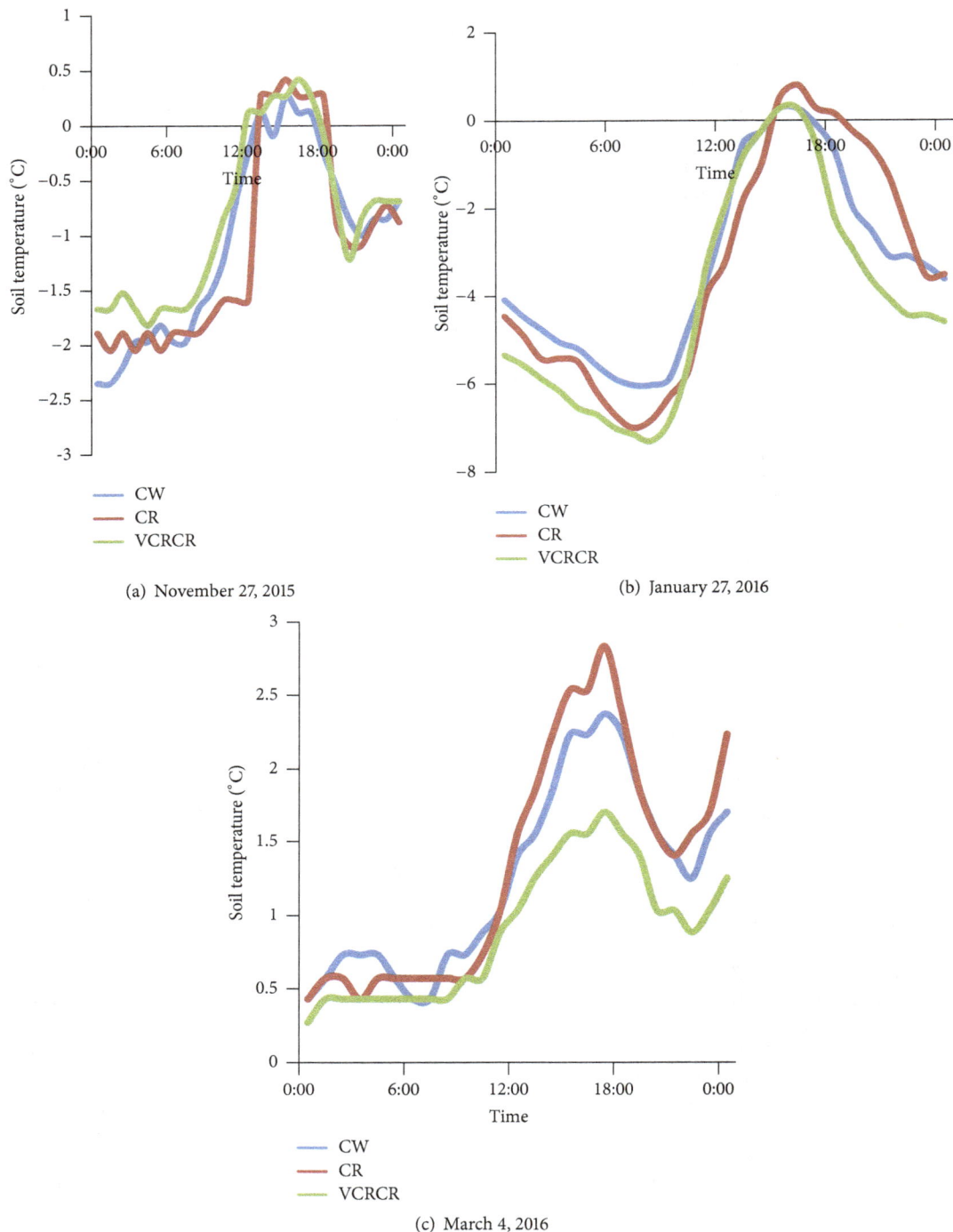

(a) November 27, 2015

(b) January 27, 2016

(c) March 4, 2016

FIGURE 5: Temporal soil temperature changes on interrow soil surface under different postsowing compaction treatments.

TABLE 1: Shoot number under different treatments on November 27, 2015, and March 4, 2016 (thousands/ha).

Date	Treatments		
	CW	CR	VCRCR
27-Nov	8688a	8146a	8350a
4-Mar	6893a	6038b	6329ab
Winter survival	0.79a	0.74b	0.76ab

Means in the same rows followed by the same letter are not significantly different.

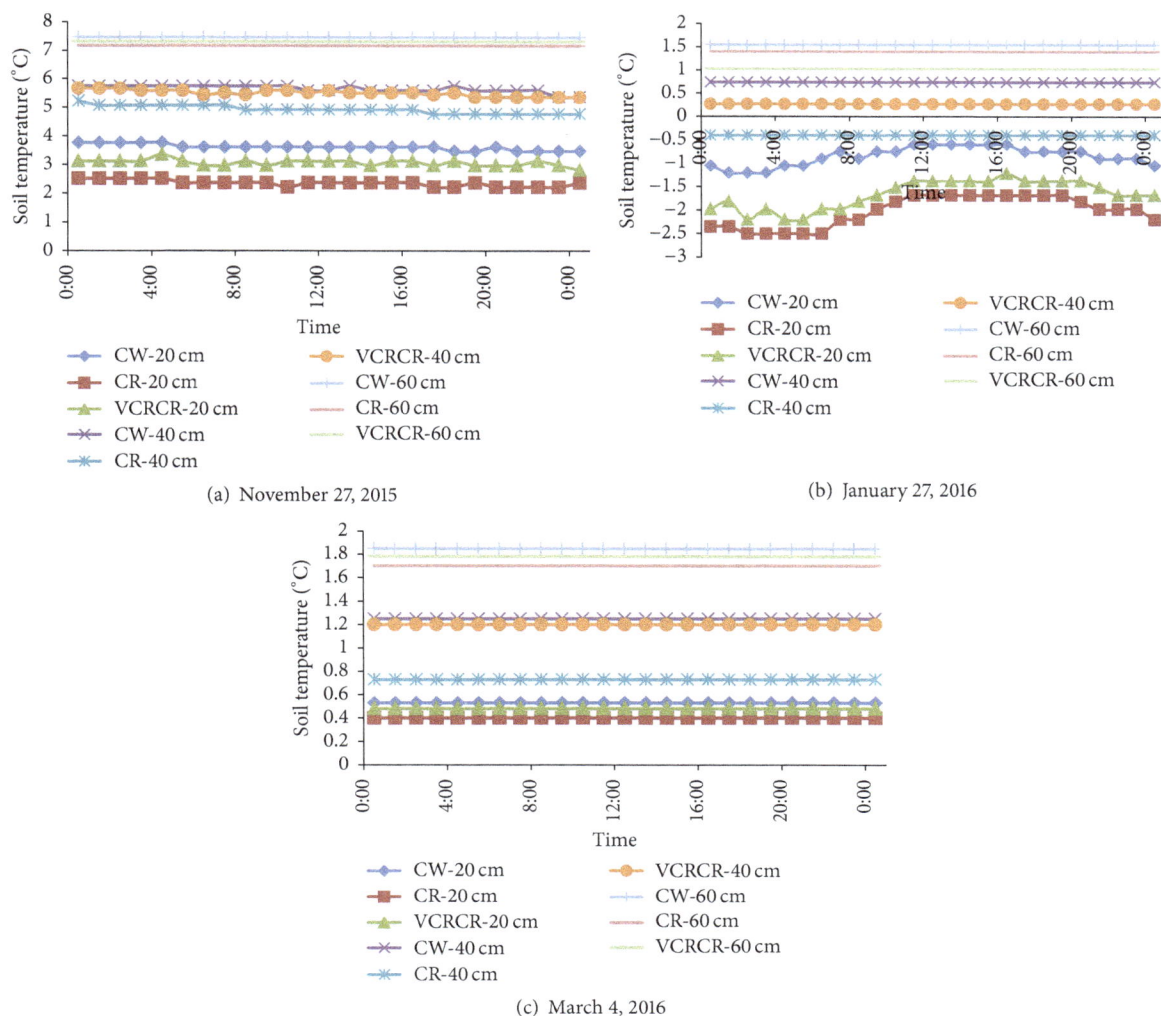

(a) November 27, 2015

(b) January 27, 2016

(c) March 4, 2016

FIGURE 6: Temporal soil temperature changes in different soil depth under different postsowing compaction treatments during wintering period.

TABLE 2: Shoot length and fresh and dry weight per 10 shoots under different treatments on March 4, 2016.

Test item	Treatment		
	CW	CR	VCRCR
Shoot length/mm	366.30a	334.50b	348.80ab
Shoot fresh weight per 10 shoots/g	36.06a	32.91b	35.43a
Shoot dry weight per 10 shoots/g	8.95a	7.83b	8.91a

Means in the same row followed by the same letter are not significantly different.

showed that seedlings and wheel compaction had a function to increase soil temperature (there were no compaction and seedlings on the interrow soil surface under CW treatment). The CW treatment had higher soil temperature and shoot winter survival (Table 1), length, and fresh and dry weight (Table 2) compared to the CR or the VCRCR treatments at the end of wintering period. This indicates that improvement in soil temperature was a factor contributing to the superior cold and frost tolerance capacity during wintering period [32–34].

Our data indicate that postsowing compaction with the CW treatment could improve soil temperature and winter wheat cold and frost tolerance capacity.

5. Conclusions

We measured how soil temperature was affected by different postsowing compaction methods during the wintering period in the North China Plain. To determine the resistance of wheat frost damage response to soil temperature as affected by postsowing compaction during wintering period, shoot growth parameters including shoot number, length, and fresh and dry weight at the beginning and end of wintering period were measured. We found that, during the wintering period, the trend of soil temperature change was similar to air temperature, but the gap between maximum and minimum soil temperature was smaller compared to air temperature. Data indicated that the adoption of CW significantly improved a wide range of soil temperature on the intrarow and deeper depth soil layer and thus contributed to higher shoot number, length, and fresh and dry weight.

The resultant positive changes in soil temperature and shoot growth parameters reduced the shoot death caused by frost damage and improved shoot winter survival under the CW treatment, making it a viable option for improving wheat production in the North China Plain. Since physicochemical presowing seed treatments have been reported to be highly efficient in enhancing the germination, growth, and productively of crops, further experiments for the synthetic action of postsowing compaction and presowing seed treatments are needed to confirm the effects of enhancing crop growth characteristics and productivity.

Conflicts of Interest

The authors declare that they have no conflicts of interest.

Acknowledgments

This work was financed by the National Key Research and Development Plan (2016YFD0200607), Anhui Science and Technology Major Project (15czz03134), and Postdoctoral Science Foundation of Beijing Academy of Agriculture and Forestry Sciences of China (2014002).

References

[1] G. Vico, V. Hurry, and M. Weih, "Snowed in for survival: quantifying the risk of winter damage to overwintering field crops in northern temperate latitudes," *Agricultural and Forest Meteorology*, vol. 197, pp. 65–75, 2014.

[2] P. Peltonen-Sainio, K. Hakala, and L. Jauhiainen, "Climate-induced overwintering challenges for wheat and rye in northern agriculture," *Acta Agriculturae Scandinavica Section B: Soil and Plant Science*, vol. 61, no. 1, pp. 75–83, 2011.

[3] J. Levitt, *Chilling, Freezing, and High Temperature Stresses*, New York Academic Press, New York, NY, USA, 1980.

[4] B. Holmer, "Fluctuations of winter wheat yields in relation to length of winter in Sweden 1866 to 2006," *Climate Research*, vol. 36, no. 3, pp. 241–252, 2008.

[5] A. K. Bergjord and A. O. Skjelvå, "Water soluble carbohydrates and growth potential of winter wheat as influenced by weather conditions during winter," *Acta Agriculturae Scandinavica Section B: Soil and Plant Science*, vol. 61, no. 6, pp. 523–534, 2011.

[6] E. Reinsdorf and H.-J. Koch, "Modeling crown temperature of winter sugar beet and its application in risk assessment for frost killing in central Europe," *Agricultural and Forest Meteorology*, vol. 182-183, pp. 21–30, 2013.

[7] R. Licker, C. J. Kucharik, T. Doré, M. J. Lindeman, and D. Makowski, "Climatic impacts on winter wheat yields in Picardy, France and Rostov, Russia: 1973–2010," *Agricultural and Forest Meteorology*, vol. 176, pp. 25–37, 2013.

[8] D. Z. Skinner and K. A. Garland-Campbell, "The relationship of LT50 to prolonged freezing survival in winter wheat," *Canadian Journal of Plant Science*, vol. 88, no. 5, pp. 885–889, 2008.

[9] M. S. Li, D. L. Wang, X. L. Zhong et al., "Current situation and prospect of research on frost of whiter wheat," *Journal of Natural Disasters*, vol. 14, no. 4, pp. 72–78, 2005.

[10] A. V. Malyshev and H. A. L. Henry, "Frost damage and winter nitrogen uptake by the grass *Poa pratensis* L.: consequences for vegetative versus reproductive growth," *Plant Ecology*, vol. 213, no. 11, pp. 1739–1747, 2012.

[11] M. Weih and P. S. Karlsson, "The nitrogen economy of mountain birch seedlings: implications for winter survival," *Journal of Ecology*, vol. 87, no. 2, pp. 211–219, 1999.

[12] E. W. R. Barlow, L. Boersma, and J. L. Young, "Photosynthesis, transpiration, and leaf elongation in corn seedlings at suboptimal soil temperatures," *Agronomy Journal*, vol. 69, no. 1, pp. 95–100, 1977.

[13] J. M. Walker, "One-degree increments in soil temperatures affect maize seedling behavior," *Soil Science Society of America Journal*, vol. 33, no. 5, pp. 729–736, 1969.

[14] H. M. Salem, C. Valero, M. Á. Muñoz, M. G. Rodríguez, and L. L. Silva, "Short-term effects of four tillage practices on soil physical properties, soil water potential, and maize yield," *Geoderma*, vol. 237-238, pp. 60–70, 2015.

[15] M. A. Licht and M. Al-Kaisi, "Strip-tillage effect on seedbed soil temperature and other soil physical properties," *Soil and Tillage Research*, vol. 80, no. 1-2, pp. 233–249, 2005.

[16] J. Tong, Q. Zhang, L. Guo et al., "Compaction performance of biomimetic press roller to soil," *Journal of Bionic Engineering*, vol. 12, no. 1, pp. 152–159, 2015.

[17] P. A. O. Odjugo, "The effect of tillage systems and mulching on soil microclimate, growth and yield of yellow yam (Dioscorea cayenensis) in Midwestern Nigeria," *African Journal of Biotechnology*, vol. 7, no. 24, pp. 4500–4507, 2008.

[18] J. A. Andrade, C. A. Alexandre, and G. Basch, "Effects of soil tillage and mulching on thermal performance of a Luvisol topsoil layer," *Folia Oecologica*, vol. 37, no. 1, pp. 1–7, 2010.

[19] J. M. Reichert, L. E. A. S. Suzuki, D. J. Reinert, R. Horn, and I. Håkansson, "Reference bulk density and critical degree-of-compactness for no-till crop production in subtropical highly weathered soils," *Soil and Tillage Research*, vol. 102, no. 2, pp. 242–254, 2009.

[20] J. M. Reichert, C. M. P. Bervald, M. F. Rodrigues, O. R. Kato, and D. J. Reinert, "Mechanized land preparation in eastern Amazon in fire-free forest-based fallow systems as alternatives to slash-and-burn practices: hydraulic and mechanical soil properties," *Agriculture, Ecosystems and Environment*, vol. 192, pp. 47–60, 2014.

[21] M. Iqbal, Z. ul Haq, A. Malik, Ch. M. Ayoub, Y. Jamil, and J. Nisar, "Pre-sowing seed magnetic field stimulation: a good option to enhance bitter gourd germination, seedling growth and yield characteristics," *Biocatalysis and Agricultural Biotechnology*, vol. 5, pp. 30–37, 2016.

[22] M. Iqbal, Z. U. Haq, Y. Jamil, and M. R. Ahmad, "Effect of presowing magnetic treatment on properties of pea," *International Agrophysics*, vol. 26, no. 1, pp. 25–31, 2012.

[23] M. Iqbal and R. A. Khera, "Adsorption of copper and lead in single and binary metal system onto *Fumaria indica* biomass," *Chemistry International*, vol. 1, pp. 157b–163b, 2015.

[24] Z. ul Haq, Y. Jamil, S. Irum, M. A. Randhawa, M. Iqbal, and N. Amin, "Enhancement in the germination, seedling growth and yield of radish (*Raphanus sativus*) using seed pre-sowing magnetic field treatment," *Polish Journal of Environmental Studies*, vol. 21, no. 2, pp. 369–374, 2012.

[25] G. O. Awe, J. M. Reichert, and O. O. Wendroth, "Temporal variability and covariance structures of soil temperature in a sugarcane field under different management practices in southern Brazil," *Soil and Tillage Research*, vol. 150, pp. 93–106, 2015.

[26] T. R. Karl, R. W. Knight, and N. Plummer, "Trends in high-frequency climate variability in the twentieth century," *Nature*, vol. 377, no. 6546, pp. 217–220, 1995.

[27] Q. Hu and S. Feng, "A daily soil temperature dataset and soil temperature climatology of the contiguous United States," *Journal of Applied Meteorology*, vol. 42, no. 8, pp. 1139–1156, 2003.

[28] T. R. Karl, R. W. Knight, D. R. Easterling, and R. G. Quayle, "Indices of climate change for the United States," *Bulletin of the American Meteorological Society*, vol. 77, no. 2, pp. 279–292, 1996.

[29] B. Shi, C. S. Tang, L. Gao, H. T. Jiang, and C. Liu, "Difference in shallow soil temperature at urban and rural areas," *Journal of Engineering Geology*, vol. 20, pp. 58–65, 2012.

[30] C.-S. Tang, B. Shi, L. Gao, J. L. Daniels, H.-T. Jiang, and C. Liu, "Urbanization effect on soil temperature in Nanjing, China," *Energy and Buildings*, vol. 43, no. 11, pp. 3090–3098, 2011.

[31] X. Liao, Z. Su, G. Liu, L. Zotarelli, Y. Cui, and C. Snodgrass, "Impact of soil moisture and temperature on potato production using seepage and center pivot irrigation," *Agricultural Water Management*, vol. 165, pp. 230–236, 2016.

[32] S. Y. Chen, X. Y. Zhang, D. Pei, H. Y. Sun, and S. L. Chen, "Effects of straw mulching on soil temperature, evaporation and yield of winter wheat: field experiments on the North China Plain," *Annals of Applied Biology*, vol. 150, no. 3, pp. 261–268, 2007.

[33] X.-L. Wang, F.-M. Li, Y. Jia, and W.-Q. Shi, "Increasing potato yields with additional water and increased soil temperature," *Agricultural Water Management*, vol. 78, no. 3, pp. 181–194, 2005.

[34] Q. Hu and G. Buyanovsky, "Climate effects on corn yield in Missouri," *Journal of Applied Meteorology*, vol. 42, no. 11, pp. 1626–1635, 2003.

Effect of Fungicide Applications on Grain Sorghum (*Sorghum bicolor* L.) Growth and Yield

Dan D. Fromme,[1] Trey Price,[2] Josh Lofton,[3] Tom Isakeit,[4] Ronnie Schnell,[5] Syam Dodla,[6] Daniel Stephenson,[1] W. James Grichar,[7] and Keith Shannon[1]

[1]*LSU AgCenter, 8208 Tom Bowman Drive, Alexandria, LA 71302, USA*
[2]*LSU AgCenter, 212A Macon Ridge Road, Winnsboro, LA 71295, USA*
[3]*Oklahoma State University, Stillwater, OK 74078, USA*
[4]*Department of Plant Pathology and Microbiology, Texas A&M University, College Station, TX 77843, USA*
[5]*Department of Soil and Crop Science, Texas A&M University, College Station, TX 77843, USA*
[6]*LSU AgCenter, 262 Research Station Drive, Bossier City, LA 71112, USA*
[7]*Texas A&M AgriLife Research and Extension Center, Corpus Christi, TX 78406, USA*

Correspondence should be addressed to W. James Grichar; w-grichar@tamu.edu

Academic Editor: Kent Burkey

Field studies were conducted in the upper Texas Gulf Coast and in central Louisiana during the 2013 through 2015 growing seasons to evaluate the effects of fungicides on grain sorghum growth and development when disease pressure was low or nonexistent. Azoxystrobin and flutriafol at 1.0 L/ha and pyraclostrobin at 0.78 L/ha were applied to the plants of two grain sorghum hybrids (DKS 54-00, DKS 53-67) at 25% bloom and compared with the nontreated check for leaf chlorophyll content, leaf temperature, and plant lodging during the growing season as well as grain mold, test weight, yield, and nitrogen and protein content of the harvested grain. The application of a fungicide had no effect on any of the variables tested with grain sorghum hybrid responses noted. DKS 53-67 produced higher yield, greater test weight, higher percent protein, and N than DKS 54-00. Results of this study indicate that the application of a fungicide when little or no disease is present does not promote overall plant health or increase yield.

1. Introduction

Fungicides are a vital solution to the effective control of plant diseases which are estimated to cause yield reductions of almost 20% in major food and cash crops worldwide [1]. In the past few years, there has been increased controversy over whether fungicides should be applied to field crops in the absence of disease [2, 3]. Despite a lack of scientific evidence supporting "plant health," the US Environmental Protection Agency has granted a supplemental label for the use of pyraclostrobin (Headline®, BASF Corp., Research Triangle Park, NC 27709) fungicide for "plant health" [4], which may be misleading. Fungicides have been shown to have effects on crop growth and physiology by various disruptions such as growth reduction, perturbation in the development of reproductive organs, alteration of nitrogen, and/or carbon metabolism leading to a lower nutrient availability for plant growth. The sensitivity of some plant species may depend on the developmental stage (e.g., more sensitive to the treatments at young stages or during critical events such as reproduction) or the type of pesticides used [1].

Grain sorghum in the US is primarily grown on dryland hectares in the "sorghum belt" which stretches from South Dakota to southern Texas. In 2015, grain sorghum was planted on 3.4 million hectares with an average yield of 4409 kg/ha [5]. Fungicides have been used in grain sorghum to manage a number of foliar diseases in the southern US including anthracnose (*Colletotrichum graminicola* [Ces.] G. W. Wilson [syn. *C. sublineolum* P. Henn., in Kabat. & Bubak]), gray leaf spot (*Cercospora sorghi* Ellis & Everh.), target leaf spot (*Bipolaris sorghicola* [Lefebvre & Sherwin] Alcorn), and zonate leaf spot (*Gloeocercospora sorghi* D. Bain & Edgerton ex Deighton) [6, 7]. Older fungicides used to manage grain sorghum diseases had multisite modes of action and were contact fungicides which remain on the leaf surface and were easily washed off [8]. With improved chemistry, systemic fungicides are now available and these types of fungicides are absorbed by the leaves and move within the treated plant [8]. Systemic fungicides allow growers to properly manage grain sorghum diseases now more than ever before.

The quinone outside inhibitor (QoI) class of fungicides, commonly referred to as strobilurins, are a relatively newer class of fungicides that have been commercially available in Europe and the US since the mid-to-late 1990s [9]. This class of fungicides disrupt electron transport in the mitochondria and diminish adenosine triphosphate (ATP) production [9], which effectively prevents spore germination and reduces mycelial growth in ascomycetes, basidiomycetes, deuteromycetes, and oomycetes [10, 11]. This mode of action is also purported to induce nonfungicidal, physiological changes within the plant, such as greater chlorophyll retention, increased water and nitrogen use efficiency, and delayed senescence [12–14].

Three of the newer fungicides labeled for use in grain sorghum are azoxystrobin, flutriafol, and pyraclostrobin. Azoxystrobin and pyraclostrobin are strobilurin-type fungicides that have shown activity against many different fungal pathogens in soybean (*Glycine max* L.), peanut (*Arachis hypogaea* L.), and various other crops [15–18]. Pyraclostrobin is rapidly absorbed by leaf tissue and has demonstrated translaminar movement through layers of the leaf; however, the material is not redistributed throughout the plant like a true systemic fungicide [9, 19]. Selected rates of pyraclostrobin have superior activity against the peanut leafspots, caused by *Cercospora arachidicola* and *C. personatum*, as well as soil-borne diseases such as southern stem rot (*Sclerotium rolfsii* Sacc.) [20].

Azoxystrobin is reported to have activity against both *Rhizoctonia solani* and *Pythium* spp. [21–23]. Uptake of azoxystrobin into leaves is a gradual process; for example, 1 to 3% of the applied material is absorbed into a grape leaf within 24 hr of foliar application [9]. Strobilurins like azoxystrobin move across the leaf surface and into the waxy cuticle of the leaf (locally systemic) and may even move into the cuticle of the underside of the leaf (translaminar activity) [24]. Some of the material also may move into the xylem and be transported upwards; however, the plant does not transport much, if any, fungicide down to the roots [9].

Flutriafol is a systemic demethylation inhibitor (DMI) fungicide that is labeled as a curative or a preventative treatment in grain sorghum [25]. The material inhibits the specific enzyme, C14-demethylase, a fungal cytochrome P450, which plays a role in sterol production [25]. Sterols are a requirement for fungal membrane structure and function and are essential for the development of functional cell walls [24, 25]. It was first registered by the EPA for use on apples (*Malus domestica* Borkh) and soybeans in 2010. In 2012, the fungicide was approved for use on corn (*Zea mays* L.) [25]. Recently, flutriafol received approval for use on cotton (*Gossypium hirsutum* L.) to manage *Phymatotrichopsis* root rot of cotton [26].

There is very limited information available on the response of grain sorghum to fungicides, especially to document whether a fungicide application causes nonfungicidal physiological changes and yield increases in years of little disease development. However, growers continue to inquire about the use of fungicides in grain sorghum production and their value. Therefore, the objectives of this research were (1) to assess the effects of fungicide applications applied to grain sorghum on seed yield when foliar disease incidence was low or nonexistent and (2) to determine if a fungicide application can be associated with nonfungicidal physiological changes and yield components or grain composition under field conditions.

2. Materials and Methods

2.1. Research Sites. Grain sorghum fungicide studies were conducted in 2013 along the upper Texas Gulf Coast near Wharton on the Michael Beard Farm (29.296036N, 96.221369W) and in 2014 (31.17540N, 92.40517W) and 2015 (31.17302N, 92.40922W) in central Louisiana at the Rapides Parish-Dean Lee Research and Extension Center near Alexandria, Louisiana. Cotton was planted prior to grain sorghum at the two locations in all three years. Soil type at the Wharton location was a Lake Charles clay (fine, smectitic, hyperthermic, Typic Hapluderts) with a pH of 6.5 while soils at both Alexandria locations was a Coushatta silt loam (fine-silty, mixed, superactive, thermic Fluventic Eutrudepts-Coushatta silt loam) with a pH range of 6.5 to 6.8. Conventional tillage systems were used at both locations and both sites were maintained under rainfed conditions. Fertilizer at the Wharton location included 125 kg/ha of N, 44 kg/ha of P, 17 kg/ha of K, and 8 kg/ha of Zn while at Alexandria 168 kg/ha of N, 34 kg/ha of P, 67 kg/ha of K, and 2 kg/ha of Zn were applied in both years. Plots were maintained weed-free at the Wharton County location throughout the growing season using a preemergence (PRE) application of a premix of dimethenamid P plus atrazine (Guardsman Max®, BASF Corporation, Research Triangle Park, NC 27709) at 3.2 L/ha plus dimethenamid (Outlook®, BASF Corporation, Research Triangle Park, NC 27709) at 0.13 kg/ha while at the Alexandria location, S-metolachlor (Medal II®, Syngenta Crop Protection, Greensboro, NC 27419) at 1.2 L/ha plus atrazine at 3.55 L/ha were applied PRE. No insecticides were applied at the Wharton location; however, at Alexandria in 2014, two applications of sulfoxaflor (Transform®, Dow AgroSciences, Indianapolis, IN 46268) were applied at 110 ml/ha. In 2015, sulfoxaflor was applied at 73 ml/ha followed by flupyradifurone (Sivanto, Bayer CropSciences LP, Research Triangle Park, NC 27709) at 365 ml/ha. In both years, these insecticides were applied to control the sugarcane aphid (*Melanaphis sacchari* L.). Prior to harvest, glyphosate (Roundup WeatherMaxx®, Monsanto Co., St. Louis, MO 63167) at 2.24 L/ha was applied at 30% physiological maturity to hasten grain drydown and aid in plant desiccation for ease of harvest. Plots were hand harvested and mechanically threshed at Wharton (July 9, 2013) while a small plot combine was used at the Alexandria location (August 18, 2014; August 5, 2015). Final yields were adjusted to 14% moisture.

TABLE 1: Rainfall amounts during the 2013 through 2015 growing season at each test location.

| Month | Wharton, Texas (2013) | | Alexandria, Louisiana | | |
	Monthly	30 yr average	2014	2015	30 yr average
			Mm		
March	12.7	80.8	49.0	180.9	131.9
April	116.8	83.8	96.5	229.6	116.1
May	66.0	119.4	110.5	160.8	119.9
June	20.3	126.5	111.0	137.9	136.9
July	33.0	83.1	132.1	46.5	112.0
August	38.1	74.9	215.4	41.7	103.9
Total	286.9	568.5	714.5	797.4	720.7

2.2. Grain Sorghum Hybrids and Fungicide Treatments. Grain sorghum hybrids (DKS 54-00 and DKS 53-67) were those commonly used in production fields in south-central Texas and central Louisiana. Grain sorghum seed was planted at the rate of 198,000 seed/ha with either a 2- or 4-row cone planter. Planting dates were March 8, 2013, at the Wharton County location and April 4, 2014, and April 1, 2015, at the Alexandria location. Treatments consisted of a factorial arrangement of the two grain sorghum hybrids with four fungicide treatments (nontreated, azoxystrobin and flutriafol at 1.0 L/ha and pyraclostrobin at 0.78 L/ha). Only azoxystrobin included Agridex (Helena Chemical Co., 6075 Poplar Ave., Memphis, TN 38119), a crop oil concentrate, at 1.0% v/v. Each study was replicated four times.

2.3. Plots and Rainfall. Individual plots consisted of four rows (102 cm centers at Wharton and 97 cm centers at Alexandria) by 15.2 m long with 4 reps at each location. The four rows of each plot in each study were sprayed with fungicide and data (including yield) were collected from the middle two rows. Rainfall for the upper Texas Gulf Coast (Wharton, TX) in 2013 can be best described as well below average for all months of the growing season with the exception of April (Table 1). Seasonal rainfall at the Alexandria location in 2014 can be described as below average during the early part of the growing season (March through April) but near or above average during the rest of the growing season. The 2015 growing season can be described as above normal rainfall early (March through June) but below average during July and August (Table 1).

2.4. Fungicide Application. At the Wharton County location, fungicides were applied with a Lee® Spider sprayer equipped with one 8003 XR flat fan nozzle (TeeJet Spraying Systems Co., Wheaton, IL 60188) per row, while at Alexandria, fungicides were applied with a CO_2-propellant backpack sprayer equipped with one 8001 flat fan spray nozzle (TeeJet Spraying Systems Co., Wheaton, IL 60188) per row. At all locations, fungicides were applied in 140 L of water/ha at a pressure of 315 kPa at the Wharton County location or 504 kPa at Alexandria. Fungicides were applied at 25% bloom at each location which was 84 (May 31), 80 (June 23), and 83 (June 23) days after planting (DAP) in 2013, 2014, and 2015, respectively.

2.5. Leaf Chlorophyll Concentration. Grain sorghum leaf chlorophyll concentrations were determined using a SPAD 502 chlorophyll meter (Konica Minolta; Ramsey, NJ 07446). Readings were taken on the flag leaf between the mid-vein and leaf margin and reported values are the average of readings collected from ten plants in the center two rows of each plot at solar noon. Measurements were taken on a weekly basis beginning the day before fungicide application until black layer was reached.

2.6. Leaf Temperature. Leaf temperature readings were taken using a Raytek ST Pro temperature gun (Raytek Corp.; Santa Cruz, CA 95061). Readings were taken at the same location and time as leaf chlorophyll measurements on ten plants and reported values are the average of the readings.

2.7. Seed N and Protein, Stalk Lodging, and Grain Mold Ratings. Seed N and protein were determined by the high temperature combustion process [27–29] (Texas A&M Soil, Water and Forage Testing Lab; College Station, TX 77843). Stalk lodging data was collected the day prior to harvest and was calculated based on the total number of plants lodged per plot divided by the total number of plants per plot. Grain mold was determined on a scale of 1 to 5 with 1 = seed bright with no mold or discoloration and 5 = seed covered entirely with mold and was deteriorated and looked dead.

2.8. Data Analysis. The treatment design was a factorial arrangement using a randomized complete block design with fungicides and grain sorghum hybrids as factors. An analysis of variance was performed using the ANOVA procedure for SAS (SAS Institute, 1998, SAS user's guide, SAS Institute, Cary, NC) to evaluate the significance of fungicide and grain sorghum hybrid on overall plant health which included leaf chlorophyll concentration and temperature, plant lodging, grain mold, grain test weight and yield, and percent nitrogen and protein in the grain. The Fishers Protected LSD at the 0.05 level of probability was used for separation of mean differences.

3. Results and Discussion

Across years at both locations, very few or no foliar disease symptoms were present among the two grain sorghum hybrids (data not shown) even though rainfall was near normal or above normal at Alexandria (Table 1). Also, across all variables tested, differences between the untreated check and any fungicide treatment were noted only in 2015 and there was not a fungicide by hybrid interaction for any variables tested (data not shown). Grain sorghum hybrid response to the several test variables was noted (Table 2). This indicates that grain sorghum was only slightly affected by the application of a fungicide. This is contrary to industry advertisements claiming otherwise.

3.1. Leaf Chlorophyll Content. The effect of fungicide was only noted in 2015 at 13 weeks after planting when the untreated and flutriafol treatments produced the highest readings and pyraclostrobin the lowest (Table 2). Also, hybrid response was noted in 2013 at 13 weeks after planting when DKS 53-67 resulted in higher readings than DKS 54-00 and also in 2013 and 2014 at 15 weeks after planting when different results were noted. In 2013, DKS 53-67 resulted in the highest reading while in 2014 the opposite was found. Henry et al. [30] noted no difference in leaf chlorophyll readings with nine different fungicides on soybean.

3.2. Leaf Temperature Readings. Only in 2015 at week 11 was a response noted with fungicides while hybrid differences were noted in 2013 and 2014 (Table 2). In 2015 at week 11, leaf temperature was the greatest with pyraclostrobin while azoxystrobin produced the lowest temperature. Hybrid differences in leaf temperature were noted in 2013 and 2014 but not 2015 (Table 2). Results were inconsistent in both years as in some instances DKS 53-67 produced higher leaf temperatures and in other instances DKS 54-00 produced higher temperatures.

3.3. Test Weight. Only a hybrid response was noted in 2013 and 2014 with DKS 53-67 producing higher test weight than DKS 54-00 in both years.

3.4. Yield. A response to fungicides was noted only in 2015 (Table 2). Flutriafol produced the highest yield and azoxystrobin and pyraclostrobin the lowest. However, no fungicide treatment was different from the untreated check. In 2013 and 2014, DKS 53-67 outyielded DKS 54-00. Swoboda and Pedersen [31] noted, in a 2-year study on soybean, that fungicides had no effect on yield. Grichar [15] noted that the use of fungicides on soybean along the upper Texas Gulf Coast under little or no disease pressure resulted in few increases in yield and subsequent increases in net returns. In several instances, the use of fungicides resulted in a decrease in net returns, especially in a year with below normal rainfall. However, Paul et al. [32] reported that, in corn, generally the mean yield was higher in plots treated with fungicides than in the nontreated plots.

3.5. Protein Content. Only a hybrid response was noted with DKS 53-67 having a higher protein content than DKS 54-00 in both 2013 and 2015 with no differences noted in 2014 (Table 2). Swoboda and Pedersen [31] noted no differences in protein content among fungicide treatments in soybean.

3.6. Seed N Content. As was seen with protein content, only a hybrid response was noted. In all three years, DKS 53-67 had a higher N content than DKS 54-00.

3.7. Grain Mold. In 2015, the untreated check resulted in a higher grain mold rating than any of the fungicide treatments while in 2014 a hybrid difference was noted as DKS 54-00 had a higher rating than DKS 53-67 (Table 2).

3.8. Lodging. No lodging was noted with any hybrid or fungicide treatment (data not shown).

The results from the various variables tested in this study fail to show a consistent overall improvement in plant health and other advantages to using a fungicide under low disease pressure. Other studies in various crops have also failed to indicate that the application of a fungicide would improve overall plant health and result in an increase in yield. The prophylactic use of fungicides may confer risks beyond economic losses. QoI or strobilurin fungicides are classified by the Fungicide Resistance Action Committee (FRAC) as high-risk for resistance development [30]. Over 40 pathogens have been reported as being resistant to QoI fungicides worldwide [33]. Recently, isolates of *Cercospora sojina* Hara on soybean were confirmed as resistant to strobilurin fungicides even though these fungicides have only been widely used in soybean for a few years [34]. Swoboda and Pedersen [31] concluded that under the low disease incidence noted in Iowa a low probability exists that the use of a fungicide on soybean would increase yield by mechanisms other than disease control. Wrather et al. [35] reported that foliar applications of azoxystrobin may be useful for the management of some foliar soybean diseases, but azoxystrobin may increase the percent of *Phomopsis* spp. seed infection. It was felt that azoxystrobin may interfere with the plant's natural defense mechanism to *Phomopsis* spp., or it may protect the plant from other diseases, thus extending the life of the plant so that *Phomopsis* spp. has more time to move from the pod into the seed. Spokas and Jacobson [36] reported no long-term negative impacts on the soil system or strawberry (*Fragaria* x ananassa) production as a consequence of strobilurin usage. In one of two years, they reported a yield boost on strawberries as a result of fungicide usage.

Paul et al. [32] stated that unless a corn crop is at risk of developing fungal disease, farmers would be smart to skip fungicide treatments that promise increased yields. They reported that fungicides used in fields where conditions were optimal for fungal diseases improved yields and paid for themselves. Some studies in wheat (*Triticum aestivum*) have indicated that QoI and DMI fungicides may delay senescence causing a "greening effect" by reducing oxygen in the leaves thereby protecting plants from toxic reactive oxygen species [37]. Others have hypothesized that the greening effect in wheat is a result of the inhibition of ethylene production by the QoI fungicides [13]. Other research has found no effect of the QoI or DMI fungicides on wheat senescence, biomass, or yield [38].

In conclusion, in the absence of any disease pressure, the application of a fungicide may not improve grain sorghum plant health or increase yield and may ultimately reduce net

TABLE 2: Fungicide and hybrid response to the different variables for the 2013 through 2015 growing seasons[a].

Year, fungicide, hybrid	Leaf chlorophyll concentration		Leaf temperature (°C) Weeks after planting					Test weight	Yield Kg/ha	Protein	N %	Grain mold
	13	15	11	12	13	14	15					
2013												
Fungicide												
Untreated	48.7a	39.7a	31.9a	31.6a	31.3a	33.3a	34.3a	64.5a	6391a	11.7a	1.9a	1.0a
Azoxystrobin	47.6a	39.6a	31.9a	31.4a	31.0a	33.2a	33.6a	64.0a	6297a	12.0a	1.9a	1.0a
Flutriafol	48.1a	41.0a	31.3a	30.7a	30.9a	33.7a	31.1a	64.0a	6387a	11.5a	1.8a	1.0a
Pyraclostrobin	49.2a	40.5a	31.9a	30.6a	31.5a	32.9a	33.6a	64.1a	6394a	11.9a	1.9a	1.0a
Hybrid												
53-67	49.1a	43.4a	31.3a	32.5a	31.0a	31.3b	31.2b	65.2a	6610a	12.1a	1.9a	1.0a
54-00	47.7b	37.0b	32.2a	29.6b	31.4a	35.2a	36.6a	63.1b	6125b	11.4b	1.8b	1.0a
2014												
Fungicide												
Untreated	46.8a	42.3a	31.2a	31.6a	32.3a	31.7a	32.1a	62.1a	8978a	10.0a	1.6a	2.0a
Azoxystrobin	45.6a	42.4a	31.7a	31.4a	32.4a	31.4a	32.1a	61.7a	8897a	9.6a	1.5a	1.9a
Flutriafol	46.5a	41.9a	31.1a	30.7a	32.0a	30.9a	32.1a	63.1a	9069a	9.8a	1.6a	1.9a
Pyraclostrobin	46.8a	42.1a	31.1a	31.3a	31.9a	31.2a	31.9a	63.0a	9279a	10.3a	1.6a	1.9a
Hybrid												
53-67	46.5a	40.4b	32.2a	31.9a	31.4b	30.3b	31.6b	64.1a	9365a	10.2a	1.6a	1.6b
54-00	46.3a	43.9a	30.4b	30.6b	32.9a	32.3a	32.6a	60.8b	8746b	9.7a	1.5b	2.3a
2015												
Fungicide												
Untreated	47.7a	44.2a	29.2bc	28.8a	33.2a	30.9a	31.0a	65.5a	7058ab	9.0a	1.4a	1.5a
Azoxystrobin	46.5ab	45.4a	29.1c	28.7a	33.3a	31.2a	31.4a	64.3a	6520b	8.4a	1.4a	1.0b
Flutriafol	47.6a	44.3a	29.8ab	26.7a	32.4a	30.5a	31.1a	64.2a	7499a	8.6a	1.4a	1.0b
Pyraclostrobin	45.1b	44.9a	30.1a	28.6a	31.8a	30.1a	31.3a	63.9a	6600b	8.8a	1.4a	1.0b
Hybrid												
53-67	46.7a	44.7a	29.6a	27.7a	32.7a	30.7a	31.2a	64.5a	6919a	9.4a	1.5a	1.1a
54-00	46.7a	44.7a	29.6a	28.8a	32.7a	30.7a	31.2a	64.5a	6919a	8.0b	1.3b	1.1a

[a]Means followed by the same letter are not significantly different at $P = 0.05$.

[b]Grain mold score: 1 to 5 with 1: seed bright with no mold or discoloration and 5: seed covered entirely with mold and deteriorated and looking dead.

returns because of the added cost of a fungicide and application. However, if conditions are conducive to disease development, the application of a fungicide may prove beneficial.

Competing Interests

The authors declare that there is no conflict of interests regarding the publication of this manuscript.

References

[1] A.-N. Petit, F. Fontaine, P. Vatsa, C. Clément, and N. Vaillant-Gaveau, "Fungicide impacts on photosynthesis in crop plants," *Photosynthesis Research*, vol. 111, no. 3, pp. 315–326, 2012.

[2] P. Vincelli and D. Hershman, "Update on fungicides for 'plant health'," *Kentucky Pest News*, no. 1212, 2009, http://www.uky.edu/Ag/kpn/kpn_09/pn_090915.html.

[3] D. Brown-Rytlewski and P. Vincelli, *Letter from Universities Regarding the Strobilurin, Pyraclostrobin (Headline), Supplemental Label*, U.S. Environmental Protection Agency, Washington, DC, USA, 2009, http://pmep.cce.cornell.edu/profiles/fungnemat/febuconazole-sulfur/pyraclostrobin/pyraclos_let_0209.pdf.

[4] Anonymous, *Headline Fungicide Supplemental Label*, BASF Corporation, Florham Park, NJ, USA, 2008.

[5] Anonymous, "All about sorghum," United sorghum checkoff program, http://www.sorghumcheckoff.com/all-about-sorghum.

[6] C. Hollier, "Commercial crop production field crops-rain sorghum," Louisiana Field Crops, 2015, http://www.lsuagcenter.com/NR/rdonlyres/E8519934-CB7D-4785-970A-709C01CC03B0/101294/10GrainSorghum2015.pdf.

[7] F. Gould and C. Gautreaux, "Odds are low for grain sorghum disease, but still scout," LSU AgCenter, 2015, http://www.lsuagcenter.com/topics/crops/soybeans/soybean_grain_promotion_board_reports/odds-are-low-for-grain-sorghum-disease-but-still-scout.

[8] J. Harrington, "Foliar fungicides for soybeans increase yield potential," Plant Health Progress, 2009, http://www.plantmanagementnetwork.org/pub/php/news/2009/FoliarFungicides2/.

[9] D. W. Bartlett, J. M. Clough, J. R. Godwin, A. A. Hall, M. Hamer, and B. Parr-Dobrzanski, "The strobilurin fungicides," *Pest Management Science*, vol. 58, no. 7, pp. 649–662, 2002.

[10] R. Stierl, M. Merk, W. Schrof, and W. S. Butterfield, "Activity of the new BASF strobilurin fungicide, BAS 500 F, against *Septoria tritici* on wheat," in *Proceedings of the British Crop Protection Council Conference*, pp. 859–864, Brighton, UK, 2000.

[11] W. Venancio, M. Rodrigues, E. Begliomini, and N. Souza, "Physiological effects of strobilurin fungicide on plants," *Ponta Grossa*, vol. 9, pp. 59–68, 2003.

[12] J. Glaab and W. M. Kaiser, "Increased nitrate reductase activity in leaf tissue after application of the fungicide Kresoxim-methyl," *Planta*, vol. 207, no. 3, pp. 442–448, 1999.

[13] K. Grossmann and G. Retzlaff, "Bioregulatory effects of the fungicidal strobilurin kresoxim-methyl in wheat (*Triticum aestivum*)," *Pesticide Science*, vol. 50, no. 1, pp. 11–20, 1997.

[14] H. Kohle, K. Grossmann, T. Jabs et al., "Physiological effects of the strobilurin fungicide F 500 on plants," in *Proceedings of the Modern Fungicides and Antifungal Compounds III*, Thuringia, Germany, 2002.

[15] W. J. Grichar, "Soybean (*Glycine max* L.) response to fungicides in the absence of disease pressure," *International Journal of Agronomy*, vol. 2013, Article ID 561370, 5 pages, 2013.

[16] D. Mueller, "Evaluation of foliar fungicides for management of soybean rust," Integrated Crop Management IC-498, no. 3, 2007.

[17] A. K. Culbreath, T. B. Brenneman, and R. C. Kemerait, "Management of early leaf spot of peanut with pyraclostrobin as affected by rate and spray interval," *Plant Health Progress*, 2002.

[18] A. K. Hagan, H. L. Campbell, K. L. Bowen, and L. Wells, "Impact of application rate and treatment interval on the efficacy of pyraclostrobin in fungicide programs for the control of early leaf spot and southern stem rot on peanut," *Peanut Science*, vol. 30, no. 1, pp. 27–34, 2003.

[19] R. Stierl, E. J. Butterfield, H. Koehle, and G. Lorenz, "Biological characterization of the new strobilurin fungicide BAS 500 F," *Phytopathology*, vol. 90, p. S74, 2000.

[20] H. E. Portillo, R. R. Evans, J. S. Barnes, and R. E. Gold, "F500, a new broad-spectrum fungicide for control of peanut diseases," *Phytopathology*, vol. 91, p. S202, 2001.

[21] W. J. Grichar, B. A. Besler, and A. J. Jaks, "Use of azoxystrobin for disease control in texas peanut," *Peanut Science*, vol. 27, no. 2, pp. 83–87, 2000.

[22] M. Mihajlovic, E. Rekanovic, J. Hrustic et al., "In vitro and in vivo toxicity of several fungicides and Timorex gold biofungicide to *Pythuim aphanidermatum*," *Pesticides & Phytomedicine*, vol. 28, no. 2, pp. 117–123, 2013.

[23] C. E. Windels and J. R. Brantner, "Early-season application of azoxystrobin to sugarbeet for control of *Rhizoctonia solani* AG 4 and AG 2-2," *Journal of Sugarbeet Research*, vol. 42, no. 1, pp. 1–16, 2005.

[24] H. Balba, "Review of strobilurin fungicide chemicals," *Journal of Environmental Science and Health - Part B Pesticides, Food Contaminants, and Agricultural Wastes*, vol. 42, no. 4, pp. 441–451, 2007.

[25] Flutriafol, "New use review," 2012, http://www.mda.state.mn.us/chemicals/pesticides/regs/~/media/Files/chemicals/reviews/nur-flutriafol.ashx.

[26] T. Isakeit, "Using Topguard to control root rot: pay attention to that label," Texas Row Crops Newsletter, 2015, http://agrilife.org/texasrowcrops/2015/02/27/using-topguard-to-control-root-rot-pay-attention-to-that-label/.

[27] B. H. Sheldrick, "Test of the LECO CHN-600 determinator for soil carbon and nitrogen analysis," *Canadian Journal of Soil Science*, vol. 66, no. 3, pp. 543–545, 1986.

[28] R. A. Sweeney, "Generic combustion method for determination of crude protein in feeds: collaborative study," *Journal of the Association of Official Analytical Chemists*, vol. 72, no. 5, pp. 770–774, 1989.

[29] D. W. Nelson and L. E. Sommers, "Determination of total nitrogen in plant material," *Agronomy Journal*, vol. 65, no. 1, pp. 109–112, 1973.

[30] R. S. Henry, W. G. Johnson, and K. A. Wise, "The impact of a fungicide and an insecticide on soybean growth, yield, and profitability," *Crop Protection*, vol. 30, no. 12, pp. 1629–1634, 2011.

[31] C. Swoboda and P. Pedersen, "Effect of fungicide on soybean growth and yield," *Agronomy Journal*, vol. 101, no. 2, pp. 352–356, 2009.

[32] P. A. Paul, L. V. Madden, C. A. Bradley et al., "Meta-analysis of yield response of hybrid field corn to foliar fungicides in the U.S. corn belt," *Phytopathology*, vol. 101, no. 9, pp. 1122–1132, 2011.

[33] FRAC, "List of plant pathogenic organisms resistant to disease control agents," 2011, http://www.frac.info/docs/default-source/publications/list-of-resistant-plant-pathogens/list-of-resistant-plant-pathogenic-organisms---february-2013.pdf?sfvrsn=4.

[34] C. A. Bradley, "Frogeye leaf spot pathogen with reduced sensitivity to fungicides found in Tennessee soybean field," University Illinois Extension Bulletin 172, 2010.

[35] J. A. Wrather, J. G. Shannon, W. E. Stevens, D. A. Sleper, and A. P. Arelli, "Soybean cultivar and foliar fungicide effects on *Phomopsis* sp. seed infection," *Plant Disease*, vol. 88, no. 7, pp. 721–723, 2004.

[36] K. Spokas and B. Jacobson, *Impacts of Strobilurin Fungicides on Yield and Soil Microbial Processes for Minnesota Strawberry Production*, Minnesota Fruit and Vegetable Growers Association, St. Cloud, Minn, USA, 2010.

[37] Y.-X. Wu and A. Von Tiedemann, "Physiological effects of azoxystrobin and epoxiconazole on senescence and the oxidative status of wheat," *Pesticide Biochemistry and Physiology*, vol. 71, no. 1, pp. 1–10, 2001.

[38] J. R. Bertelsen, E. De Neergaard, and V. Smedegaard-Petersen, "Fungicidal effects of azoxystrobin and epoxiconazole on phyllosphere fungi, senescence and yield of winter wheat," *Plant Pathology*, vol. 50, no. 2, pp. 190–205, 2001.

3

Evaluation of the Effect of Irrigation and Fertilization by Drip Fertigation on Tomato Yield and Water Use Efficiency in Greenhouse

Wang Xiukang[1] and Xing Yingying[1,2]

[1]College of Life Science, Yan'an University, Yan'an, Shaanxi 716000, China
[2]Key Laboratory of Agricultural Soil and Water Engineering in Arid and Semiarid Areas of Ministry of Education, Northwest A&F University, Yangling, Shaanxi 712100, China

Correspondence should be addressed to Wang Xiukang; wangxiukang@126.com

Academic Editor: Othmane Merah

The water shortage in China, particularly in Northwest China, is very serious. There is, therefore, great potential for improving the water use efficiency (WUE) in agriculture, particularly in areas where the need for water is greatest. A two-season (2012 and 2013) study evaluated the effects of irrigation and fertilizer rate on tomato (*Lycopersicum esculentum Mill.*, cv. "Jinpeng 10") growth, yield, and WUE. The fertilizer treatment significantly influenced plant height and stem diameter at 23 and 20 days after transplanting in 2012 and 2013, respectively. As individual factors, irrigation and fertilizer significantly affected the leaf expansion rate, but irrigation × fertilizer had no statistically significant effect on the leaf growth rate at 23 days after transplanting in 2012. Dry biomass accumulation was significantly influenced by fertilizer in both years, but there was no significant difference in irrigation treatment in 2012. Our study showed that an increased irrigation level increased the fruit yield of tomatoes and decreased the WUE. The fruit yield and WUE increased with the increased fertilizer rate. WUE was more sensitive to irrigation than to fertilization. An irrigation amount of 151 to 208 mm and a fertilizer amount of 454 to 461 kg·ha^{-1} (nitrogen fertilizer, 213.5–217 kg·ha^{-1}; phosphate fertilizer, 106.7–108 kg·ha^{-1}; and potassium fertilizer, 133.4–135.6 kg·ha^{-1}) were recommended for the drip fertigation of tomatoes in greenhouse.

1. Introduction

Technologies such as drip irrigation and fertigation can improve WUE and decrease salinization while maintaining or increasing yields [1]. Fertigation is an agricultural water management technology that supplies water and fertilizer simultaneously in a drip irrigation system, feeding a crop by injecting soluble fertilizers into water and then transporting them into the root zone [2]. Fertigation, which can improve the efficiency of irrigation water and fertilizer, is a new fertilization method of precision agriculture [3]. In the late 1970s, the use of fertigation technology was widespread in China, particularly in the North and Northwest regions, where water shortage is very serious [4]. In drip fertigation systems, which combine drip irrigation with fertilizer application, the fruit yield of tomato was 20–30% higher in drip fertigation than

in furrow irrigation [5]. It is well known that water and fertilizer are the two main factors limiting vegetable and crop production in arid and semiarid regions [6–8].

Tomato is one of the most popular and widely grown vegetables in the world. The first reason for this is that tomatoes are beneficial to our heath and are good sources of provitamins, β carotene, and vitamin C. The second reason is that tomatoes are particularly rich sources of lycopene, which is a very powerful antioxidant and helps prevent the development of many forms of cancer [9–12]. Hence, this vegetable is gaining importance in both developing and developed countries, and efforts are being made to improve the quality and quantity of tomato production [13–15]. Of course, water supply is important for tomato yield quantity and quality. Increasing the water supply increases fruit yield but significantly reduces the brix, lycopene, and total polyphenol contents of fruits; the

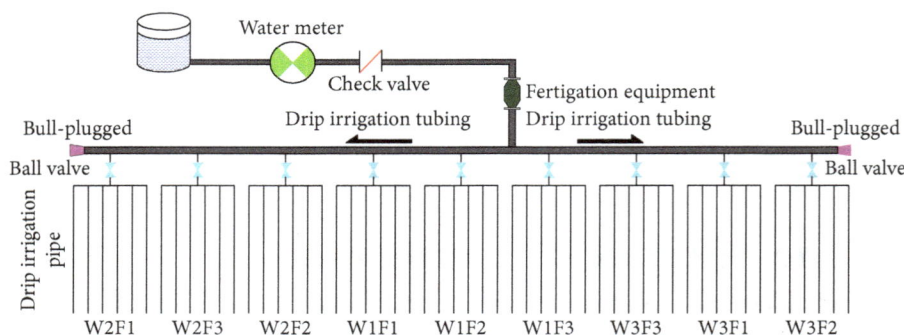

FIGURE 1: The layout of experiment included water sources, water meter, check valve, fertigation equipment, and ball valve and drip irrigation pipe positions of different treatments in greenhouse.

ascorbic acid content is significantly higher under optimum water supply conditions [16–18]. Water stress is one of the most important environmental factors that regulate plant growth and development and limit plant production [19].

Tomato responds well to fertilizer application and is reported to be a heavy feeder of nitrogen (N), phosphorus (P), and potassium (K) fertilizer [5]. It is partially compatible with drip irrigation, which is a very efficient use of water and nutrients [20, 21]. Previous studies have suggested that the N use efficiency and N agronomic efficiency decreased with increases in fertilizer N rate and that the N rate of 271 kg·ha^{-1} produced the highest marketable yield and 265 kg·ha^{-1} produced the optimum economic yield [22, 23]. Nitrogen, phosphorus, and potassium are essential for tomato production [24], and the recommended balanced rates of fertilization include twice as much N as P and K [25]. According to the literature, the application of deficit irrigation at the seedling stage may not significantly influence the total yield of greenhouse tomato [26]. The excessive use of irrigation water leads to low water productivity and deterioration [27]. To maximize tomato water productivity, priority must be given to the efficient use of water, both to improve yields and to control water use by minimizing nonbeneficial water use [28].

Therefore, the development of efficient agricultural water use is not only necessary but also feasible, being critical to improve the fruit yield and WUE. To obtain high yields and maximum profits in commercial tomato production, the optimal management of both fertilizer and water is required. Previous studies have focused on the influence of irrigation amount and fertilizer rate on tomato growth, fruit yield, and quality. Meanwhile, it is necessary to select an optimal combination of irrigation and fertilization to improve agricultural water and fertilizer management practices. Therefore, the aim of this study is to explore the effect of irrigation and fertilization on the growth, yield, and quality of tomato with fertigation by drip irrigation and to make recommendations regarding the strategies for growing greenhouse tomatoes.

2. Materials and Methods

2.1. Experimental Site and Treatments. The tomatoes plant (*Lycopersicum esculentum Mill.*, cv. "Jinpeng 10") was used for

our experiment in the greenhouse, located at the Key Laboratory of Agricultural Soil and Water Conservation Engineering in Arid Areas (34°20′N, 108°04′E, and altitude 521 m), Shaanxi Province, China. The topsoil (0–80 cm) has a bulk density of 1.42 g·cm^{-3}, a pH of 8.12, and a field capacity of 25% cm^3·cm^{-3}, an organic matter content of 13.8 g·kg^{-1}, a total nitrogen content of 0.82 g·kg^{-1}, an available phosphorus content of 13.2 g·kg^{-1}, an available potassium content of 105.8 g·kg^{-1}, and an available nitrogen content of 74.12 mg·kg^{-1}.

In this experiment, nine treatments were designed with three different irrigation levels (W1: 100% ET_c; W2: 75% ET_c; W3: 50% ET_c) and fertilizer levels (F1: N240-$P_2O_5$120-K_2O150 kg·ha^{-1}; F2: N180-$P_2O_5$90-K_2O112.5 kg·ha^{-1}; F3: N120-$P_2O_5$60-K_2O75 kg·ha^{-1}). The experiment was organized using a randomized block design with three replications; each plot was 6 m long, 1.25 m wide, and 22.5 (3 × 6 × 1.25 = 22.5) m^2 in area. Total 9 divided and ridged experimental plots were divided by a water barrier sheet. A mosaic column emitter type drip irrigation tape (Hebei Green Water Conservancy Engineering Co., Ltd., Shijiazhuang, Hebei Province, China) was used in this experimental irrigation system, with an external diameter of 16 mm, drip tape emitter spacing of 30 cm, a head flow of 2 L h^{-1}, and a drip irrigation operating pressure of 0.3 MPa (Figure 1).

2.2. Crop Management, Harvesting, and Measurements. The tomato seedlings were transplanted on 21 Mar 2012 and 31 Mar 2013. The furrow-film mulch was cultivated by the local traditional planting patterns and calendars using tomato ridging in a tube with a two-line spaced layout; the tubes were placed 50 cm apart, with a 45 cm planting distance and 78 plants in each experimental plot. Drip fertigation was performed using a fertilizer of urea (46% N), diammonium phosphate (44% P_2O_5), and potassium chloride (60% K_2O). This fertilizer was applied five times at the recovering stage, the blossoming and bearing fruits stage, the first fruit enlargement period, the second fruit enlargement period, and the third fruit enlargement period, and the fertilization ratio was 1 : 1 : 2 : 2 : 2 for those applications.

The plant height in each treatment was measured every 15 to 20 days after transplanting. The height of 3 randomly selected plants from each experimental unit was measured three times per month from the soil level to the growing point.

Changes in stem diameter were continuously recorded during the treatment period using a shrinkage-type microdisplacement detector (Portable Battery Internal Resistance Tester, JZ-1A, Peking, 2010). All of the measurements were recorded three times and the pattern of response was similar in all.

The number of leaves longer than 20 mm was determined once every 2 weeks, and the maximal leaf width was measured for every leaf. The leaf area was estimated by multiplying the product of leaf length and leaf width by a conversion factor estimated from the destructive sampling.

Plants were harvested in three replicates and separated into roots, leaves, fruits, and stem. The plant parts were dried in an open-air draught oven at 75°C for 72 h to estimate the dry weight.

Ripe tomatoes were harvested and fresh total yield and total number of tomatoes from all of the plants in each plot were determined. The fruit yield was measured throughout the crop. Fruits were harvested twice a week for a period of 9 weeks and were separated into marketable and total yields.

2.3. Irrigation Management.

Irrigation treatments were initiated using the surface drip irrigation system during transplanting, and the irrigation amount was 40 mm. Irrigation was applied using a subsurface drip system based on the daily crop evapotranspiration (ET_c), which was calculated as a product of the reference evapotranspiration (ET_0) and the stage-specific crop coefficients (K_c).

The FAO 56 Penman-Monteith method is recommended as the standard method for ET_0 estimation [29, 30]. Fernández et al. reported the FAO 56 Penman-Monteith equation with a fixed aerodynamic resistance of 295 s m^{-1} can better estimate daily ET_0 in greenhouse [31]:

$$ET_0 = \frac{0.48\Delta\left(R_n - G\right) + \gamma\left(628/\left(T + 273\right)\right)\left(e_s - e_a\right)}{\Delta + 628\gamma}, \quad (1)$$

where R_n is the net radiation (MJ m^{-2} d^{-1}), G is the soil heat flux (MJ m^{-2} d^{-1}), Δ is the slope of the saturated vapour pressure curve (kPa °C^{-1}), γ is the psychometric constant (kPa °C^{-1}), e_s is saturation vapour pressure (kPa), e_a is actual vapour pressure (kPa), and $e_s - e_a$ (VPD) is the vapour pressure deficit (kPa). The calculation procedures of parameters $R_n, G, e_s, e_a, \Delta, \gamma$, and T were described in FAO 56 [31, 32]. The average daily environmental condition at different growth stages of tomato inside the greenhouse and the seasonal variation of daily ET_0 are calculated using (1) (Figure 2).

The K_c values were as follows: $K_{c\,ini} = 0.5$, $K_{c\,mid} = 0.85$, and $K_{c\,end} = 0.6$ [32]. The irrigation amounts of the W1, W2, and W3 treatments were, respectively, 262.00, 206.50, and 151.00 mm in 2012 and 279.80, 219.85, and 159.85 mm in 2013. The irrigation frequency and total amount of water applied during the full irrigation treatment are given in Figure 3.

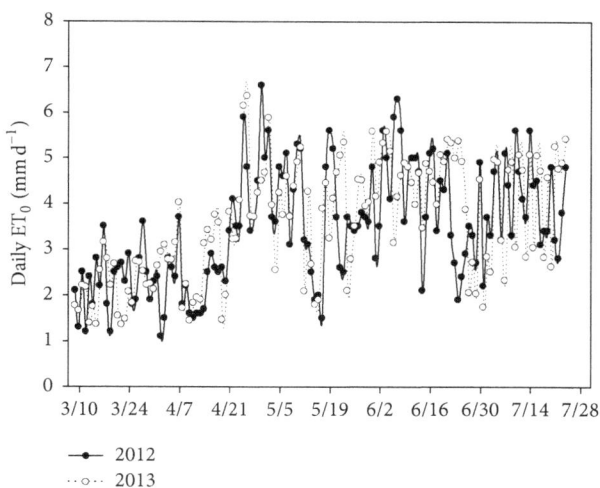

FIGURE 2: Daily variation of the reference evapotranspiration (ET_0) against days for the tomatoes growing seasons of 2012 and 2013.

The water use efficiency (WUE) was determined using the following equation [33, 34]:

$$WUE = \frac{Y}{ET_c} \times 100, \quad (2)$$

where WUE is measured in kg·m^{-3}, Y is the total fruit yield (t·ha^{-1}), and ET_c is the crop water consumption (mm).

2.4. Statistical Analysis.

Analysis of variance was conducted on the plant height, stem diameter, dry biomass accumulation, and distribution in different organs using a two-way analysis of variance (SAS GLM procedure version 9.2, SAS Institute Ltd., North Carolina, USA). Duncan's multiple range tests were considered significant when $p < 0.05$.

3. Results

3.1. Plant Height, Stem Diameter, and Leaf Growth Rate.

The effects of irrigation amount and fertilizer rate on plant height at the whole growth stages are shown in Table 1. In 2012, the results show that the single factors of irrigation amount or fertilizer application rate very significantly affected plant height at 23 days after transplanting, and the interaction between irrigation and fertilization had an obvious effect. The highest plant height was 40.8 cm in the W2F1 treatment, which was significantly higher than that of the other plants. At 37 days after transplanting, the average added values of plant height in W3 (40.8 cm) were 6.9% and 10.3% higher than those in the W1 and W2 treatments in 2012. During the recovering stage, the rate of plant height increase was faster, and, with the advancement of the blossoming and bearing fruits stage, the rate of increase decreased by 53 days after transplanting. The average plant height in W3 was higher than that in the W1 treatment, but there was no significant plant height difference among the irrigation treatments 70 days after transplanting. In addition, irrigation × fertilizer had no statistically significant effect on plant height.

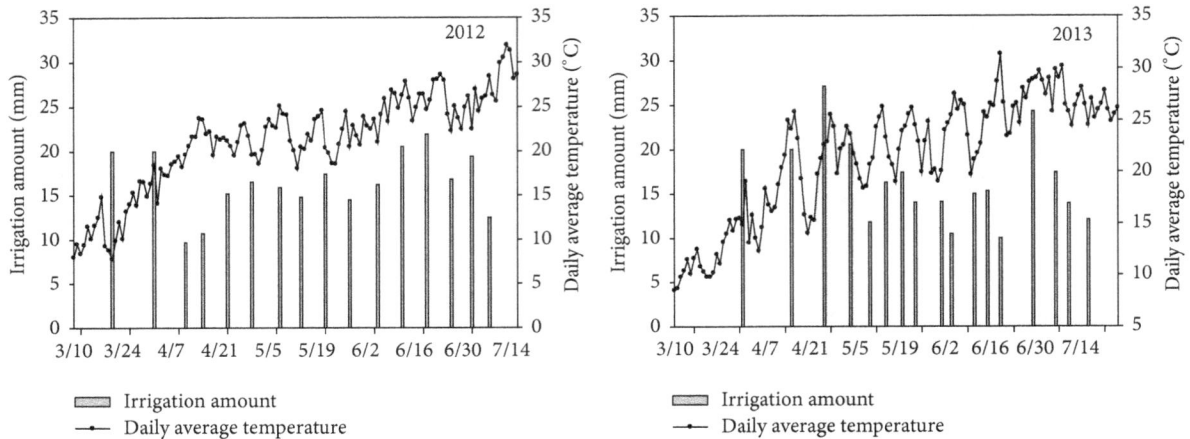

FIGURE 3: In tomato growing season, the distribution of daily average temperature, irrigation interval, and numbers were recorded in the study years.

TABLE 1: The effects of irrigation amount and fertilizer rate on plant height of tomato (cm).

Treatments	2012				2013			
	23 D	37 D	53 D	70 D	20 D	40 D	60 D	80 D
W1F1	35^{cd}	71.2^{cd}	102^d	125.8^b	23.9^{bc}	70.9^{ab}	93.9^{bcd}	112.5^{ab}
W1F2	35.7^{cd}	72.5^{bcd}	109^{abc}	134.1^{ab}	24.1^{bc}	68.6^{abc}	101.5^a	117.4^a
W1F3	34.3^d	70.3^{cd}	105^{bcd}	130^{ab}	23.2^{bc}	67.8^{abc}	88^{cde}	112^{ab}
W2F1	40.8^a	76.5^{ab}	110.5^{ab}	140.3^a	29.7^a	68.4^{abc}	94.1^{bcd}	111.5^{ab}
W2F2	37.2^{bc}	72.5^{bcd}	103.5^{cd}	131.5^{ab}	30^a	74.1^a	87.2^{de}	110.4^{ab}
W2F3	34.3^d	68.3^d	100.5^d	127.7^{ab}	26.1^b	62^{cd}	86.2^e	110.2^{ab}
W3F1	38^b	73^{abcd}	113.5^a	135.1^{ab}	23.1^{bc}	63.8^{bcd}	97.5^{ab}	113.3^{ab}
W3F2	35.5^{cd}	78^a	115^a	136.9^{ab}	24.6^{bc}	64.3^{bcd}	95.2^{abc}	113^{ab}
W3F3	34.1^d	73.8^{abc}	111^{ab}	132.1^{ab}	22.8^c	59.7^d	92.1^{bcde}	107^b
				p value of significance test				
Irrigation	**	NS	***	NS	***	**	*	NS
Fertilizer	***	*	NS	NS	*	*	*	NS
Irrigation × fertilizer	*	NS	*	NS	NS	NS	NS	NS

D is the days after transplanting and columns with the same letter represent values that are significant at the 5% probability level. "***" means $p < 0.001$, "**" means $0.001 < p < 0.01$, "*" means $0.01 < p < 0.05$, and "NS" means $p > 0.05$.

The plant stem plays a very important role in plant anchorage and in the movement and transport of water, solutes, and nutrients. Importantly, the stem functions in photosynthesis and nutrient storage. The effects of irrigation and fertilizer on tomato stem diameter at the whole growth stages are shown in Table 2. At 23 days after transplanting, the highest stem diameter was 8.9 mm in 2012 and 6.8 mm in 2013. The fertilizer treatment very significantly influenced the tomato stem diameter, irrigation × fertilizer significantly influenced the stem diameter, and irrigation treatment had no statistically significant effect in 2012. In 2013, the single factors of irrigation or fertilizer very significantly ($p < 0.01$) affected the stem diameter, but irrigation × fertilizer had no statistically significant effect. In addition, the rate of stem diameter increase decreased obviously at 37 days after transplanting, and the added values of stem diameter ranged from 0.9 to 2.2 mm at the blossoming and bearing fruits stage. The average stem diameter in F3 was significantly lower than that

in the F1 and F2 treatments, and the fertilizer treatment very significantly influenced the tomato stem diameter. By 53 days after transplanting, the rate of stem diameter increase had undergone a steady decline, and the added value ranged from 0.6 to 1.1 mm. The influence of stem diameter on fertilization was greater than that on irrigation throughout the entire stage.

The major function of leaves is to take in carbon dioxide for photosynthesis, the process of converting light energy into chemical energy. The effects of irrigation and fertilization on the leaf growth rate at the whole growth stages are shown in Figure 4. The overall pattern of change in the leaf growth rate was represented by positive-negative and single-peak curves in both years. The single factors of irrigation or fertilizer very significantly affected the leaf expansion rate, but irrigation × fertilizer had no statistically significant effect on the leaf growth rate at 23 days after transplanting. The highest leaf expansion rate was 4.5 cm$^2 \cdot$leaf$^{-1} \cdot$day^{-1} at 23–37 days after

TABLE 2: The effects of irrigation and fertilizer on stem diameter tomato (mm).

Treatment	2012				2013			
	23 D	37 D	53 D	70 D	20 D	40 D	60 D	80 D
W1F1	8.46^{ab}	10.66^a	10.98^a	12.28^{ab}	7.85^a	8.74^{bc}	10.95^{ab}	12.18^{ab}
W1F2	7.66^{cd}	9.58^{bc}	10.1^{abc}	11.3^{bcd}	7.28^b	9.14^{ab}	11.68^a	12.88^a
W1F3	7.45^d	8.57^{cde}	9.6^{bcd}	11.01^{cd}	6.84^{bc}	7.8^d	10.55^{ab}	11.26^{bc}
W2F1	8.9^a	10.15^{ab}	10.74^{ab}	12.47^a	7.29^b	8.89^{bc}	9.91^{bc}	11.33^{bc}
W2F2	8.13^{bcd}	9.16^{bcde}	10.11^{abc}	11.6^{abc}	7.09^{bc}	9.59^a	11.37^a	11.43^{bc}
W2F3	6.49^e	8.44^{de}	9.26^{cd}	10.39^d	6.8^{bc}	7.37^d	9.01^c	10.45^c
W3F1	8.39^{abc}	9.33^{bcd}	9.98^{abcd}	11.86^{abc}	6.62^{cd}	7.87^d	9.07^c	11.78^b
W3F2	8.03^{bcd}	8.93^{cde}	9.53^{cd}	10.84^{cd}	6.24^{de}	8.44^c	9.43^c	10.54^c
W3F3	5.94^e	8.21^e	8.91^d	10.29^d	5.96^e	6.65^e	7.93^d	9.25^d
				p value of significance test				
Irrigation	NS	*	NS	NS	***	***	***	***
Fertilizer	***	***	**	***	**	***	***	***
Irrigation × fertilizer	**	NS	NS	NS	NS	NS	NS	NS

D is the days after transplanting and columns with the same letter represent values that are significant at the 5% probability level. "$***$" means $p < 0.001$, "$**$" means $0.001 < p < 0.01$, "$*$" means $0.01 < p < 0.05$, and "NS" means $p > 0.05$.

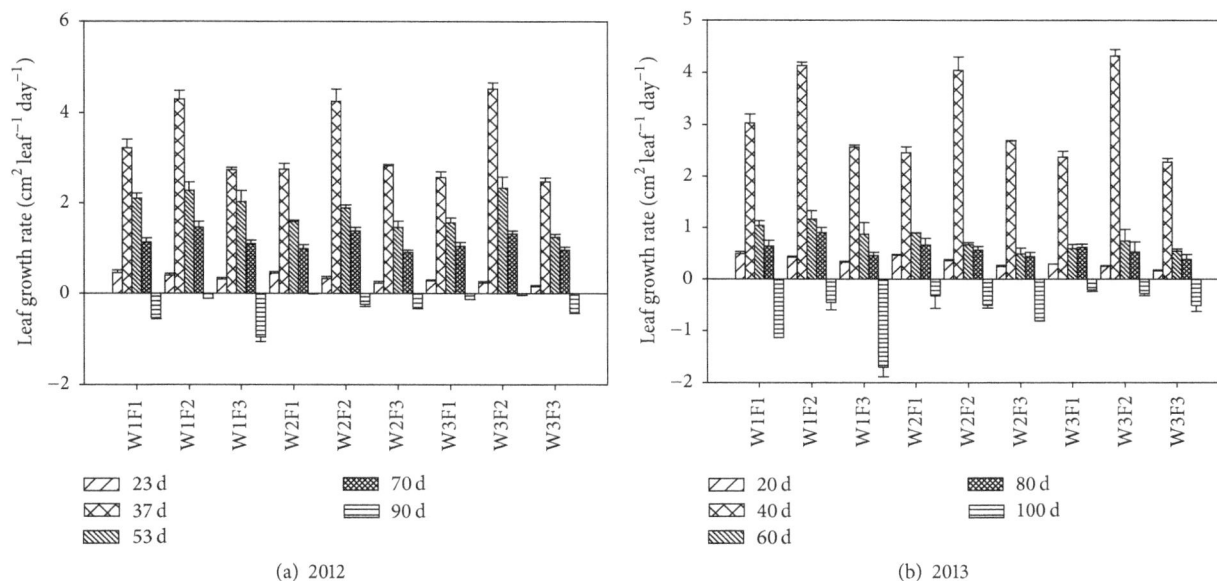

FIGURE 4: The effects of irrigation and fertilizer on tomato leaf growth rate.

transplanting. The highest average leaf expansion rate was produced in the F2 treatment, and the fertilizer amount very significantly affected the leaf expansion rate in 2012. The leaf area continued to increase, but the rate of increase decreased in 37–53 days after transplanting.

3.2. Dry Biomass Accumulation. The effects of irrigation and fertilizer on tomato dry biomass accumulation and the distribution in different organs are shown in Table 3. The highest dry biomass accumulation in the W1F1 treatment was 12 t·ha^{-1}, which was significantly higher than that in the other treatments; the fertilizer treatment significantly influenced dry biomass accumulation, but the irrigation treatment had no statistically significant effect in either year. The average

root dry biomass in W1 (221.9 kg·ha^{-1}) was 7% and 20.4% higher than that in the W2 and W3 treatments and the same as the stem and fruit dry biomass. There was no significant difference between the W1 and W2 treatments, and the average dry biomass accumulation in W1 was 6% and 6% higher than W2 in 2012 and 2013, respectively. With the same irrigation amount, the dry biomass accumulation in the F2 treatment was higher than that in the F1 and F3 treatments.

3.3. Tomato Yield. The interactions between irrigation and fertilizer treatments were important for tomato yield, and the single factors of irrigation or fertilizer very significantly ($p < 0.01$) affected the fruit yield in two consecutive years (Figure 5). The highest fruit yield was 96.7 t·ha^{-1} in the W1F1

TABLE 3: Effects of irrigation and fertilizer on tomato dry biomass accumulation and distribution in different organs.

Year	Treatment	Dry biomass accumulation (kg·hm^{-2})					Distribution in different organs (%)			
		Fruit	Stem	Leaf	Root	Total	Fruit	Stem	Leaf	Root
2012	W1F1	6219[a]	2922[a]	2635[a]	247[a]	12023[a]	51.7[bcd]	24.3[ab]	21.9[ab]	2.1[a]
	W1F2	5828[b]	2785[ab]	2552[a]	223[b]	11389[b]	51.2[d]	24.4[ab]	22.4[ab]	2[a]
	W1F3	5100[d]	2660[ab]	2172[b]	195[c]	10127[c]	50.4[d]	26.3[a]	21.4[bc]	1.9[ab]
	W2F1	5615[bc]	2504[bc]	2573[a]	230[b]	10922[b]	51.4[cd]	22.9[bc]	23.6[a]	2.1[a]
	W2F2	5475[c]	2221[cd]	2309[b]	215[b]	10221[c]	53.6[bc]	21.7[c]	22.6[ab]	2.1[a]
	W2F3	5030d[e]	1993[de]	1720[c]	174[d]	8917[e]	56.4[a]	22.3[bc]	19.3[de]	2[a]
	W3F1	5119[d]	2231[cd]	2247[b]	197[c]	9794[cd]	52.3[bcd]	22.8[bc]	23[ab]	2[a]
	W3F2	4984[de]	2221[cd]	1849[c]	194[c]	9247[de]	53.9[b]	24[abc]	20[cd]	2.1[a]
	W3F3	4722[e]	1876[e]	1460[d]	140[e]	8198[f]	57.6[a]	22.9[bc]	17.8[e]	1.7[b]
					p value of significance test					
	Irrigation	**	*	*	**	**	**	*	*	**
	Fertilizer	***	NS	*	*	*	***	NS	*	*
	Irrigation × fertilizer	***	NS	*	NS	NS	***	NS	*	NS
2013	W1F1	5100[a]	1818[c]	1973[a]	254[c]	9145[a]	55.8[abc]	19.9[d]	21.6[bcd]	2.8[b]
	W1F2	4781[ab]	2106[a]	2031[a]	330[a]	9248[a]	51.7[c]	22.8[abc]	22[bc]	3.6[a]
	W1F3	4090[bc]	1663[d]	1965[a]	209[d]	7927[bc]	51.5[c]	21[bcd]	24.8[a]	2.6[b]
	W2F1	4461[abc]	1477[e]	1757[b]	217[d]	7913[bc]	56.3[ab]	18.7[d]	22.2[b]	2.7[b]
	W2F2	4300[bc]	1960[b]	1809[b]	306[b]	8375[b]	51.3[c]	23.4[ab]	21.6[bcd]	3.7[a]
	W2F3	3707[c]	1772[cd]	1394[d]	193[d]	7066[d]	52.4[bc]	25.1[a]	19.7[d]	2.7[b]
	W3F1	4087[bc]	1692[d]	1498[cd]	197[d]	7473[cd]	54.6[bc]	22.7[abc]	20.1[cd]	2.6[b]
	W3F2	3945[c]	1502[e]	1607[c]	151[e]	7206[cd]	54.7[bc]	20.9[cd]	22.3[b]	2.1[c]
	W3F3	3740[c]	1168[f]	1243[e]	139[e]	6290[e]	59.4[a]	18.6[d]	19.8[d]	2.2[c]
					p value of significance test					
	Irrigation	***	*	*	**	**	***	*	*	**
	Fertilizer	***	NS	*	*	*	***	NS	*	*
	Irrigation × fertilizer	***	NS	*	NS	NS	***	NS	*	NS

Columns with the same letter represent values that are significant at the 5% probability level. "∗∗∗" means $p < 0.001$, "∗∗" means $0.001 < p < 0.01$, "∗" means $0.01 < p < 0.05$, and "NS" means $p > 0.05$.

treatment, which was 9.7% and 17.7% higher than that in W2F1 and W3F1 in 2012. The same result was produced in 2013, and the highest fruit yield was 97.1 t·ha^{-1} in the W1F1 treatment, which was 12.5% and 19.9% higher than that in W2F1 and W3F1. In both years, the results indicated that the increased irrigation level and fertilizer rate increased the fruit yield of tomatoes. The mean fruit yield of the W1 treatment was 88.9 t·ha^{-1} in 2012, which was 6% and 13.5% higher than that in W2 and W3. The mean fruit yield in the F1 treatment was 87.9 t·ha^{-1} in 2012, which was 3.9% and 12.4% higher than that in F2 and F3. The results indicated that fruit yield had a slightly higher sensitivity to treatment with irrigation than to that with fertilization.

3.4. Water Use Efficiency.

The effects of irrigation and fertilizer on the WUE are shown in Figure 6. The irrigation treatment significantly affected the WUE. The results showed a significant negative correlation between WUE and irrigation amount. The highest WUE was obtained in the W3F1 treatment and was 45 kg·m^{-3} and 47.7 kg·m^{-3} in 2012 and 2013,

respectively. When the irrigation amount decreased, the WUE increased. The average WUE in the W3 treatment was 27.3% and 18.7% higher than that in the W1 and W2 treatments in 2012, and the same result was observed in 2013. There was a positive correlation between the WUE and fertilizer amount; the average WUE in the F3 treatments was 14.8% and 10.7% higher than that in the F1 and F2 treatments in 2012. The results indicated that the WUE was more sensitive to irrigation than fertilizer.

3.5. Recommended Levels of Irrigation and Fertilization.

The regression model was used to predict the effect of an unknown dependent variable on the fruit yield and WUE, given the values of the independent variables of irrigation amount and fertilizer level. The fertilizer and irrigation supply affected the fruit yield and WUE, with a very significant interaction between them in both years. In the two successive growing seasons, the extreme calculation results showed that the fruit yield peaked at the maximal irrigation amount, while the WUE peaked at the minimal irrigation amount.

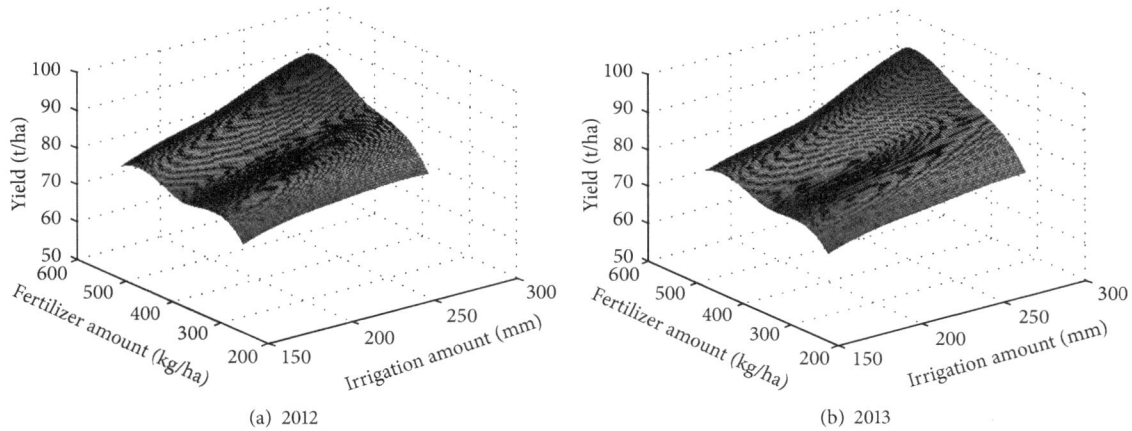

(a) 2012   (b) 2013

FIGURE 5: The effects of irrigation and fertilizer on tomato fruit yield in 2012 and 2013.

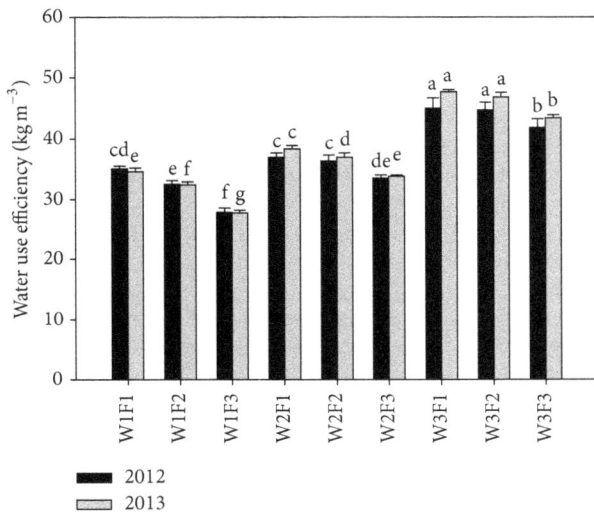

FIGURE 6: The effects of irrigation and fertilizer on water use efficiency in 2012 and 2013; the same letter represents values that are significant at the 5% probability level.

Therefore, the tomato yield and WUE cannot simultaneously reach their maxima.

When the WUE was maximal, irrigation was minimal, and the fertilizer amount was close to maximal. Therefore, it is necessary to perform further studies on the input of irrigation and fertilizer, which affect fruit yield and WUE.

In our study, a multiple regression analysis was used to develop a hypothesis in which fruit yield and WUE were equally important ($\lambda_1 = \lambda_2 = 0.5$). The relationships among irrigation, fertilizer, and the weighted fruit yield and WUE were determined using the following equations:

$$2012: Y = 1.18 - 4.83 \times 10^{-3}W + 7.62 \times 10^{-4}F + 6.80$$
$$\times 10^{-6}W^2 - 1.42 \times 10^{-6}F^2 + 3.62$$
$$\times 10^{-6}WF$$

$$2013: Y = 1.33 - 7.11 \times 10^{-3}W + 1.18 \times 10^{-3}F + 1.21$$
$$\times 10^{-5}W^2 - 1.75 \times 10^{-6}F^2 + 3.02$$
$$\times 10^{-6}WF,$$

$$(3)$$

where Y is the optimization value considering the fruit yield and WUE (t·ha^{-1}), W is the irrigation amount (mm), and F is the fertilizer amount (kg·ha^{-1}).

The optimal value of the target function was calculated by MATLAB. The irrigation amount and fertilizer amount were 151 mm and 453.6 kg·ha^{-1} (nitrogen, phosphorus, and potassium fertilizers were 213.5, 106.7, and 133.4 kg·ha^{-1}, resp.) in 2012, respectively. The irrigation amount and fertilizer amount were 207.8 mm and 461.08 kg·ha^{-1} (nitrogen, phosphorus, and potassic fertilizers were 217, 108, and 135.6 kg·ha^{-1}, resp.) in 2013, respectively.

4. Discussion

The relationships between the growth indexes and the irrigation amount and fertilizer level were statistically analyzed, and the positive correlation was significant. In two consecutive years, the plant height and leaf growth rate were higher in W2F1 than in the other treatments at 23 days after transplanting, which may be due to water consumption. The results were the same as those of Zhu et al. [35], who reported that higher levels of irrigation can inhibit plant height increase. It is of great importance to study the growth and soil water required for tomato dry matter accumulation under irrigation and fertilization strategies with drip irrigation, as too much water can cause excessive vegetative growth; the most important thing is the proper proportion of irrigation and fertilizer [36]. There was a positive correlation between dry matter accumulation and fertilizer amount, and fertilizer treatment was more sensitive to dry matter accumulation than irrigation treatment. During the growth period, the leaf growth rate increased rapidly at first, and then the rate of increase decreased and the growth plateaued. There was no significant difference between the irrigation treatment and leaf growth rate, and the same result was obtained as with dry matter accumulation. Crop evapotranspiration is closely related to crop growth and water consumption during crop growth, which is irreversible. The result was same as that of Wang et al. [37], the tomato dry matter accumulation was mainly due to irrigation and fertilization, and there was no significant difference in the dry matter accumulation at the whole growth stage. Fertilization was more sensitive to fruit yield than irrigation and the irrigation treatment was more sensitive than the interaction between irrigation and fertilization [38].

The results showed that irrigation and fertilization had significant effects on tomato yield, and the effects of the interaction between irrigation and fertilizer were very significant. The same results were obtained in greenhouse that there was a significant difference in the tomato yield in the irrigation and fertilization treatments, due to irrigation, fertilizer, and the interaction between the two factors [39]. However, in this experiment, the interaction between irrigation and fertilization was more sensitive than the single factors of fertilization or irrigation; the reason for this result may be that the effect of the interaction between water and fertilizer was obvious and could be used in further studies on the effect of the interaction between irrigation and fertilization on tomato growth.

The results showed that different irrigation and fertilization supplies significantly affected the WUE and that the effect of irrigation treatment on the WUE was significant. The influence of the interaction between irrigation and fertilization on the WUE was not significant; however, the effect of irrigation treatment on the WUE was greater than the effect of fertilization. Javanmardi and Kubota [40] and Maria do Rosário et al. [41] reported that fertilizer improved the WUE; under the same irrigation level, fertilization could effectively improve the WUE. This result is consistent with our experimental results. There are two reasons explaining why fertilization improved the WUE. One reason might be that fertilization can promote tomato root growth and development, thereby improving the root system's capacity to absorb water and nutrients [42]. Another reason may be that tomato water consumption improved during growth, causing the roots near soil water movement to increase the efficiency of soil water absorption and to further improve the WUE of the soil.

5. Conclusions

Generally, it is difficult to obtain the maximal WUE and the maximum yield simultaneously. Reducing the amount of irrigation water will result in higher WUE; based on this characteristic, the highest WUE and fruit yield cannot occur at the same time. According to this characteristic, tomatoes require fertilizer and water at the same time. The tomato yield and WUE are equally important and the tomato yield and WUE coefficients are each 0.5. An irrigation amount of 151.1 to 207.8 mm and a fertilizer amount of 453.6 to 461.1 kg·ha^{-1} for greenhouse tomato surface drip fertigation are recommended (nitrogen fertilizer, 213.5–217 kg·ha^{-1}; phosphate

fertilizer, 106.7–108 kg·ha^{-1}; and potassium fertilizer, 133.4–135.6 kg·ha^{-1}).

Competing Interests

The authors declare that there is no conflict of interests regarding the publication of this paper.

Acknowledgments

This research was supported by Specialized Research Fund for the Doctoral Program of Yan'an University (205040119, 205040123), Shaanxi Province High-Level University Construction Special Fund Projects of Ecology (2012SXTC03), and Special Scientific Research Project in Shaanxi Province Department of Education (16JK1853).

References

[1] S. Phuntsho, H. K. Shon, S. Hong, S. Lee, and S. Vigneswaran, "A novel low energy fertilizer driven forward osmosis desalination for direct fertigation: evaluating the performance of fertilizer draw solutions," *Journal of Membrane Science*, vol. 375, no. 1-2, pp. 172–181, 2011.

[2] J. Hagin and A. Lowengart, "Fertigation for minimizing environmental pollution by fertilizers," *Fertilizer Research*, vol. 43, no. 1–3, pp. 5–7, 1996.

[3] T. A. Howell, "Enhancing water use efficiency in irrigated agriculture," *Agronomy Journal*, vol. 93, no. 2, pp. 281–289, 2001.

[4] J. Li, J. Zhang, and L. Ren, "Water and nitrogen distribution as affected by fertigation of ammonium nitrate from a point source," *Irrigation Science*, vol. 22, no. 1, pp. 19–30, 2003.

[5] S. S. Hebbar, B. K. Ramachandrappa, H. V. Nanjappa, and M. Prabhakar, "Studies on NPK drip fertigation in field grown tomato (*Lycopersicon esculentum* Mill.)," *European Journal of Agronomy*, vol. 21, no. 1, pp. 117–127, 2004.

[6] Z. Hou, P. Li, B. Li, J. Gong, and Y. Wang, "Effects of fertigation scheme on N uptake and N use efficiency in cotton," *Plant and Soil*, vol. 290, no. 1-2, pp. 115–126, 2007.

[7] K. O. Burkey, F. L. Booker, W. A. Pursley, and A. S. Heagle, "Elevated carbon dioxide and ozone effects on peanut: II. Seed yield and quality," *Crop Science*, vol. 47, no. 4, pp. 1488–1497, 2007.

[8] E. Baldi, G. Marcolini, M. Quartieri, G. Sorrenti, and M. Toselli, "Effect of organic fertilization on nutrient concentration and accumulation in nectarine (*Prunus persica* var. *nucipersica*) trees: the effect of rate of application," *Scientia Horticulturae*, vol. 179, pp. 174–179, 2014.

[9] G. Evgenidis, E. Traka-Mavrona, and M. Koutsika-Sotiriou, "Principal component and cluster analysis as a tool in the assessment of tomato hybrids and cultivars," *International Journal of Agronomy*, vol. 2011, Article ID 697879, 7 pages, 2011.

[10] I. K. Arah, H. Amaglo, E. K. Kumah, and H. Ofori, "Preharvest and postharvest factors affecting the quality and shelf life of harvested tomatoes: a mini review," *International Journal of Agronomy*, vol. 2015, Article ID 478041, 6 pages, 2015.

[11] W. You and A. V. Barker, "Effects of soil-applied glufosinate-ammonium on tomato plant growth and ammonium accumulation," *Communications in Soil Science and Plant Analysis*, vol. 35, no. 13-14, pp. 1945–1955, 2004.

[12] S. De Pascale, A. Maggio, V. Fogliano, P. Ambrosino, and A. Ritieni, "Irrigation with saline water improves carotenoids content and antioxidant activity of tomato," *The Journal of Horticultural Science and Biotechnology*, vol. 76, no. 4, pp. 447–453, 2001.

[13] G. Farré, G. Sanahuja, S. Naqvi et al., "Travel advice on the road to carotenoids in plants," *Plant Science*, vol. 179, no. 1-2, pp. 28–48, 2010.

[14] H. Zhao, Y.-C. Xiong, F.-M. Li et al., "Plastic film mulch for half growing-season maximized WUE and yield of potato via moisture-temperature improvement in a semi-arid agroecosystem," *Agricultural Water Management*, vol. 104, pp. 68–78, 2012.

[15] G. Mahajan and K. G. Singh, "Response of Greenhouse tomato to irrigation and fertigation," *Agricultural Water Management*, vol. 84, no. 1-2, pp. 202–206, 2006.

[16] F. Favati, S. Lovelli, F. Galgano, V. Miccolis, T. Di Tommaso, and V. Candido, "Processing tomato quality as affected by irrigation scheduling," *Scientia Horticulturae*, vol. 122, no. 4, pp. 562–571, 2009.

[17] J. P. Mitchell, C. Shennan, S. R. Grattan, and D. M. May, "Tomato fruit yields and quality under water deficit and salinity," *Journal of the American Society for Horticultural Science*, vol. 116, no. 2, pp. 215–221, 1991.

[18] V. Dewanto, X. Wu, K. K. Adom, and R. H. Liu, "Thermal processing enhances the nutritional value of tomatoes by increasing total antioxidant activity," *Journal of Agricultural and Food Chemistry*, vol. 50, no. 10, pp. 3010–3014, 2002.

[19] H. Gulen and A. Eris, "Effect of heat stress on peroxidase activity and total protein content in strawberry plants," *Plant Science*, vol. 166, no. 3, pp. 739–744, 2004.

[20] K. Liu, T. Q. Zhang, C. S. Tan, and T. Astatkie, "Responses of fruit yield and quality of processing tomato to drip-irrigation and fertilizers phosphorus and potassium," *Agronomy Journal*, vol. 103, no. 5, pp. 1339–1345, 2011.

[21] A. S. Isah, E. B. Amans, E. C. Odion, and A. A. Yusuf, "Growth rate and yield of two tomato varieties (*Lycopersicon esculentum* mill) under green manure and NPK fertilizer rate Samaru northern guinea savanna," *International Journal of Agronomy*, vol. 2014, Article ID 932759, 8 pages, 2014.

[22] K. Liu, T. Q. Zhang, and C. S. Tan, "Processing tomato phosphorus utilization and post-harvest soil profile phosphorus as affected by phosphorus and potassium additions and drip irrigation," *Canadian Journal of Soil Science*, vol. 91, no. 3, pp. 417–425, 2011.

[23] T. Q. Zhang, C. S. Tan, K. Liu, C. F. Drury, A. P. Papadopoulos, and J. Warner, "Yield and economic assessments of fertilizer nitrogen and phosphorus for processing tomato with drip fertigation," *Agronomy Journal*, vol. 102, no. 2, pp. 774–780, 2010.

[24] H. Kuşçu, A. Turhan, and A. O. Demir, "The response of processing tomato to deficit irrigation at various phenological stages in a sub-humid environment," *Agricultural Water Management*, vol. 133, pp. 92–103, 2014.

[25] F. He, Q. Chen, R. Jiang, X. Chen, and F. Zhang, "Yield and nitrogen balance of greenhouse tomato (*Lycopersicum esculentum* Mill.) with conventional and site-specific nitrogen management in Northern China," *Nutrient Cycling in Agroecosystems*, vol. 77, no. 1, pp. 1–14, 2007.

[26] J. Chen, S. Kang, T. Du, R. Qiu, P. Guo, and R. Chen, "Quantitative response of greenhouse tomato yield and quality to water deficit at different growth stages," *Agricultural Water Management*, vol. 129, pp. 152–162, 2013.

[27] J. Zheng, G. Huang, D. Jia et al., "Responses of drip irrigated tomato (*Solanum lycopersicum* L.) yield, quality and water productivity to various soil matric potential thresholds in an arid region of Northwest China," *Agricultural Water Management*, vol. 129, pp. 181–193, 2013.

[28] C. Patanè and S. L. Cosentino, "Effects of soil water deficit on yield and quality of processing tomato under a Mediterranean climate," *Agricultural Water Management*, vol. 97, no. 1, pp. 131–138, 2010.

[29] M. T. Castellanos, M. J. Cabello, M. C. Cartagena, A. M. Tarquis, A. Arce, and F. Ribas, "Nitrogen uptake dynamics, yield and quality as influenced by nitrogen fertilization in 'Piel de sapo' melon," *Spanish Journal of Agricultural Research*, vol. 10, no. 3, pp. 756–767, 2012.

[30] R. Qiu, J. Song, T. Du et al., "Response of evapotranspiration and yield to planting density of solar greenhouse grown tomato in northwest China," *Agricultural Water Management*, vol. 130, pp. 44–51, 2013.

[31] M. D. Fernández, S. Bonachela, F. Orgaz et al., "Erratum to: measurement and estimation of plastic greenhouse reference evapotranspiration in a Mediterranean climate," *Irrigation Science*, vol. 29, no. 1, pp. 91–92, 2011.

[32] R. G. Allen, L. S. Pereira, D. Raes, and M. Smith, *Crop Evapotranspiration—Guide-lines for Computing Crop Water Requirements*, FAO Irrigation and drainage paper no. 56, FAO, Rome, Italy, 1998.

[33] Z. Wang, Z. Liu, Z. Zhang, and X. Liu, "Subsurface drip irrigation scheduling for cucumber (*Cucumis sativus* L.) grown in solar greenhouse based on 20 cm standard pan evaporation in Northeast China," *Scientia Horticulturae*, vol. 123, no. 1, pp. 51–57, 2009.

[34] F. Wang, S. Kang, T. Du, F. Li, and R. Qiu, "Determination of comprehensive quality index for tomato and its response to different irrigation treatments," *Agricultural Water Management*, vol. 98, no. 8, pp. 1228–1238, 2011.

[35] G.-L. Zhu, Y.-Y. Hu, and Q.-R. Wang, "Nitrogen removal performance of anaerobic ammonia oxidation co-culture immobilized in different gel carriers," *Water Science & Technology*, vol. 59, no. 12, pp. 2379–2386, 2009.

[36] W. Xiukang, L. Zhanbin, and X. Yingying, "Effects of mulching and nitrogen on soil temperature, water content, nitrate-N content and maize yield in the Loess Plateau of China," *Agricultural Water Management*, vol. 161, pp. 53–64, 2015.

[37] C. Wang, W. Liu, Q. Li et al., "Effects of different irrigation and nitrogen regimes on root growth and its correlation with above-ground plant parts in high-yielding wheat under field conditions," *Field Crops Research*, vol. 165, pp. 138–149, 2014.

[38] R. B. Singandhupe, G. G. S. N. Rao, N. G. Patil, and P. S. Brahmanand, "Fertigation studies and irrigation scheduling in drip irrigation system in tomato crop (*Lycopersicon esculentum* L.)," *European Journal of Agronomy*, vol. 19, no. 2, pp. 327–340, 2003.

[39] A. Ozbahce and A. F. Tari, "Effects of different emitter space and water stress on yield and quality of processing tomato under semi-arid climate conditions," *Agricultural Water Management*, vol. 97, no. 9, pp. 1405–1410, 2010.

[40] J. Javanmardi and C. Kubota, "Variation of lycopene, antioxidant activity, total soluble solids and weight loss of tomato during postharvest storage," *Postharvest Biology and Technology*, vol. 41, no. 2, pp. 151–155, 2006.

[41] G. O. Maria do Rosário, A. M. Calado, and C. A. M. Portas, "Tomato root distribution under drip irrigation," *Journal of the American Society for Horticultural science*, vol. 121, no. 4, pp. 644–648, 1996.

[42] A. M. K. Nassar, S. Kubow, and D. J. Donnelly, "High-throughput screening of sensory and nutritional characteristics for cultivar selection in commercial hydroponic greenhouse crop production," *International Journal of Agronomy*, vol. 2015, Article ID 376417, 28 pages, 2015.

4

Impact of Forage Fertilization with Urea and Composted Cattle Manure on Soil Fertility in Sandy Soils of South-Central Vietnam

Keenan C. McRoberts,[1] Quirine M. Ketterings,[2] David Parsons,[3] Tran Thanh Hai,[4] Nguyen Hai Quan,[4] Nguyen Xuan Ba,[4] Charles F. Nicholson,[5] and Debbie J. R. Cherney[6]

[1]Department of Animal Science, Cornell University, 149 Morrison Hall, Ithaca, NY 14853, USA
[2]Department of Animal Science, Cornell Nutrient Management Spear Program, Cornell University, 323 Morrison Hall, Ithaca, NY 14853, USA
[3]School of Land and Food, University of Tasmania, Private Bag 98, Hobart, TAS 7001, Australia
[4]Hue University of Agriculture and Forestry, 102 Phung Hung Street, Hue, Vietnam
[5]Department of Supply Chain and Information Systems, The Pennsylvania State University, 467 Business Building, University Park, PA 16802, USA
[6]Department of Animal Science, Cornell University, 329 Morrison Hall, Ithaca, NY 14853, USA

Correspondence should be addressed to Keenan C. McRoberts; kcm45@cornell.edu

Academic Editor: Glaciela Kaschuk

Increased production in smallholder beef systems requires improved forage management. Our objective was to evaluate the effects of composted cattle manure and mineral nitrogen (urea) application on soil fertility and partial nutrient balances in plots established to *Brachiaria* cv. Mulato II in south-central coastal Vietnam from 2010 to 2013. A randomized complete block design was implemented on six farms (blocks), with five rates of composted cattle manure (0, 4, 8, 12, and 24 Mg DM/ha per yr) and three urea rates (0, 60, and 120 kg N/ha per yr) in a factorial design. Soil was analyzed before and after the experiment. Compost increased soil pH, organic matter, Ca, Mg, and Mn. The effect of compost and urea applications on postexperiment soil fertility depended on preexperiment soil fertility for K, P, S, Mg, Zn, Mn, Cu, and organic matter, suggesting that the ability to maintain soil fertility depends on the interaction between soil organic and inorganic amendments and existing soil fertility. Highest farm yields were also achieved on farms with higher preexperiment soil fertility levels. Negative partial nutrient balances for N, P, and K suggest that yields will not be sustainable over time even for the highest fertilization inputs used in this experiment.

1. Introduction

Smallholder crop-livestock farms (≤5 head of cattle/household) in south-central coastal Vietnam contribute to beef supply for urban areas [1]. Semi-intensive (grazing and stall feeding) and extensive (grazing of communal land) cattle management systems are dominant in the region. Supplementary stall feeding, with a basal diet of rice straw, peanut straw, and cultivated forage (e.g., *Pennisetum purpureum*) or cut-and-carry native grasses and legumes from private and communal land, complements supervised or unsupervised grazing on communal (and private) lands [1]. Progressive farmers have transitioned toward semi-intensive management of backyard plots to increase animal productivity (daily gain) using supplementary forages (native grasses and legumes and cultivated improved forages) and concentrates and reducing energy expenditures incurred from walking to open grazing areas. Key to this transition is cultivation of high-yielding, high nutritive value forages that are well adapted to local climatic and soil conditions [1–4]. Farmers also benefit from reduced labor requirements for cattle production when forages are cultivated near the household [3, 5].

Adoption of *Brachiaria* spp. grasses in tropical Latin America has been important to productivity and economic gains in dairy and beef systems [6] and can be managed under grazing or cut-and-carry systems. Similar benefits are emerging in smallholder crop-livestock systems in Vietnam and other southeast Asian countries due to adoption of *Brachiaria* spp. hybrids such as Mulato (*B. brizantha* × *B. ruziziensis*) and Mulato II (*B. ruziziensis* × *B. decumbens* × *B. brizantha*) and other improved forages [2, 4, 7]. *Brachiaria* spp. hybrids are popular among farmers in Vietnam due to high palatability, high leaf to stem ratios, easy establishment, and good yield potential relative to other forages [2, 4].

Sandy soils in Vietnam, similar to other tropical systems, pose a unique set of fertility constraints [8], including nutrient deficiencies and low soil organic matter (OM) concentrations [9]. Low organic matter contributes to low cation exchange capacity (CEC), low water holding capacity, and high potential nutrient losses due to runoff, leaching, volatilization, and denitrification. Data supporting forage management decisions in sandy, tropical, rain-fed systems such as those located in the south-central Vietnam coast are inadequate [7, 10]. Vietnamese farmers identify fertilization management of forages as a critical decision impacting productivity and soil fertility over time. Palm et al. [11], Zingore et al. [12], Goyal et al. [13], and Kaur et al. [14] suggest that a combination of locally available organic nutrient sources and fertilizer might be needed to enhance productivity in the tropics. Farmers in south-central coastal Vietnam typically use composted cattle manure combined with occasional inorganic fertilizer applications (urea and/or NPK blends) and generally follow a "more is better" approach to composted manure applications based on plant color and vigor.

The objective of this study was to assess the effects of urea and composted cattle manure on soil fertility and partial nutrient balances in a multiyear (29 mo) on-farm field experiment. We report on soil fertility and nutrient balances under various nutrient management scenarios.

2. Methods

2.1. Cát Trinh Commune Characteristics.
The experiment was undertaken from 2010 to 2013 in Cát Trinh Commune, Phù Cát District, Bình Định Province in south-central coastal Vietnam. The district capital of Ngô Mây (14°0′2″N, 109°2′38″E) borders Cát Trinh Commune to the southwest. Sandy soils in the commune belong to the Arenosols Group [21]. Annual rainfall in the region is about 1,200 mm, with over 70% falling during the rainy season (September to December). The range in average monthly temperature is from 23 to 31°C. Farms were selected in regions where the Australian Centre for International Agricultural Research-funded Project SMCN/2007/109, entitled "Sustainable and Profitable Crop and Livestock Systems for South-Central Coastal Vietnam," was taking place. Project presence simplified access to commune farms (including

permission requirements with local and regional authorities) and permitted adequate experiment tracking due to partnerships with rural development professionals and extension educators operating in the commune. This experiment responded directly to problems identified by local farmers and revealed during baseline project surveys [1, 22].

2.2. Farm Selection and Experiment Preparation.
Recent field histories for the six farms participating in the experiment included (1) *Pennisetum purpureum* Schumach. cultivation, (2) vacant with household waste and ash accumulation, (3) cassava and eggplant production, (4) cassava production, (5) peanut production, and (6) vacant lot. A range of backyard sandy soils in Vietnam and potentially in other tropical regions were represented in initial soil fertility conditions (Table 1). Preexperiment (2010) soil samples at a depth of 15 to 25 cm indicated low organic matter, low nutrient concentrations, and acidic soils (Table 1). Soil textural classes consisted of sands and loamy sands with very low cation exchange capacity (Table 1).

Experimental areas were prepared using animal traction and handheld hoes in July 2010 according to farmer practice. Weeding was done using handheld hoes as needed throughout the experiment (typically one to two times per month). A woven wire fence (1.2 m high) was installed around each area to inhibit animal entry. No pesticides were applied during the experiment.

Starter fertilizer (20-20-15; N-P_2O_5-K_2O) was applied at 175 kg/ha before transplanting *Brachiaria* cv. Mulato II (*B. ruziziensis* × *B. decumbens* × *B. brizantha*) seedlings into experimental plots in September 2010 and a second time after transplanting in October 2010. Mulato II plant establishment density in each 2 × 2 m plot was 20 plants, derived from 50 cm row spacing and 40 cm plant spacing. Replanting was completed during the first two months after establishment (September and October 2010) to obtain desired density (50,000 plants/ha). One-meter buffer strips and outer borders separated plots and contained single row of Mulato II with 40 cm spacing between plants.

2.3. Composted Cattle Manure Treatment Preparation, Sampling, and Analyses.
Preexperiment compost samples contained 50% DM and 1 g/kg N, which were used to define treatment application rates. Farmer practice in the region dictated compost preparation. Experimental compost treatments were supplied entirely by farm 3 to ensure consistent composition. Compost preparation included daily removal of cattle manure and rice straw refusals from cattle pens into an uncovered pile. Compost was turned at least twice prior to use during the approximately 45 d process. Experimental treatment composition was determined by collecting three representative subsamples (150 g each) of the composted cattle manure during each treatment application period (Table 2). The subsamples were dried to stable weight at 60°C in a forced-air oven and ground to pass a 2 mm screen in a Retsch cutting mill (Germany). Brookside Laboratories Inc. (New Bremen, OH) analyzed samples as described in

TABLE 1: Preexperiment soil fertility levels at 15 to 25 cm depth on six farms in Cát Trinh Commune.

Parameter[c]	Overall mean	Overall SD	Farm means ($n = 16$)					
	($n = 96$)		Farm 1	Farm 2	Farm 3	Farm 4	Farm 5	Farm 6
CEC (cmol$_+$/kg)[a]	1.3	0.60	2.4	1.4	1.3	0.9	0.9	1.0
pH (1:1 soil:water)	5.67	0.658	5.98	5.53	6.23	5.76	4.63	5.88
OM (g/kg)[b]	3.5	2.1	5.2	4.5	2.1	1.6	5.1	2.5
S (mg/kg)	9.1	4.47	10.1	9.9	6.3	5.6	17.0	5.6
P (mg/kg)	37	27.3	67	41	68	22	14	12
Ca (mg/kg)	176	68.3	269	204	190	126	116	151
Mg (mg/kg)	29	23.3	71	19	28	17	21	16
K (mg/kg)	36	36.2	91	41	32	11	31	12
Na (mg/kg)	23	13.1	43	20	19	19	21	17
B (mg/kg)	0.18	0.19	0.25	0.38	0.24	0.01	0.15	0.04
Fe (mg/kg)	116	75.2	180	142	153	28	169	27
Mn (mg/kg)	7	7.5	14	7	16	1	2	5
Cu (mg/kg)	1.6	1.29	3.3	2.3	1.6	0.6	1.1	0.6
Zn (mg/kg)	2.7	5.43	4.4	7.1	1.5	0.4	1.0	1.6
Al (mg/kg)[d]	280	179.7	375	346	235	116	540	72
NO$_3$-N (mg/kg)	3	2.6	5	5	1	1	5	2
NH$_4$-N (mg/kg)	8	4.1	5	15	7	6	8	7
Ca:Mg	7.8	3.79	4.2	12.5	7.2	7.7	5.6	9.4

[a] CEC: cation exchange capacity determined by summation of cations Ca, Mg, K, Na, and H, according to Ross [15].
[b] Soil organic matter (OM) was determined by loss-on-ignition at 360°C [16].
[c] Elements are Mehlich-3 extractable nutrients [17].
[d] Aluminum toxicity might be an issue due to low pH (<5) on farm 5.

Peters et al. [18]. Total N was determined using an Elementar Vario Max (Elementar Analysensysteme, Hanau, Germany), and minerals were determined using a CEM Mars Express microwave (CEM Corporation, Matthews, NC) with digest analyzed in a Thermo Scientific iCAP 6500 inductively coupled plasma-atomic emission spectrometer (Thermo Electron Corp., Waltman, MA). Ammonium-N and NO$_3$-N were determined using a Lachat QuickChem 8000 flow injection calorimetric analyzer (Lachat Instruments, Loveland, CO). Organic matter (OM) was estimated by loss-on-ignition at 550°C for 2 h [18] and mineral matter was calculated as 100 − %OM. Average compost treatment composition was 50% DM, 261 g OM/kg DM, 12 g N/kg DM, 2733 mg P/kg DM, and 6750 mg K/kg DM (Table 2).

2.4. Experimental Design.

Fifteen treatments were derived from the factorial combination of five compost rates (0, 4, 8, 12, and 24 Mg DM/ha per yr) and three urea rates (0, 60, and 120 kg N/ha per yr). Treatments were implemented in a randomized complete block design on six farms (blocks). A sixteenth treatment added 80 kg K$_2$O/ha per yr to the highest compost rate in 2011 and also included the highest urea rate from January 2012 onward. Buffer strips received 4 Mg DM/ha per yr compost.

Preexperiment compost samples determined field application rates to achieve desired compost N rates (0, 40, 80, 120, and 240 kg N/ha per yr). Actual N rates (Table 3) differed due to variation in composted manure DM and N concentrations over time. Treatments were selected to

balance for Mulato II N requirement (240 kg N/ha per yr) with the highest compost treatment. Annual treatments were divided into six parts and applied at approximately 2 mo intervals immediately following each harvest. During the first year, treatments were surface-applied, while second-year treatments were incorporated in the top 15 cm of soil using handheld hoes to reduce potential for nutrient transfer across plots and to reduce nutrient loss. Initial compost and urea treatments were applied in November 2010 (two months after transplanting) after all Mulato II plants were cut uniformly at 25 cm above ground level to initiate the experiment.

2.5. Sampling Protocols

2.5.1. Soil Samples. Preexperiment soil samples ($n = 96$) were collected in August 2010 using a handheld trowel at a depth of 15 to 25 cm below ground level (Table 1) and postexperiment samples in May 2013 at two depths (0 to 15 cm and 15 to 25 cm). In each plot, three samples were gathered on a diagonal line across the plot and combined into a single composite sample for each depth. Samples were air-dried, litter was removed, and soil was passed through a 2 mm sieve. Brookside Laboratories Inc. (New Bremen, OH) determined soil pH (1:1 H$_2$O) [23], using an AS-3000 Dual pH Analyzer, soil OM by loss-on-ignition at 360°C [16], Mehlich-3 [17] extractable P, K, Ca, Mg, S, Na, B, Mn, Cu, Zn, and Al in a Thermo Scientific iCAP 6500 inductively coupled plasma-atomic emission spectrometer (Thermo Electron Corp., Waltman, MA), inorganic N (NH$_4$-N and NO$_3$-N via 1 N KCl cadmium reduction) using a Flow

TABLE 2: Means and standard deviations for composition of composted cattle manure from the supplying farm in Cát Trinh Commune on a dry matter basis.

Parameter	Mean[c]	SD[c]
Dry matter (%)	50	8.4
Organic matter (g/kg)[a]	261	99
Mineral matter (g/kg)[a]	739	99
C:N[b]	12.8	2.3
Total N (g/kg)	12	3
Organic N (g/kg)	11	3
P (mg/kg)	2733	1302
K (mg/kg)	6750	4105
Ca (mg/kg)	9988	3581
Mg (mg/kg)	3088	1094
Na (mg/kg)	955	661
S (mg/kg)	2090	516
B (mg/kg)	14	4
Fe (mg/kg)	4266	2196
Mn (mg/kg)	791	129
Cu (mg/kg)	27	25
Zn (mg/kg)	110	25

[a]Organic matter was estimated by loss-on-ignition at 550°C for 2 h [18] and mineral matter was calculated as 100 − %OM.
[b]1.724 factor was applied to convert organic matter (OM) to C, based on the assumption that loss-by-ignition = OM and OM contains 58% organic carbon (Factor of Wolff, as cited in Pribyl [19]).
[c]Means and standard deviations (SD) were calculated from 14 treatment applications, and each treatment application was an average of 3 subsamples.

TABLE 3: Actual nitrogen (N) treatment rates were combinations of urea N and N in composted cattle manure.

Compost rate[a] (Mg DM/ha per yr)	Urea rate[a] (kg N/ha per yr)		
	0	60	120
0	0	60	120
4 ± 0.7	46 ± 12.4	106 ± 12.4	166 ± 12.4
8 ± 1.3	92 ± 24.9	152 ± 24.9	212 ± 24.9
12 ± 2.0	138 ± 37.3	198 ± 37.3	258 ± 37.3
24 ± 4.0	276 ± 74.7	336 ± 74.7	396 ± 74.7

[a]Error is standard deviation and exists due to variation in compost composition over time during the experiment.

Injection Analyzer (FIAlab Instruments Inc., Bellevue, WA) [24], and cation exchange capacity by summation of cations Ca, Mg, K, Na, and H [15]. Values below the limit of detection were replaced by *limit of detection/2* to enable statistical analysis.

2.5.2. Forage Sampling. Forage plots were harvested every 36 d with SD ± 8.6 d (24 total harvest events) during the experiment, starting in January 2011. Farms were harvested in the same order and within two to four days for each harvest event. All fresh forage biomass was weighed for each plot and small subsamples (20 to 50 g) were collected to determine dry matter concentration. Every second harvest,

larger subsamples (200 to 300 g) were collected instead of small subsamples to evaluate forage nutritive value. Samples were dried to stable weight at 60°C in a forced-air oven to determine dry matter concentration. Samples were ground to pass a 4 mm screen in a Retsch cutting mill (Haan, Germany), and those that had not incurred damage from heat or water during the subsequent transportation and storage process were reground to pass a 1 mm screen in a UDY Cyclone Mill (UDY Corp., Fort Collins, CO). September 2011 and December 2012 samples were analyzed by Dairyland Laboratories Inc. (Arcadia, WI) to determine nutritive value using near infrared spectroscopy [25] with a Foss model 5000 (Foss-NIR System, Silver Spring, MD). Measured parameters included N, P, K, Mg, and Ca. Nutrient uptake in September 2011 and December 2012 was used to calculate partial nutrient balances.

2.5.3. Partial Nutrient Balance and Nitrogen Recovery Calculations. Partial nutrient balances for N, P, and K were calculated as *inputs-outputs*, where *inputs* included nutrients applied in fertilizer and compost treatments and *outputs* were nutrients accumulated in harvested forage. Nutrient outputs were calculated by multiplying back-transformed mean treatment yields (geometric means) by nutrient concentration for each plot in September 2011 samples and December 2012 samples. Nutrient outputs for farm partial nutrient balances were determined using plot yields for September 2011 and December 2012 (in lieu of overall treatment yields) to capture farm-to-farm yield variation. Nitrogen recovery was calculated according to the difference method of Jokela and Randall [20] as % N recovery = [(N uptake in treatment$_i$− N uptake in zero N control)/total N applied in treatment$_i$] × 100. Mean treatment yields for the experiment and September 2011 and December 2012 plot nutrient concentration were used to calculate N uptake.

2.6. Composted Manure Incubation Test. An incubation study was conducted with a representative composted manure sample to test if compost addition resulted in a release of nitrate to the soil or a reduction in nitrate due to immobilization. A New York State fine-loamy soil with a relatively high soil nitrate content (pH = 7.6, OM = 64 g/kg, and Morgan NO_3-N = 17 mg/kg) (100 g) was amended with 1 g or 10 g representative compost (4 replicates) in 450 mL plastic incubation cups with perforated lids. Cups were arranged in a completely randomized design and incubated at 23°C in the dark. Soil moisture was maintained between 70% and 75% of field capacity during the incubation. After incubation, samples were dried to stable weight at 50°C in a forced-air oven. Subsamples were analyzed for KCl extractable nitrate [26] in an EasyChem Analyzer (Systea Scientific, Oak Brook, IL).

2.7. Statistical Methods. The effect of compost and urea treatments on soil fertility factors was evaluated in PROC MIXED [27], with fixed effects of compost, urea, compost × urea, and preexperiment soil fertility, and block as a random effect. Interactions between preexperiment soil fertility

TABLE 4: Soil chemical models in 0 to 15 cm and 15 to 25 cm strata selected from candidate fixed effects of compost (C), urea (U), compost × urea (C × U), preexperiment soil fertility for each parameter sampled at 15 to 25 cm depth (PSF), and significant interactions, with block included as a random effect.

Parameter[b]	0 to 15 cm depth		15 to 25 cm depth	
	Model effect	Farm (%)	Model effect	Farm (%)[c]
S (mg/kg)	C × U × PSF[a]	82	C × U × PSF[a]	85
Mg (mg/kg)	C × U × PSF[a]	87	C × U × PSF[a]	89
Zn (mg/kg)	C × U × PSF[a]	28	C × U × PSF[a]	16
Mn (mg/kg)	C × PSF[a]	87	C × U × PSF[a]	49
OM (g/kg)	C × U × PSF[a]	65	U × PSF[a]	77
K (mg/kg)	C × PSF[a]	51	C × PSF[a]	60
Cu (mg/kg)	C × U × PSF[a]	62	PSF	53
CEC (cmol$_+$/kg)	PSF	31	C × U × PSF[a]	46
P (mg/kg)	PSF	52	C × U × PSF[a]	28
NH$_4$-N (mg/kg)	PSF	66	PSF C × U[a]	42
pH (1:1 soil:water)	C PSF	88	PSF	75
Fe (mg/kg)	PSF	88	PSF	70
Al (mg/kg)	PSF	62	PSF	58
NO$_3$-N (mg/kg)	PSF	43	PSF	46
Ca (mg/kg)	C × U[a]	75	PSF	49
Na (mg/kg)	PSF	47	C × U[a]	49
B (mg/kg)	NS	—	NS	—

[a]Where interactions were significant, only highest level interactive effect or effects are shown, although lower level terms were also in the model.
[b]Each row represents a single model for each soil depth.
[c]Percentage of residual variance explained by farm in each model is indicated.

factors and main effects were also evaluated to determine if soil fertility responses depended on initial soil fertility levels. Mean differences among treatments were declared at $P \leq 0.05$ using Tukey's statistic to control the familywise error rate for multiple comparisons. One-tailed t-tests were used to assess paired mean differences between preexperiment and postexperiment soil fertility data, with Bonferroni correction used to identify significant mean differences.

Partial nutrient balance means for the compost × urea interaction were extracted from mixed models with fixed effects of compost, urea, compost × urea, and harvest period (September 2011 or December 2012), and included block as a random effect. Models estimating farm partial nutrient balances contained fixed effects of compost, urea, and harvest period (September 2011 or December 2012) and block (farm). Nitrogen recovery means for the compost × urea interaction for each harvest period were extracted from mixed models with fixed effects of compost, urea, and compost × urea, and included block as a random effect.

3. Results

3.1. Soil Fertility

3.1.1. Preexperiment and Postexperiment Soil Fertility. Preexperiment farm soil fertility levels were consistent with yield differences observed in the experiment (i.e., preexperiment soil nutrient concentrations were higher for high-yielding farms). All treatment means ($n = 6$ for each compost × urea interaction rate) were numerically higher in postexperiment

samples in the 15 to 25 cm stratum than preexperiment samples at the same depth for OM and NH$_4$-N, and most means were numerically higher after the experiment for Na and Fe (Table 8). More than half of postexperiment treatment means were also numerically higher than preexperiment for Cu and Mn. All treatment means were numerically lower in postexperiment samples at the 15 to 25 cm stratum than preexperiment samples at the same depth for S, P, K, and NO$_3$-N, and most were lower postexperiment for pH, Mg, Al, Ca, and CEC. Significant mean difference (Bonferroni correction) was detected only in soil Na for compost applied at 12 Mg DM/ha per yr without urea, which was higher after the experiment (Table 8).

Models estimating soil fertility parameters in the 15 to 25 cm stratum (*postexperiment level – preexperiment level*) indicated that compost impacted differences for soil Al ($P = 0.0189$, ordered 12, 24, 4, 8, and 0, with 12 > 8 and 0) (Table 9). Urea application decreased soil OM ($P = 0.0036$, ordered 0, 60, and 120, with 0 and 60 > 120) and tended to affect Mg ($P = 0.0898$, ordered 0, 60, and 120, no significant differences) (Table 4). A compost × urea interaction was detected for K ($P = 0.0413$, no significant differences), with a tendency to impact pH ($P = 0.0595$), Zn ($P = 0.0816$), and NH$_4^+$ ($P = 0.0985$).

Preexperiment soil factors (PSF) most highly associated with yield were Mn ($r = 0.48$), P ($r = 0.44$), Fe ($r = 0.33$), exchange capacity ($r = 0.32$), and Ca ($r = 0.32$). Many preexperiment soil fertility factors were significant in multivariate models with compost and urea main effects (Table 10). Models selected from main effects, PSFs,

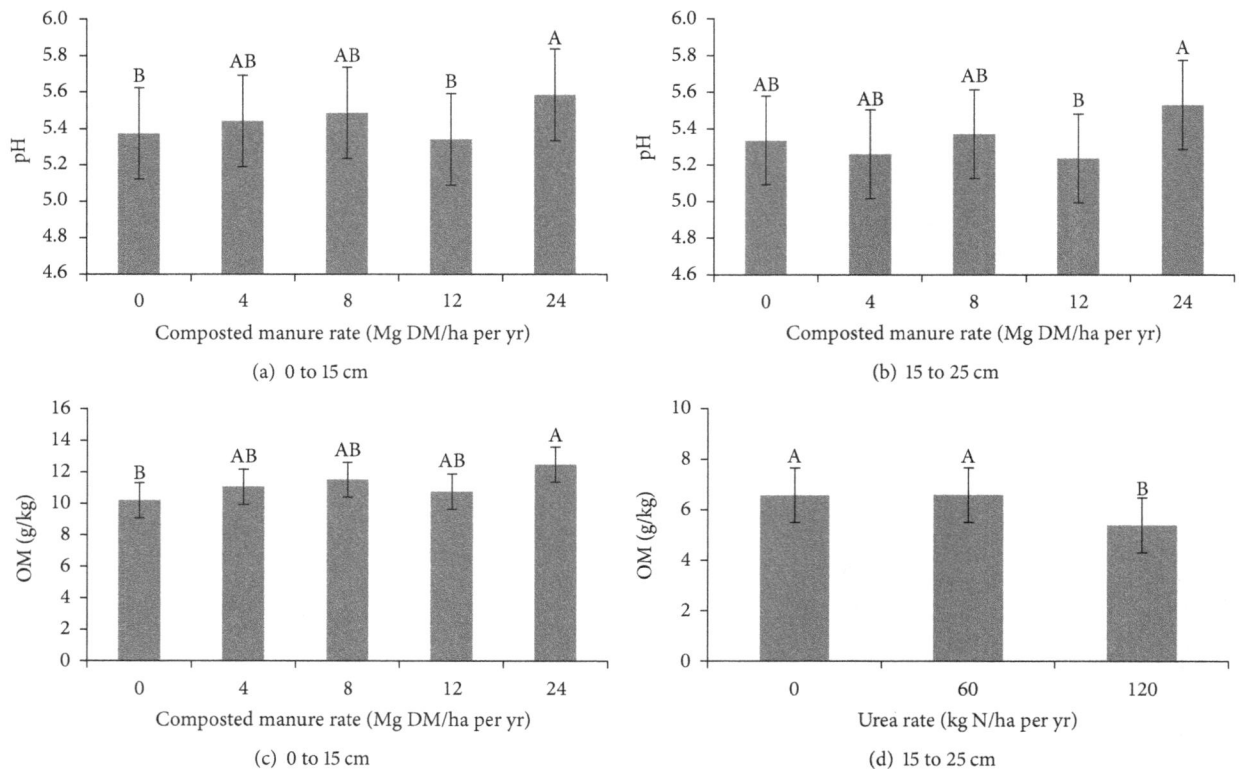

FIGURE 1: Postexperiment pH least squares means for composted manure rates in the 0 to 15 cm ((a), $P = 0.0071$) and 15 to 25 cm strata ((b), $P = 0.0460$) and soil organic matter (OM) least square means for compost treatments in the 0 to 15 cm stratum ((c), $P = 0.0087$) and urea treatments in the 15 to 25 cm stratum ((d), $P = 0.0035$). pH models contained preexperiment soil pH and OM models contained preexperiment OM (Table 10). Error bars are ±1 SE of the mean. Bars connected by the same letter are not significantly different ($P \leq 0.05$).

and significant interactions indicated the importance of preexperiment soil fertility on treatment responses for some soil fertility parameters (Table 4). The effect of compost and urea (main effects) on postexperiment soil S, K, Mg, Zn, Mn, and OM depended on their respective PSFs for both soil depths. Compost and urea effects on soil Cu depended on Cu PSF in the 0 to 15 cm stratum, while only Cu PSF emerged from the 15 to 25 cm stratum. The effect of compost and urea on soil CEC and P depended on PSF in the 15 to 25 cm stratum, while only PSF mattered in the 0 to 15 cm stratum. The effect of compost and urea on soil pH at 0 to 15 cm and NH_4-N at 15 to 25 cm did not depend on PSF, although it was in the model. Only PSF emerged from models estimating Fe, Al, and NO_3-N at both depths, CEC, P, NH_4-N, and Na at 0 to 15 cm, and Ca, pH, and Cu at 15 to 25 cm. Preexperiment soil fertility was not significant for Ca at 0 to 15 cm or for Na at 15 to 25 cm. Farm explained greater than 40% of residual variation in soil fertility models at both depths for all factors except P, Zn, and CEC.

Compost addition increased soil pH in both soil strata (Figure 1, Table 10), although pH for the 12 Mg DM/ha per yr compost rate was lower than the 24 Mg DM/ha per yr rate. Postexperiment soil pH was highest where compost had been applied, although pH decreased numerically in most plots relative to preexperiment levels (Table 8). Preexperiment soil pH (1:1 soil:water) was higher than postexperiment pH for

all treatments in the 15 to 25 cm stratum, which suggests minor soil acidification during the experiment. The difference between post- and preexperiment pH was less for the highest compost treatment than 12 Mg DM/ha per yr. Postexperiment pH was impacted by compost treatments. Soil pH in the 0 to 15 cm stratum for the highest compost rate (pH = 5.58) was higher than without compost (pH = 5.37) and 12 Mg DM/ha per yr (pH = 5.33) (Figure 1). Response to compost in the 15 to 25 cm stratum indicated that pH for the highest compost rate (pH = 5.53) was higher than for the 12 Mg DM/ha per yr (pH = 5.24) treatment. Overall, the deeper layer was slightly more acidic than the topsoil layer (Figure 1).

Compost increased soil OM in the 0 to 15 cm stratum (Figure 1, Table 10). Urea application decreased OM in the 15 to 25 cm stratum (Figure 1, Table 10). Organic matter for the highest urea rate was lower than the intermediate rate and plots not receiving urea, possibly explained by high mineralization rates in this stratum combined with limited OM addition to the soil in the form of plant litter. Organic matter in the 15 to 25 cm stratum for treatment plots increased by 0.8 g/kg to 3.8 g/kg from preexperiment to postexperiment levels (Table 8).

Consistent with the OM effect, postexperiment Ca in the 0 to 15 cm stratum increased steadily with added compost (Figure 2) and likely contributed to buffering capacity (Figure 2, Table 10). Calcium was higher for the highest

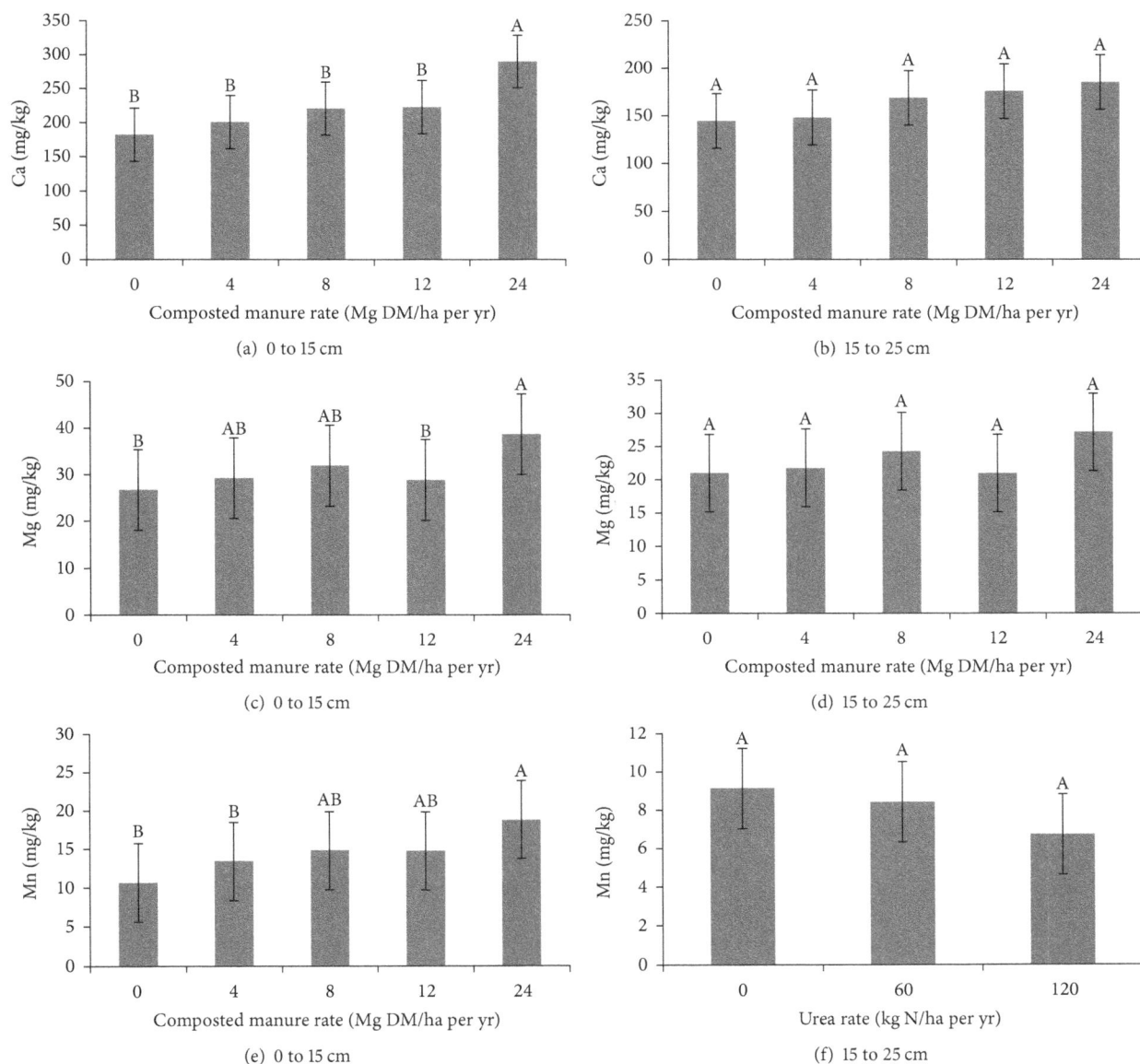

FIGURE 2: Postexperiment soil Ca least squares means for composted manure rates in the 0 to 15 cm ((a), $P < 0.0001$) and 15 to 25 cm strata ((b), $P = 0.1873$), Mg least squares means in the 0 to 15 cm ((c), $P = 0.0015$) and 15 to 25 cm ((d), $P = 0.0542$) strata, and Mn least squares means for compost treatments in the 0 to 15 cm stratum ((e), $P = 0.0002$) and for urea treatments in the 15 to 25 cm stratum ((f), $P = 0.2015$). Error bars are ± 1 SE of the mean. Bars connected by the same letter are not significantly different ($P \leq 0.05$). Models contain preexperiment soil fertility factors (Table 10).

compost rate than all other compost treatments. Compost × urea interaction was also observed (results not shown), but the interaction response was inconsistent. Calcium also increased numerically with applied compost in the 15 to 25 cm stratum.

Soil Mg increased with compost treatments in the 0 to 15 cm stratum, and a trend was observed in the 15 to 25 cm stratum (Figure 2, Table 10). Similar to soil pH, Mg for the highest compost rate was numerically higher than all other treatments and soil Mg tended to increase with compost application (excepting the 12 Mg DM/ha per yr plots).

Soil Mn in the 0 to 15 cm stratum increased with compost addition, and Mn for the highest compost rate was higher than 4 Mg DM/ha per yr and plots not receiving compost (Figure 2, Table 10). Similar to OM, Mn at 15 to 25 cm depth decreased with added urea, although the effect was not significant ($P = 0.2015$).

Soil Na responses in the 15 to 25 cm stratum were probably due to an outlying treatment response. The compost × urea interaction effects for soil Na and soil NH_4-N in this stratum were inconsistent (results not shown). Preexperiment soil Al levels tended to be higher than postexperiment

TABLE 5: Nitrate-N extracted from compost-amended soil, control, and nonincubated soil in a laboratory incubation study.

Treatment (g dry compost added to 100 g dry soil)	N treatment equivalent (kg N/ha)	Soil nitrate-N[a] (mg/kg)	SE
10	2,735	161.5 a	3.36
1	273.5	15.7 b	3.36
0	0	19.5 b	3.36
Nonincubated soil	—	15.2 b	3.36

[a]Treatments connected by the same letter are not significantly different ($P \leq 0.05$).

TABLE 6: *Brachiaria* cv. Mulato II dry matter yield and N, P, and K balances by farm.

Farm	Yield[a] (Mg DM/ha/yr)	N balance[bc] (kg N/ha per yr)	P balance[bc] (kg P/ha per yr)	K balance[bc] (kg K/ha per yr)
3	26.2	−653 e	−113 d	−783 e
1	16.1	−342 d	−71 c	−610 d
2	14.8	−325 d	−57 c	−388 c
5	13.3	−150 c	−21 b	−186 b
6	8.8	−56 b	−9 ab	−107 ab
4	6.6	53 a	7 a	−38 a

[a]Farm yields are overall experiment means (geometric means).
[b]Farm N, P, and K balances were calculated using dry matter yield and plot nutrient concentrations for the September 2011 and December 2012 samples.
[c]Means not connected by the same letter in each column are significantly different ($P \leq 0.05$).

levels in the 15 to 25 cm stratum (Table 8). Soil Al for compost applied at 12 Mg DM/ha per yr was higher (lower mean difference) than 8 Mg DM/ha per yr and without compost (Table 9). This suggests soil Al depletion during the experiment, especially for plots receiving no or a low rate of compost. Compost effect on Al was not detected in statistical analyses of postexperiment data.

Observed soil fertility effects could be a function of pre- or postharvest sampling time. Preexperiment samples were collected in August 2010 and postexperiment samples in early May 2013. These soil sampling time frames were both during the dry season but rainfall was slightly higher leading up to August 2010 sampling than May 2013. Soil moisture near sampling times (not measured) could have affected mineralization rates and outcomes.

3.1.2. Incubation Study Results. The compost incubation study confirmed that addition of composted cattle manure at a high rate resulted in rapid N mineralization and nitrification as expected for material with C : N of 13 : 1 (Table 5). The incubation study did not provide evidence of N immobilization, consistent with the relatively low C : N ratio of the compost. However, at the lower application rate, no increase in soil nitrate was determined. The latter supports the hypothesis that the lack of a yield response to compost addition in the farm experiments is due to an insufficient amount applied and not due to immobilization.

3.2. Partial Nutrient Balances and Nitrogen Recovery. Nutrient concentration across the two sampling dates averaged 24 g N/kg DM, 3.8 g P/kg DM, 21.2 g K/kg DM, 3.5 g Mg/kg DM, and 6.9 g Ca/kg DM. Average partial N, P, and K

balances were negative on five of the six farms for N and P and negative on all six farms for K (Table 6). Overall partial N and P balances on lower-yielding farms were positive or less negative (farms 4, 5, and 6) than for higher-yielding farms (Table 6). Farm-to-farm variation explained a large proportion of residual variance in partial nutrient balances (39% in N balance model, 27% in P balance model, and 52% in K balance model).

Farms with better soil fertility indicators yielded higher, but this also resulted in more negative partial N balances (Table 6). Variable farm plot histories were also responsible for differences in farm yields. Partial N balance (Table 7) was most negative for treatments containing low N inputs and progressively became less negative with additional N applications as urea and compost. Only the highest compost rate achieved partial N balances that were close to zero or positive (Table 7). Partial P and K balances by farm followed a similar pattern to N, but all P and K balances were negative (Table 7). Treatments receiving higher rates of compost were less negative or positive in P balance. Potassium balances were very negative, impacted more by yield than by compost rates, reflecting the fact that compost K inputs (0, 26.6, 53.2, 79.9, and 159.7 kg K/ha per yr) were low relative to K removal with harvested forage. Dry season partial nutrient balances were higher due to lower DM yields (not shown).

Greatest N recovery occurred in plots receiving the highest rate of urea (ranging from 5% to 42%) (Table 7). Highest overall N recovery was observed with 60 kg N/ha per yr urea × 4 Mg DM/ha per yr compost (79% and 112%, depending on nutritive value period) due to high dry matter yield for this treatment. Farm-to-farm variation explained 38% of residual variation in N recovery for September 2011 samples and 65% for December 2012.

TABLE 7: Dry matter yield, partial N, P, and K balances, and nitrogen recovery, by composted manure and urea treatment combinations.

Compost rate	Urea rate	Yield	N balance[ab]	P balance[ab]	K balance[ab]	Nitrogen recovery[bc]	
						Sept 2011	Dec 2012
(Mg DM/ha per yr)	(kg N/ha per yr)	(Mg DM/ha per yr)	(kg N/ha per yr)	(kg P/ha per yr)	(kg K/ha per yr)	(%)	(%)
24	120	18.3	−67 bc	−5 b	−241 bc	34.0 ab	42.2 ab
24	60	12.5	32 a	17 a	−123 a	3.0 b	−7.8 b
24	0	11.7	−8 ab	20 a	−104 a	−8.5 b	−12.4 b
12	120	15.4	−113 cd	−25 d	−258 cd	22.4 ab	23.0 b
12	60	15.7	−168 def	−25 d	−247 bc	21.4 ab	32.7 ab
12	0	12.2	−153 de	−13 c	−192 b	−10.7 b	−20.4
8	120	15.3	−155 de	−36 e	−263 cd	21.0 ab	30.3 ab
8	60	13.3	−160 de	−28 d	−230 bc	0.5 b	−0.6 b
8	0	12.8	−208 efg	−26 d	−216 bc	−13.1 b	−13.0 b
4	120	14.1	−181 def	−42 fg	−248 bc	23.4 ab	17.9 b
4	60	16.7	−307 hi	−52 h	−316 d	78.5 a	112.2 a
4	0	12.8	−264 ghi	−38 ef	−240 bc	1.1 b	−10.8 b
0	120	13.6	−202 efg	−49 h	−280 cd	10.6 ab	5.0 b
0	60	12.5	−236 fgh	−47 gh	−276 cd	−13.5 b	−42.2 b
0	0	12.6	−312 i	−47 gh	−240 bc	—	—

[a]Balances were calculated using overall dry matter yield treatment means and plot treatment nutrient concentrations from September 2011 and December 2012 samples.

[b]Means not connected by the same letter in each column are significantly different ($P \leq 0.05$).

[c]Nitrogen recovery was calculated according to Jokela and Randall [20] as % N recovery = [(N uptake in treatment$_i$ − N uptake in zero N control)/total N applied in treatment$_i$] × 100. Nitrogen recovery was based on overall dry matter yield treatment means and application rates in September 2011 and December 2012.

4. Discussion

4.1. Soil Fertility. Mehlich-3 soil test guidelines for agronomic crops in the subtropics (Florida) suggest medium levels at 26 to 40 mg/kg P, 26 to 40 mg/kg K, and 11 to 23 mg/kg Mg [28]. Although local calibration experiments should be conducted, these results suggest the potential for P and K deficiency in our experiment (medium range), while Mg is sufficiently high. Alternative interpretation guidelines for forage crops in temperate US regions suggest that P, K, and Mg levels are below optimum for forage crops (optimum range is 30 to 50 mg/kg for P, 100 to 200 mg/kg for K, and 120 to 180 mg/kg for Mg) [29]. Variability in soil fertility among farms was high as indicated by large standard deviations for most elements; preexperiment soil nutrient concentrations were higher for high-yielding farms. The interaction of preexperiment soil fertility with main compost and urea effects for OM and some macro- and micronutrients in compost (P, K, Mg, S, Zn, Mn, and Cu) suggests that the ability to maintain soil fertility is dependent on the interaction between soil organic and inorganic amendments and existing soil fertility. These observations support the hypothesis that soils are deficient in multiple nutrients and that compost supplies most soil nutrients but insufficient N, while urea application can address the N shortfall.

Soil OM increase during the experiment also took place in plots that did not receive compost, indicating that litter from aboveground and belowground plant biomass may be accumulating over time. Combination of OM from compost and from Mulato II leaf and root residues probably contributed to higher topsoil OM. Compost may play an important role in maintaining or perhaps increasing soil OM over time. Soil Mn was most highly correlated with yield and there were possible Mn deficiencies on some farms with low preexperiment Mn levels (plots at or below 1 mg Mn/kg) that were ameliorated by compost mineralization over time, further supporting the hypothesis that compost addition increases soil fertility status.

4.2. Partial Nutrient Balances. Farms represented a range in tropical sandy soils in the region. Preexperiment soil fertility and field histories differed substantially among participating farms. These conditions, combined with variation in rainfall and irrigation, resulted in large yield differences among farms. In general, preexperiment soil fertility was better for the three farms (farms 1, 2, and 3) with the highest yields and most negative N, P, and K balances. However, residual soil nutrients are unlikely to entirely account for the observed negative balances in Arenosols with low CEC (1.3 cmol$_+$/kg with SD ± 0.57) and low organic matter (0.35% with SD ± 0.21) (Table 1). Negative balances suggest that yields will not be sustainable over time even for the highest input levels used in this experiment. For example, DM yield of 20 Mg/ha per yr with 24 g N/kg DM removes 480 kg N/ha per yr through forage mass. Supplying 480 kg N/ha per yr with the cheapest commercial N fertilizer available in Vietnam (urea, 46% N, $0.56 USD/kg in 2013) would require 1043 kg urea at a total cost of $584 USD/ha per yr. This calculation assumes no losses and application at the crop removal rate, assuming limited soil N supply. Supply of other nutrients in addition to

TABLE 8: Soil fertility means before and after the experiment at the 15 to 25 cm depth for compost × urea treatment combinations. Elements are Mehlich-3 extractable nutrients. Data from before the experiment are included as a treatment baseline. Means calculated in each cell are an average of 6 farms. The only pair that differed significantly based on a paired one-tailed t-test, using the Bonferroni-adjusted P value cutoff of 0.003, is in bold.

Parameter	Compost treatment (Mg DM/ha per yr) / urea treatment (kg N/ha per yr)														
	0	4	8	12	24	0	4	8	12	24	0	4	8	12	24
	0	0	0	0	0	60	60	60	60	60	120	120	120	120	120
Before pH (1:1 soil:water)	5.37	5.98	5.63	5.87	5.68	5.52	5.27	5.83	5.83	5.63	5.72	5.50	5.82	5.75	5.45
After pH (1:1 soil:water)	5.05	5.30	5.47	5.48	5.67	5.58	5.05	5.48	5.37	5.50	5.18	5.32	5.32	5.10	5.32
Before CEC (cmol$_+$/kg)	1.19	1.16	1.19	1.22	1.53	1.52	1.20	1.40	1.29	1.43	1.33	1.17	1.35	1.28	1.23
After CEC (cmol$_+$/kg)	0.922	1.10	1.13	1.16	1.47	1.69	1.05	1.26	1.26	1.26	1.14	1.04	1.31	1.44	1.25
Before OM (g/kg)	4.07	2.58	2.87	3.05	4.15	4.68	3.28	3.35	3.13	3.87	3.98	3.47	3.15	4.22	3.15
After OM (g/kg)	6.82	5.5	5.97	6.65	7.23	6.62	7.12	6.13	6.1	7.48	4.8	4.98	5.53	6.32	5.58
Before S (mg/kg)	9.67	8.00	9.33	7.50	9.50	9.67	9.33	8.67	8.67	10.30	9.83	9.17	8.17	8.50	9.00
After S (mg/kg)	8.67	6.83	7.17	7.33	7.67	6.67	8.50	7.00	7.67	7.50	8.33	7.33	7.67	7.67	7.33
Before P (mg/kg)	36.5	37.7	36.5	27.2	39.0	50.8	33.2	46.2	34.2	39.7	36.2	37.3	42.8	29.3	37.3
After P (mg/kg)	25.3	18.5	24.2	23.5	36.5	29.0	27.7	26.5	22.0	24.2	23.2	20.7	25.3	17.0	24.2
Before Ca (mg/kg)	165	159	165	178	207	198	170	193	159	201	174	162	179	165	150
After Ca (mg/kg)	119	156	154	163	220	181	141	168	185	174	136	141	187	176	169
Before Mg (mg/kg)	25.8	22.0	23.5	22.0	30.2	31.8	22.8	31.3	33.2	29.2	34.3	23.5	30.7	27.7	34.2
After Mg (mg/kg)	20.8	20.2	23.8	24.8	26.7	24.0	21.7	23.5	21.5	26.5	20.0	19.7	25.8	16.2	30.5
Before K (mg/kg)	24.3	35.5	32.7	23.2	47.3	64.8	25.7	29.5	48.3	32.5	26.5	29.5	38.7	33.5	38.2
After K (mg/kg)	12.3	10.7	10.8	11.8	10.2	16.0	12.2	12.0	11.0	12.5	11.0	11.5	11.8	10.7	12.0
Before Na (mg/kg)	21.3	20.7	20.5	**18.8**	28.8	21.2	20.3	22.5	21.3	23.7	24.2	20.8	24.5	32.5	22.7
After Na (mg/kg)	28.2	29.7	29.3	**25.0**	27.5	42.3	30.5	28.2	28.0	32.0	26.3	32.0	28.8	28.3	27.8
Before B (mg/kg)	0.135	0.170	0.228	0.208	0.243	0.203	0.135	0.148	0.100	0.262	0.230	0.075	0.203	0.127	0.138
After B (mg/kg)	0.262	0.115	0.125	0.090	0.170	0.290	0.223	0.100	0.207	0.128	0.300	0.235	0.143	0.183	0.313
Before Fe (mg/kg)	132	123	116	97.8	116	153	106	127	115	132	100	128	107	109	114
After Fe (mg/kg)	138	112	120	132	148	154	125	135	127	125	108	122	152	120	121
Before Mn (mg/kg)	8.50	5.83	5.17	8.17	8.17	10.5	6.17	5.67	8.00	6.00	6.33	6.67	8.83	11.5	5.33
After Mn (mg/kg)	6.50	7.00	10.0	8.00	13.7	8.00	9.67	7.00	11.0	6.17	6.17	5.33	9.33	7.50	6.17
Before Cu (mg/kg)	1.74	1.11	1.19	1.24	1.76	1.45	1.20	1.50	1.69	1.50	2.05	1.94	1.17	2.39	1.56
After Cu (mg/kg)	1.81	1.65	2.02	1.67	1.73	1.73	1.74	1.59	1.44	2.05	1.99	1.69	2.24	2.02	1.68
Before Zn (mg/kg)	2.45	1.45	2.06	3.20	3.14	2.72	1.88	1.64	1.76	2.11	3.69	2.09	2.05	10.60	1.61
After Zn (mg/kg)	1.37	2.12	2.17	5.97	3.53	4.13	1.84	1.59	1.89	2.42	3.57	1.98	3.89	2.25	1.60
Before Al (mg/kg)	293	235	306	244	273	285	289	339	246	267	369	292	249	225	265
After Al (mg/kg)	250	192	245	257	254	221	267	248	230	202	313	210	245	230	242
Before NO$_3$-N (mg/kg)	3.93	2.40	2.40	2.67	5.10	3.87	3.15	1.90	3.32	2.68	1.67	2.78	3.48	4.57	2.45
After NO$_3$-N (mg/kg)	1.37	1.33	0.77	1.25	0.400	0.450	1.87	0.533	1.48	0.650	0.850	0.433	1.98	2.57	1.12
Before NH$_4$-N (mg/kg)	8.90	6.13	8.30	6.73	7.60	8.88	8.57	8.65	8.72	8.17	9.73	8.25	7.72	6.70	7.77
After NH$_4$-N (mg/kg)	13.5	9.52	9.82	10.2	10.1	10.2	12.4	9.92	10.6	10.7	11.0	10.4	13.5	10.8	10.6

N could be achieved using composted cattle manure if applied in sufficient amounts.

The 24 Mg DM/ha per yr of compost added in the experiment contained approximately 279 kg N, 66 kg P, 162 kg K, 74 kg Mg, and 240 kg Ca. Maximum N release for compost mineralization combined with urea application at removal rate could be sufficient to support a 20 Mg DM/ha per yr yield with assumed losses of 40% of total N applied. Nutrient removal estimates with a 20 Mg/ha per yr yield (75 kg P/yr, 424 kg K/yr, 70 kg Mg/yr, and 137 kg Ca/yr) suggest that only the supplies of Mg and P were adequate while K supply remained insufficient (−262 kg/ha per yr). Yet a response to added K was not observed in this experiment, suggesting some capacity of the soil to supply K despite low postexperiment soil test levels or that availability of other soil nutrients may have been more limiting than K. Grasses exhibit luxury consumption of K, although yield may remain stable with herbage K concentrations as low as 13 g K/kg DM [30].

Nutrient recycling from decaying aboveground and belowground Mulato II litter could supply additional nutrients. In a grazing system with *Brachiaria humidicola* pastures in Brazil, Boddey et al. [31] discovered litter deposits as high as 170 kg N/ha per yr (30 Mg DM/ha per yr) with low stocking rates. Magnitude of litter contributions is unknown in this

TABLE 9: Treatment P values for the difference between postexperiment soil fertility parameters and preexperiment parameters in the 15 to 25 cm stratum. Each row represents a single model. Models include block as a random effect. Elemental analyses are for Mehlich-3 extractable elements. Significant effects ($P \leq 0.05$) are in bold.

Parameter	Compost	Urea	Compost × urea
Total exchange capacity (cmol$_+$/kg)	0.9317	0.8727	0.8756
pH (1:1 soil:water)	**0.0269**	0.1749	0.0595
Organic matter (g/kg)	0.1604	**0.0036**	0.7973
S (mg/kg)	0.3240	0.4647	0.7185
P (mg/kg)	0.4661	0.2771	0.4179
Ca (mg/kg)	0.4523	0.8386	0.8740
Mg (mg/kg)	0.3694	0.0898	0.6977
K (mg/kg)	0.7658	0.4829	**0.0413**
Na (mg/kg)	0.2887	0.1182	0.4206
B (mg/kg)	0.2849	0.1422	0.5266
Fe (mg/kg)	0.3824	0.6983	0.1219
Mn (mg/kg)	0.2557	0.2068	0.4542
Cu (mg/kg)	0.1078	0.4458	0.1693
Zn (mg/kg)	0.7921	0.2485	0.0816
Al (mg/kg)	**0.0189**	0.2622	0.1371
NO$_3$-N (mg/kg)	0.4310	0.4914	0.1622
NH$_4$-N (mg/kg)	0.9296	0.3301	0.0985

TABLE 10: P values for treatment effects on soil chemical properties in 0 to 15 cm and 15 to 25 cm strata for models containing fixed effects of compost (C), urea (U), compost × urea (C × U), and preexperiment soil fertility measurement for each parameter sampled at 15 to 25 cm depth (PSF), with block as a random effect. Each row represents a single model for each soil depth. Significant values at $P \leq 0.05$ are in bold.

Parameter	C	U	C × U	PSF	C	U	C × U	PSF
		0 to 15 cm depth				15 to 25 cm depth		
CEC (cmol$_+$/kg)	0.2145	0.6271	0.3142	0.0797	0.6827	0.5997	0.4332	**0.0363**
pH (1:1 soil:water)	**0.0071**	0.1196	0.0705	**<0.0001**	**0.0460**	0.1071	0.0856	**<0.0001**
OM (g/kg)	**0.0087**	0.6440	0.0505	**<0.0001**	0.2699	**0.0035**	0.7938	**<0.0001**
S (mg/kg)	0.2249	0.9163	0.0720	0.2507	0.8146	0.7345	0.1171	0.0547
P (mg/kg)	0.6723	0.5556	0.3621	**0.0157**	0.6537	0.3633	0.4554	**0.0001**
Ca (mg/kg)	**<0.0001**	0.4688	**0.0147**	0.0718	0.1873	0.9323	0.6124	0.2071
Mg (mg/kg)	**0.0015**	0.2469	0.2219	0.0540	0.0542	0.4886	0.4991	**0.0006**
K (mg/kg)	0.1703	0.3593	0.7202	**0.0301**	0.2833	0.1041	0.3959	0.3572
Na (mg/kg)	0.0788	0.6333	0.8865	**0.0446**	0.1636	**0.0246**	**0.0201**	0.3319
B (mg/kg)	0.0954	0.4529	0.4883	0.9517	0.0788	0.2146	0.8916	0.5975
Fe (mg/kg)	0.9956	0.4733	0.2342	**0.0020**	0.4978	0.9109	0.1584	**<0.0001**
Mn (mg/kg)	**0.0002**	0.4951	0.6742	0.9859	0.5170	0.2015	0.3686	**0.0002**
Cu (mg/kg)	0.5569	0.3744	0.5829	**0.0343**	0.3963	0.8292	0.4546	**0.0003**
Zn (mg/kg)	0.8383	0.6481	0.2006	0.1849	0.9202	0.5418	0.0771	**0.0046**
Al (mg/kg)	0.6086	0.4611	0.6980	**<0.0001**	0.1240	0.2865	0.1456	**<0.0001**
NO$_3$-N (mg/kg)	0.5608	0.2398	0.2026	0.1510	0.1936	0.4103	0.2641	0.0885
NH$_4$-N (mg/kg)	0.1017	0.6409	0.1306	**0.0155**	0.9530	0.5630	**0.0386**	**0.0068**

experiment, but litter could make a large and important contribution to annual nutrient supply. Further experimentation is necessary in sandy tropical soils to obtain more accurate estimates of nutrient balances.

Reasons reported for yield persistence (despite negative balances) during postexperiment discussions with participating farmers include nutrient uptake from buffers and neighboring plots and soil nutrient reserves. Brachiaria spp. grasses are deep-rooted, which can help improve soil properties and transfer nutrients from deeper layers [32]. Assessment of the soil fertility status of deeper soil layers was not done for this study. Additionally, rising water tables during the rainy season may bring dissolved nutrients that could be used by actively growing plants and left behind in soil solution after the rainy season [33]. Furthermore, Brachiaria spp. association with N_2 fixing bacteria could supply additional N.

Biological N fixation capacity for *Brachiaria* spp. grasses has been identified using ^{15}N labeling and could supply as much as 20 to 40% of total N uptake (30 to 40 kg N/ha per yr) [34, 35]. A more minor contributor of N is OM. A hectare furrow slice (17 cm deep) contains 2,240,702 kg soil and approximately 11 mg/kg organic-bound N. Approximately 25 kg N/ha per yr could be released from topsoil OM in this study, a small amount compared to the 330 kg N/ha per yr in average crop removal across plots. Atmospheric N deposition could make a small additional contribution on a seasonal basis, perhaps as high as 10 to 15 kg N/ha per yr [36]. Magnitude, and in some cases existence, of these additional contributions is unknown but could explain relatively high Mulato II yields despite negative partial nutrient balances.

Partial nutrient balances calculated in this experiment did not account for potential nutrient losses that could make balances even more negative. Leaching losses can occur where compost or urea is applied to sandy soils. Water-logged and flooded plots during the rainy season lend favorable conditions for denitrification losses. High rainfall conditions in sandy soils may induce an annual "reset" of nutrients in soil solution, potentially inhibiting treatment response potential in this experiment. Surface applications of urea and compost (as done in experiment year one) furthermore favor losses from volatilization and runoff.

Nitrogen recovery was poor for compost application without urea primarily due to relatively low yields for compost-amended plots. In general, N recoveries were better for higher combined rates of compost and urea. In sandy, nutrient-depleted Zimbabwean soils, similar low N recovery was reported for compost (4% recovery rate) applied to corn at 17 Mg DM/ha per yr (7.8 g N/kg DM) [37]. In the study in Zimbabwe, compost combined with 40 kg N/ha per yr ammonium nitrate top-dressed to corn increased N recovery to 8%. Chikowo et al. [37] concluded that low quality composted cattle manure could not supply adequate N and that most N needs would need to be fulfilled by mineral N. Similar relatively low N recoveries were reported by Lynch et al. [38] for perennial forage production with composted dairy manure applied at 12.8 and 25.6 Mg DM/ha per yr in a temperate system (8.9% to 15.1% recovery). These studies are consistent with our findings that compost is not able to supply sufficient N for forage growth and that N recoveries improve when compost is applied together with inorganic N.

5. Conclusions

Field experiment results suggest that the ability to sustain soil fertility over time in sandy tropical soils is dependent on the interaction between existing soil fertility and soil organic and inorganic amendments. Negative partial nutrient balances for N, P, and K indicate that, for the compost and urea application rates used in this experiment with actively growing forage grass, soil degradation is occurring and yields are unlikely to be sustainable over time. Overall, compost plays a potentially important role in maintaining OM, preventing acidification, and supplying key macro- and micronutrients for plant uptake. Urea supplies readily available N, which the grass

could benefit from once other nutrient limitations had been overcome (with compost application). More broadly, these results suggest that extremely high rates of nutrient additions are required to realize the high yield potential in tropical forage systems. The consequences of high required nutrient inputs have important implications for intensification of smallholder crop-livestock systems and resulting nutrient loss potential.

Appendix

Complementary Soil Fertility Data

See Tables 8, 9, and 10.

Conflict of Interests

The authors declare that there is no conflict of interests regarding the publication of this paper.

Acknowledgments

The authors thank Françoise Vermeylen in the Cornell Statistical Consulting Unit for assistance with development of statistical models and Dr. J. H. Cherney for assistance with study design, planning, and implementation. They thank a team of institutions and individuals for supporting the experiment in Vietnam including farmers in Cát Trinh Commune, Hoang Văn Tùng at Research and Development Centre for Animal Husbandry in the Central Region, Nguyễn Văn Mười at Cát Trinh Commune, Nguyễn Tiên Thinh at Hue University of Agriculture and Forestry, and students at Hue University of Agriculture and Forestry. They acknowledge Dr. Nguyễn Hữu Văn at Hue University of Agriculture and Forestry and Jeff Corfield at Corfield Consultants for assistance with experiment design and insights about forage production systems in the experiment region. This field experiment was partially supported by an array of funding sources: Fulbright-Hays Doctoral Dissertation Research Abroad Fellowship, Richard Bradfield Research Awards, Cornell University Graduate School Travel Grants, Mario Einaudi Center for International Studies International Research Travel Grants, Gerald O. Mott Scholarship for Meritorious Graduate Students in Crop Science, Andrew W. Mellon Awards, Wilson G. Pond International Travel Award, and Clinton-Dewitt-Smith Fellowship. U.S. Department of Education Foreign Language and Area Studies Fellowships permitted acquisition of Vietnamese language proficiency, which was important to experiment participation, communication, and relationships with experiment participants.

References

[1] D. Parsons, P. A. Lane, L. D. Ngoan et al., "Systems of cattle production in South Central Coastal Vietnam," *Livestock Research for Rural Development*, vol. 25, article 25, 2013.

[2] N. X. Ba, P. A. Lane, D. Parsons et al., "Forages improve livelihoods of smallholder farmers with beef cattle in South Central

Coastal Vietnam," *Tropical Grasslands-Forrajes Tropicales*, vol. 1, no. 2, 2013.

[3] W. Stür, T. T. Khanh, and A. Duncan, "Transformation of small-holder beef cattle production in Vietnam," *International Journal of Agricultural Sustainability*, vol. 11, no. 4, pp. 363–381, 2013.

[4] N. X. Ba, N. H. Van, J. Scandrett et al., "Improved forage varieties for smallholder cattle farmers in South Central Coastal Vietnam," *Livestock Research for Rural Development*, vol. 26, article 158, 2014.

[5] H. L. P. Khanh, N. X. Ba, N. H. Van et al., "Influence of labour saving in uptake of improved forage technologies by small-holder farmers in south central Vietnam," in *Proceedings of the 16th AAAP Animal Science Congress*, pp. 1126–1129, Jogjakarta, Indonesia, November 2014.

[6] F. Holmann, L. Rivas, P. Argel, and E. Pérez, "Impact of the adoption of *Brachiaria* grasses: Central America and Mexico," *Livestock Research for Rural Development*, vol. 16, article 98, 2004.

[7] W. W. Stür, P. M. Horne, F. A. Gabunada Jr., P. Phengsavanh, and P. C. Kerridge, "Forage options for smallholder crop-animal systems in Southeast Asia: working with farmers to find solutions," *Agricultural Systems*, vol. 71, no. 1-2, pp. 75–98, 2002.

[8] H. T. T. Hoa, P. T. Cong, H. M. Tam, W. Chen, and R. Bell, "Sandy soils in South Central Coastal Vietnam: their origin, constraints and management," in *Proceedings of the 19th World Congress of Soil Science, Soil Solutions for a Changing World*, Brisbane, Australia, August 2010.

[9] C. Neve, P.-Y. Ancion, H. H. T. Thai, T. P. Khanh, C. N. Chiang, and J. E. Dufey, "Fertilization capacity of aquatic plants used as soil amendments in the coastal sandy area of Central Vietnam," *Communications in Soil Science and Plant Analysis*, vol. 40, no. 17-18, pp. 2658–2672, 2009.

[10] C. Devendra, "Crop-animal systems in Asia: implications for research," *Agricultural Systems*, vol. 71, no. 1-2, pp. 169–177, 2002.

[11] C. A. Palm, R. J. Myers, and S. M. Nandwa, "Combined use of organic and inorganic nutrient sources for soil fertility maintenance and replenishment," in *Replenishing Soil Fertility in Africa*, R. Buresh, P. Sanchez and, and F. Calhoun, Eds., pp. 193–217, 1997.

[12] S. Zingore, R. J. Delve, J. Nyamangara, and K. E. Giller, "Multiple benefits of manure: the key to maintenance of soil fertility and restoration of depleted sandy soils on African smallholder farms," *Nutrient Cycling in Agroecosystems*, vol. 80, no. 3, pp. 267–282, 2008.

[13] S. Goyal, K. Chander, M. C. Mundra, and K. K. Kapoor, "Influence of inorganic fertilizers and organic amendments on soil organic matter and soil microbial properties under tropical conditions," *Biology and Fertility of Soils*, vol. 29, no. 2, pp. 196–200, 1999.

[14] K. Kaur, K. K. Kapoor, and A. P. Gupta, "Impact of organic manures with and without mineral fertilizers on soil chemical and biological properties under tropical conditions," *Journal of Plant Nutrition and Soil Science*, vol. 168, no. 1, pp. 117–122, 2005.

[15] D. Ross, "Recommended soil tests for determining exchange capacity," in *Recommended Soil Testing Procedures for the Northeastern United States. Northeastern Regional Bulletin #493*, J. Sims and A. Wolf, Eds., pp. 62–69, Agricultural Experiment Station, University of Delaware, Newark, Del, USA, 1995.

[16] E. Schulte and B. Hopkins, "Estimation of organic matter by weight loss-on-ignition," in *Soil Organic Matter: Analysis and Interpretation*, F. Magdoff, M. Tabatabai, and E. Hanlon, Eds., pp. 21–31, Soil Science Society of America, 1996.

[17] A. Mehlich, "Mehlich 3 soil test extractant: a modification of Mehlich 2 extractant," *Communications in Soil Science and Plant Analysis*, vol. 15, no. 12, pp. 1409–1416, 1984.

[18] J. B. Peters, S. Combs, B. Hoskins et al., *Recommended Methods of Manure Analysis*, University of Wisconsin Cooperative Extension, 2003.

[19] D. W. Pribyl, "A critical review of the conventional SOC to SOM conversion factor," *Geoderma*, vol. 156, no. 3-4, pp. 75–83, 2010.

[20] W. E. Jokela and G. W. Randall, "Fate of fertilizer nitrogen as affected by time and rate of application on corn," *Soil Science Society of America Journal*, vol. 61, no. 6, pp. 1695–1703, 1997.

[21] IUSS Working Group WRB, "World reference base for soil resources 2014: international soil classification system for naming soils and creating legends for soil maps," World Soil Resources Reports 106, FAO, Rome, Italy, 2014.

[22] H. T. T. Hoa, *Survey Report for Soils Component 2*, Australian Centre for International Agricultural Development, 2009.

[23] E. McLean, "Soil pH and lime requirement," in *Methods of Soil Analysis. Part 2—Chemical and Microbiological Properties*, A. L. Page, R. H. Miller, and D. R. Keeney, Eds., pp. 199–224, Soil Science Society of America, 2nd edition, 1982.

[24] W. Dahnke and G. Johnson, "Testing soils for available nitrogen," in *Soil Testing and Plant Analysis*, R. Westerman, Ed., pp. 127–140, Soil Science Society of America, 3rd edition, 1990.

[25] G. C. Marten, J. Shenk, and F. Barton, *Near Infrared Reflectance Spectroscopy (NIRS): Analysis of Forage Quality*, vol. 643 of *Agricultural Handbook*, United States Department of Agriculture, Agricultural Research Service, 1989.

[26] G. Griffin, D. Jokela, D. Ross, D. Pettrinelli, T. Morris, and A. Wolf, "Recommended soil nitrate tests," in *Recommended Soil Testing Procedures for the Northeastern United States*, J. Sims and A. Wolf, Eds., Northeastern Regional Publication No. 493, University of Delaware Agricultural Experiment Station, 2nd edition, 1995.

[27] SAS Institute, *SAS for Windows. V. 9.3*, SAS Institute, Cary, NC, USA, 2011.

[28] G. Kidder, C. Chambliss, and R. Mylavarapu, "UF/IFAS standardized fertilization recommendations for agronomic crops," Fact Sheet SL-129, University of Florida Cooperative Extension Service, Gainesville, Fla, USA, 2002.

[29] D. Beegle, "Soil fertility management," in *The Penn State Agronomy Guide 2013-2014*, A. Kirsten, Ed., Pennsylvania State University, State College, Pa, USA, 2013.

[30] J. H. Cherney, D. J. R. Cherney, and T. W. Bruulsema, "Potassium management," in *Grass for Dairy Cattle*, J. H. Cherney and D. J. R. Cherney, Eds., pp. 137–160, CAB International, 1998.

[31] R. M. Boddey, R. MacEdo, R. M. Tarré et al., "Nitrogen cycling in *Brachiaria* pastures: the key to understanding the process of pasture decline," *Agriculture, Ecosystems & Environment*, vol. 103, no. 2, pp. 389–403, 2004.

[32] E. Amézquita, R. J. Thomas, I. M. Rao, D. L. Molina, and P. Hoyos, "Use of deep-rooted tropical pastures to build-up an arable layer through improved soil properties of an Oxisol in the Eastern Plains (Llanos Orientales) of Colombia," *Agriculture, Ecosystems & Environment*, vol. 103, no. 2, pp. 269–277, 2004.

[33] F. N. Ponnamperuma, "Effects of flooding on soils," in *Flooding and Plant Growth*, T. T. Kozlowski, Ed., Academic Press, Cambridge, Mass, USA, 1984.

[34] V. Reis, F. dos Reis Jr., D. Quesada et al., "Biological nitrogen fixation associated with tropical pasture grasses," *Australian Journal of Plant Physiology*, vol. 28, pp. 837–844, 2001.

[35] R. M. Boddey and R. L. Victoria, "Estimation of biological nitrogen fixation associated with *Brachiaria* and *Paspalum* grasses using ^{15}N labelled organic matter and fertilizer," *Plant and Soil*, vol. 90, no. 1, pp. 265–292, 1986.

[36] B. Eickhout, A. F. Bouwman, and H. Van Zeijts, "The role of nitrogen in world food production and environmental sustainability," *Agriculture, Ecosystems & Environment*, vol. 116, no. 1-2, pp. 4–14, 2006.

[37] R. Chikowo, P. Mapfumo, P. Nyamugafata, and K. E. Giller, "Maize productivity and mineral N dynamics following different soil fertility management practices on a depleted sandy soil in Zimbabwe," *Agriculture, Ecosystems & Environment*, vol. 102, no. 2, pp. 119–131, 2004.

[38] D. H. Lynch, R. P. Voroney, and P. R. Warman, "Nitrogen availability from composts for humid region perennial grass and legume-grass forage production," *Journal of Environmental Quality*, vol. 33, no. 4, pp. 1509–1520, 2004.

5

The Chilhuacle Chili (*Capsicum annuum* L.) in Mexico: Description of the Variety, Its Cultivation, and Uses

Víctor García-Gaytán,[1] **Fernando Carlos Gómez-Merino,**[2] **Libia I. Trejo-Téllez,**[1] **Gustavo Adolfo Baca-Castillo,**[1] **and Soledad García-Morales**[2]

[1]*Colegio de Postgraduados, Campus Montecillo, 56230 Texcoco, MEX, Mexico*
[2]*Colegio de Postgraduados, Campus Córdoba, 94946 Amatlán de los Reyes, VER, Mexico*

Correspondence should be addressed to Fernando Carlos Gómez-Merino; fcgmerino@gmail.com

Academic Editor: David Clay

The chilhuacle chili (*Capsicum annuum* L.) is a Mexican native variety whose production has been highly valuable because it is the main ingredient of the Oaxacan black mole, a typical Mexican dish. It is basically grown in the Cañada Region of the State of Oaxaca, Mexico, within the Tehuacán-Cuicatlán Biosphere Reserve. Importantly, it is cultivated under traditional agricultural systems, where a range of agronomic constraints associated with the production process and the incidence and severity of pests and diseases represent significant impediments that hinder the yield potential. Additionally, the genetic basis of the crop is highly restricted. Under such environmental and production conditions, the mean crop yield of chilhuacle chili can reach $1\,t\,ha^{-1}$ of dehydrated fruits, which can be used in the food, chemical, and pharmaceutical industries. In this review we summarize the current progress on chilhuacle chili cultivation and outline some crucial guidelines to improve production, as well as other research topics that need to be further addressed.

1. Introduction

Chilies belong to the genus *Capsicum* (Solanaceae family), which is one of the most cultivated groups of species in the world. These diverse species are grown worldwide for vegetable, spice, ornamental, medicinal, and lachrymator uses and are a significant source of vitamins A and C [1, 2]. *Capsicum* is native to the tropical and subtropical Americas, and the majority of the genetic diversity is concentrated in Bolivia, Peru, Brazil, and Mexico. This genus comprises over 30 species, and *C. annuum* is the most widely cultivated and economically important one. *Capsicum annuum* was domesticated in Mexico thousands of years ago and includes both sweet and spicy fruits, with a myriad of shapes, colors, and sizes [2–7].

The remnants of wild chili peppers recovered at various locations in Coxcatlán cave in the Tehuacán Valley, Mexico, and those identified in the Guilá Naquitz cave in the Oaxaca Valley, Mexico, indicate that chilies were harvested in the wild in Mexico more than 8,000 years ago, and their domestication

and cultivation for the first time in Mesoamerica occurred approximately 6,000 years ago [2, 8]. It is believed that seed dispersal was performed by wild birds, while selection and domestication by humans gave rise to various types of fruit morphology and degrees of pungency [1, 4]. The chilhuacle chili is an endemic crop of the Cañada Region in the State of Oaxaca, Mexico, the only place in the world where it is produced.

The main markets to which Mexican chilies are exported include the United States, Japan, Canada, the United Kingdom, and Germany [9]. Currently, Mexico is the second world's leading exporter of fresh chilies and the sixth largest exporter of dehydrated chilies. By entering this thriving export market, chilhuacle chili producers may find new market niches and expand their production and marketing potential. However, new technologies and innovations have to be developed to support a possible rise in production and commercialization. The locations where it is mainly cultivated are San Juan Bautista Cuicatlán, Tomellín, Valerio Trujano, Santa María Tecomavaca, San José del Chilar, and

Chilhuacle chili (*Capsicum annuum* L.)

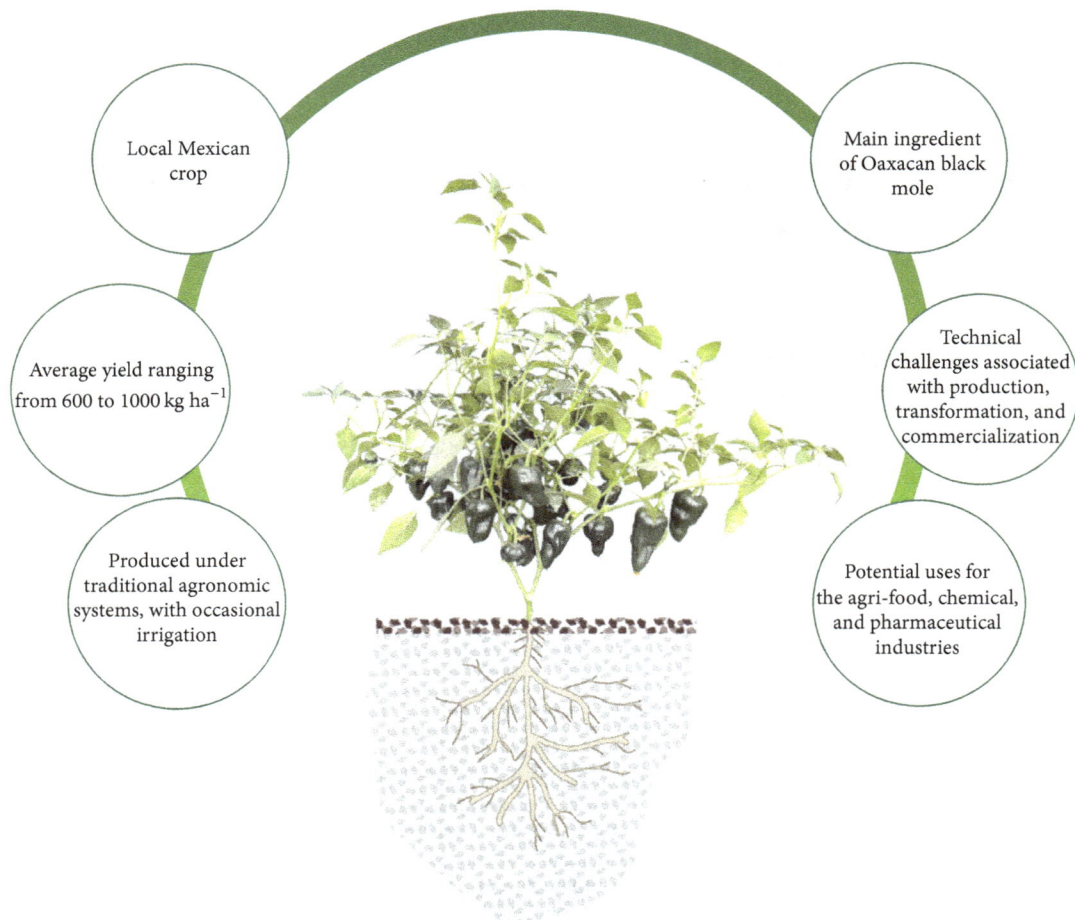

FIGURE 1: Summary of chilhuacle chili production, yield, and potential uses.

Santiago Dominguillo, all within the Cañada Region in Oaxaca, Mexico.

The importance of the chilhuacle chili lies in its international recognition as a culinary spice in the preparation of the famous "Oaxacan black mole" (Mole Negro Oaxaqueño), using dehydrated fruits, while an alternative local dish called "Texmole" is prepared with fresh chili. A graphical abstract summarizing some important issues concerning the chilhuacle chili is displayed in Figure 1.

The culinary and gastronomic value lies in the fact that this chili has unique characteristics in terms of aroma, color, and flavor, which are enhanced by the dehydration process [3]. The culinary use of the chilhuacle chili in Oaxacan regional cuisine has been recognized as a component of Mexican cuisine by the Intangible Cultural Heritage of Humanity for the Organization of the United Nations for Education, Sciences and Culture (UNESCO) [10], which emphasizes the importance of this crop. Despite its importance in Oaxacan cuisine and the demand for it, chilhuacle chili is a species that tends to be increasingly less cultivated and is in danger of extinction according to SAGARPA

Regional Offices' Joint Information System (Sistema de Información Coyuntural de las Delegaciones SAGARPA, SICDE: http://www.sicde.gob.mx) [11]. According to López-López et al. [12], new strategies for growing chilhuacle chili are needed because there are currently only a few producers who cultivate it in the state of in Oaxaca. In this state, the development and generation of technological packages are practically nonexistent, which is reflected by the scarcity of research on production systems. For these reasons, it is necessary to undertake various strategies to deepen our knowledge of this crop to provide a solid foundation that permits its sustainable utilization. This research highlights different aspects of this crop generated in the areas where it is cultivated.

2. Culinary Uses of Chilhuacle Chili by Indigenous People

The chilhuacle chili, or huacle chili, is mainly used by the Cuicatecos and Chinantecos native groups of Mexico. According to the National Commission for the Development of Indigenous Peoples (CDI: http://www.cdi.gob.mx) and the National Population Council [13], the Cuicateco population is slightly greater than 13,000, while the Chinanteco region has over 137,000 inhabitants, distributed mainly in Northern Oaxacan. The chilhuacle chili is mainly used by these towns for religious purposes at festivals such as the Day of the Dead, the Days of the Patron Saints of Towns, Christmas, and New Year's festivities, as well as family celebrations such as weddings and birthdays [14].

The chilhuacle chili provides unique flavor characteristics, and each dish is recognized by the manner and the place in which it is prepared. Oaxacan black mole is the most widely recognized culinary specialty in which this chili is used; when it is prepared with traditional methods, a wide variety of local and international ingredients, such as peanuts, walnuts, almonds, cinnamon, cloves, seedless raisins, oregano, and Oaxacan table chocolate, is used [15]. As a result of its high demand and its increasing scarcity, the price of dehydrated fruits can vary between 25 and 38 US dollars per kilogram, making it one of the most expensive ingredients in Mexican cuisine. As an alternative, and with the aim of lowering production costs, restaurants have opted to substitute the chilhuacle chili for the guajillo and ancho chilies to prepare black mole; however, this also affects the original flavor and consistency of this variant of Mexican mole.

3. Main Topics Addressed in the Literature on Chilhuacle Chili

Because of its importance, there are various publications on the culinary richness of this chili. The fruit is scarce, produced during a short period, and its Nahuatl name suggests a pre-Columbian domestication [16, 17]. There are three known types of chilhuacle chili cultivar that are differentiated by their color: the yellow, red, and black chilies. These characteristics impart a deep and intense flavor to the famous Oaxacan black mole dish [18]. The chilhuacle chili is one of the most widely consumed species (though not in large volumes) in the State of Oaxaca, in addition to other chilies, such as the agua, pasilla, guajillo, tabiche, tusta, and paradito [19]. This widespread consumption ensures that the diversity of local types of chilhuacle chili is not lost in these various localities, which promotes conservation based on consumption. Chili continues to be a catalyst in the Mexican kitchen and has been employed for thousands of years to modify the flavors of the basic diet of the country, surviving throughout the centuries in spite of the introduction and the influence of culinary traditions from other countries [20]. Generally, it is the local chilies that impart a unique characteristic to the regional food, as is the case for the use of the chilhuacle chili to prepare mole in Oaxaca or the Habanero in Yucatecan cuisine.

Other publications consider the chilhuacle chili to be a variety that is in danger of extinction, highlighting the great importance of conducting research into its production system. According to the SAGARPA Regional Offices' Joint Information System [11], Mexico is the center of origin and diversity of C. annuum species; however, there is no accurate record of chili varieties, and some of them are cultivated to only a limited extent, thus producing only low yields, a factor driving the high market price, as is the case for the chilhuacle chili. Importantly, chilhuacle chili is itself in danger of disappearing and being replaced by other types of chili. Year by year fewer farmers cultivate it, as a result of migration and farm abandonment because young people are less interested in agriculture. This phenomenon is complex, though main reasons are related to the very low profits farmers obtain from their crops. We think that technology can bring about new tools to farmers to cultivate chilhuacle chili and increase profits.

The cultivation of the chilhuacle chili, and other species, should be preserved and promoted or these chilies could disappear from the Oaxacan fields. This crop can obtain a high market value because of its desired taste and the limited extent of production. Nevertheless, there is a paucity of research on the chilhuacle chili. Among those publications that exist, the literature covers topics related to management and production strategies, culinary uses, and the recovery of genetic resources. This last section aims to emphasize the fact that chilhuacle chili is undergoing the process of extinction, demonstrating the necessity to strengthen efforts to implement strategies for its study and for the purpose of its conservation and sustainable use. In addition to exploring all biological aspects of the chilhuacle chili, technological packages for its production, processing, and sale that are based on technological innovations and improvements in organizational, commercial, and managerial strategies [21] need to be developed.

4. Origin, Taxonomy, and Botany of Chilhuacle Chili

Pungency and color of chili fruits are crucial factors in determining their quality and commercial value. Such characteristics depend on the genotype and time of planting and harvest [22]. In particular, the origin of chilhuacle chili is restricted to the Cañada Region in the Tehuacán-Cuicatlán Biosphere Reserve, which is located in southern Mexico, within the states of Puebla and Oaxaca (Figure 2). Most species that inhabit this region are restricted to the limits of this area and are therefore endemic species [23, 24].

Chilhuacle chili is a cultivar of the species C. annuum [25]. Capsicum annuum var. annuum is the domesticated form and the most important in Mexico and across the globe; it also has the highest morphological variability because it comprises the majority of the chili and pepper types grown in Mexico [26]. In Mexico, the National Commission for the Knowledge and Use of Biodiversity (CONABIO, for its acronym in Spanish) has compiled detailed taxonomic records for Capsicum annuum var. annuum [27].

The chilhuacle chili is a plant of dichotomous branching. Its root may reach depths of 70 to 120 cm. The bulk of the roots is located within a soil depth of 5–40 cm. The plants have a green cylindrical herbaceous main stem that is semiwoody at the base and slightly pubescent. The growth of the stem is limited, and the stem divides into 3 or 4 branches or secondary stems between the heights of 10 and 40 cm. The dark green leaves are lanceolate or oval, have an entire lamina

FIGURE 2: Cañada Region in the State of Oaxaca, Mexico, the only place in the world where chilhuacle chili is produced. Cañada Region is located within the Tehuacán-Cuicatlán Biosphere Reserve in southeast Mexico, within the State of Puebla and Oaxaca.

margin, and are weakly pubescent with a nonerect petiole. The flowers appear solitary in the node of branches of the stem, and each branch can have 5 to 6 or more flowers (Figure 3).

Flowers are perfect and regular and composed of 6-7 sepals partially fused together and 6-7 petals. Their position is intermediate, according to the International Plant Genetic Resources Institute [28]. The androecium is composed of 7 equal stamens, bilocular and dehiscence inwards or terminal. The ovary is superior, of 2-3 carpels with a single style and stigma (Figure 4). Chilhuacle chili flowers are self-polli-nated, though bees, wasps, and ants may contribute to cross pollination, which is common when it is cultivated under field conditions [17].

During the early stages of growth, the fruit appears round and the calyx covers much of the outside. Subsequently, the fruit acquires an elongated form, with a 4 to 8 mm thick peduncle. When the fruit acquires its final size and shape, the calyx becomes immersed in the fruit and the thickness of the peduncle decreases (Figure 5).

The fruit is a berry in the form of a capsule, with trapezoidal shape and a mean size of 10 cm in length and 8 cm in equatorial diameter. Fruits are usually consumed when they reach maturity (60 to 78 days after pollination (DAP)) and have been dehydrated. Fruit color varies according to developmental stages (Table 1, Figure 6), and this character-istic is crucial to determine the right moment of harvesting.

Changes in color of mature dehydrated fruits are depicted in Figure 7. Once the fruits have been dehydrated, seeds retain their germination capacity for 3 to 4 years [3, 29]. In order to preserve seed germination capacity, local farmers keep seeds within the dehydrated fruits, until the new production cycle starts.

5. Agronomic Practices for Chilhuacle Chili Production

In Mexico, the State of Oaxaca has one of the highest diversity levels of chilies, including chilhuacle chili [30, 31]. Under Oaxacan environmental conditions, some agronomic constraints are associated with the production process of this crop, and the incidence and severity of pests and diseases are among the most significant impediments that hinder the potential of this crop [32].

Chilhuacle chili, as well as soledad, costeño, and agua chilies, which are also native to Oaxaca, are domesticated cultivars grown under traditional agricultural systems, where no sophisticated technologies are used [33]. In the Cañada Region, seeds of this chili are germinated at the end of the dry season (i.e., May). Once seeds germinate, plantlets are transplanted to the crop fields in June and July, during the raining season. Eventually, local farmers irrigate the crop if necessary. Harvest takes place in October, once the rainy season has ended.

The greatest yield of chilhuacle chili has been obtained with a density of 53,000 plants per hectare, although the largest fruits with the greatest dry masses can be obtained at a lower density (i.e., 40,000 plants per hectare) [34]. Furthermore, pruning increases the yield per plant and per unit surface area, although the fruits tend to be smaller [34]. When nutrient solutions are applied during irrigation, the best agronomic responses are observed with the Steiner [35] and Escobar [36] nutrient solutions [29].

When grown under field conditions, this crop performs better in friable and well-drained, sandy loam soils with pH values between 6.5 and 7.5, with a minimum depth of 35 to 50 cm [37, 38], and slopes ranging from 1 to 10%. Luvisols,

FIGURE 3: Morphology of a mature chilhuacle chili plant and anatomy of fruits. (1) A chilhuacle plant during fruit development. (2) A mature fruit. (3) A stigma. (4) Stamens. (5) Petals. (6) Flowers. (7) Leaves. (8) Cross section of immature fruit showing septum and seeds. (9) Immature seeds of the fruit. (10) Mature seeds are yellow.

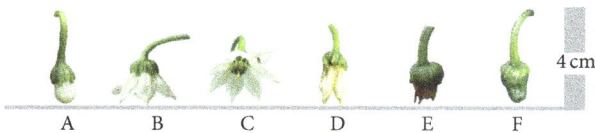

FIGURE 4: Flower developmental stages of chilhuacle chili. (A) Floral bud 1 day after pollination (DAP). (B) 2 DAP. (C) 4 DAP. (D) 5 DAP. (E) 7 DAP. (F) Fruit development 8 DAP.

Cambisols, and Phaeozems are the main types of soils where this chili variety is most often cultivated [32]. The terrains where chilhuacle chili is cultivated are located from 687 to 1085 meters above sea level (masl) [32].

Climates of the region where this variety is produced are predominantly semiarid and warm (BSh climate), with a mean annual rainfall of 450 mm, the majority of which falls between June and September, and a mean annual temperature higher than 22°C, with temperatures higher than 18°C in the coldest month [39, 40].

5.1. Seedling Production. Chilhuacle chili seedling production is accomplished by sowing the seeds in soil, with selected seeds obtained from well-selected mature chilies. The seedbed is prepared in an open (nonshaded) area, and the soil is manually tilled with a hoe or a shovel. For the seedbed, a mixture of one-third of properly composted cattle manure, another third of local soil, and a remaining third of fine sand is prepared. Planting beds of $1 \, m^2$, with 20 cm borders on the sides, are created using this substrate mixture. The seeds are broadcast-sown and covered using a broomstick. The soil should be moist for as long as the seedlings are in the seedbed, without flooding; hence, proper drainage is necessary with light and frequent irrigation. Similar practices for obtaining seedlings of other varieties of peppers native to Oaxaca, such as the soledad and agua chilies, have been described by various nurseries [41]. A solution of propamocarb chlorohydrate

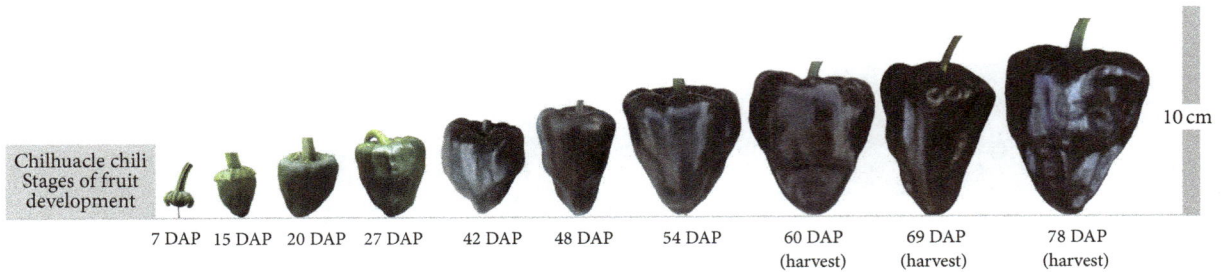

FIGURE 5: Development of the chilhuacle chili fruit. The fruit develops 7 days after pollination [DAP]. At the beginning of fruit development, the calyx covers a substantial portion of the outside of the fruit (15 DAP). Subsequently, the fruit acquires an elongated form, with a thick peduncle (20 to 27 DAP). When the fruit has achieved its final size and shape, the color is black or very deep purple and the calyx is immersed in the fruit (42 to 54 DAP). Developed fruits can be harvested between 60 and 78 DAP.

TABLE 1: Chilhuacle chili developmental stages and color changes from immature to mature fruits.

| Color scale | Developmental stage | Days after pollination (DAP) | Color | | | Description of color |
			Luminosity	Chroma	Hue	
1	Immature	15	22.8	7.0	139.2	Light green
2	Green	27	22.4	7.1	138.9	Bright green
3	Green-mature	42	22.7	7.5	141.6	Bright dark green
4	Mature	54	21.0	5.3	155.9	Completely dark green

FIGURE 6: Color changes from immature to mature fruits of chilhuacle chili: (1) Immature fruit. (2) Green fruit. (3) Green mature fruit. (4) Mature fruit.

FIGURE 7: Dehydrated fruit of chilhuacle chili. (A) A mature chilhuacle fruit 78 DAP. (B) A dehydrated fruit. (C) A dehydrated fruit, showing the apex, which consists of a thick pericarp and placental tissue to which the seeds are attached. (D) In mature dehydrated fruits, the seeds are flattened and yellow in color.

(64% aqueous solution) at a concentration of $0.5 \, \mathrm{g \, L^{-1}}$ may be applied to prevent fungal diseases when planting chilhuacle chili seedlings in polystyrene trays, with a 1 : 1 mixture of peat : perlite, with 2 seeds sown per cavity in damp substrate and irrigation to saturation [29, 42]. The success of the chili production process lies in the strength and size of the seedling used in planting, which depends on seed quality, seedbed preparation, soil disinfection, nutrient applications, and general handling in the nurseries [43]. Local farmers of the Cañada Region do not disinfect soils, though they use other sustainable practices such as crop rotation.

In our experimental conditions, seedling emergence occurred seven days after sowing, which differs from other chilies such as poblano with a germination period of 9 days [44], mirasol with 10 to 12 days, árbol and ancho chilies with 3 days, and guajillo chili with 4 days to start germination [45].

Under greenhouse conditions, transplantation of chilhuacle chili plants in polythene bags with a sand substrate was performed successfully 57 days after sowing; plants achieved a mean height of 20 cm with 2 true leaves fully formed and extended during this growth period [29]. For seedling growth and development, the choice of substrate is a key factor that contributes to seedling quality [46]. We have observed that the choice of substrate supplemented with 25% Steiner's nutrient solution [47] affects the growth and development of chilhuacle chili plantlets, which is most evident 50 days after planting (Figure 7). Plants that developed in peat were taller and had a greater number of leaves, and the root and substrate remained intact when the seedlings were removed from the seedbed (Figure 8(A)), while those cultivated in agricultural local soil were shorter than those that developed in peat, although the root ball had similar characteristics to those developed in peat (Figure 8(B)). The beneficial effect of peat on plant growth may be attributed to its biological, physical, and chemical properties that improve nutrient availability to plant roots [48], whereas the local soil conferred higher compaction and hence less water and nutrient availability to plants. Tezontle, a Mexican local volcanic gravel, is inert

FIGURE 8: Characteristic chilhuacle chili seedlings and their root balls in three different substrates 50 days after planting. (A) Seedling developed in peat. (B) Seedling developed in common soil. (C) Seedling developed in volcanic gravel. Plants were irrigated every other day with 25% Steiner's nutrient solution.

and its water retention capacity is very low [49]. In our experimental conditions, plants that developed in tezontle as substrate were shorter than those grown in peat or local soil, and the root ball had a lower transplantation value given that it disintegrated easily (Figure 8(C)).

The choice of substrates for planting is crucial because it provides appropriate conditions to the crop for root growth [46]. One of the most globally used substrates for seedling production is peat moss because its main characteristics provide excellent germination and seedling growth. However, the high cost and nonsustainable exploitation of peat are restricting its use [50], which highlights the importance of finding new local substrates that ensure excellent germination of genetic resources such as the chilhuacle chili.

5.2. Chilhuacle Chili Nutrient Management. Chilhuacle chili is mainly cultivated under rainfed conditions, eventually supplemented with irrigation systems. When irrigated, water is applied every 8 days, for a total of 17 irrigations during the crop cycle. The irrigation is applied to small surface areas, ranging from 2,500 to 5,000 m^2, and occasionally on land areas of 1 ha [29]. A chilhuacle chili fertilization program may comprise inorganic fertilizer applications at 25 and 46 days after transplantation, while a number of producers apply 200 kg of 17-17-17 (NPK) fertilizer or 200 kg of urea (equivalent to 92 kg of N ha^{-1}) for each growth cycle [32]. Interestingly, N doses applied by local Oaxacan producers of chilhuacle chili are low compared to that of the recommended fertilizer application for soledad chili (also native to the State of Oaxaca), which are 400, 300, and 300 kg ha^{-1} N, P, and K, respectively, in the municipalities of Loma Bonita, Tuxtepec, Valle Nacional, and Santa María

Jacatepec, which have sandy-crumbly and sandy-clayey soil textures [41]. Under this management regime, the mean yield ranges from 600 to 1000 kg ha^{-1} of dehydrated chili [3, 29], with N, P, K, Ca, and Mg being the most important nutrients affecting yields [51]. The fertilization regimes of other varieties of chilies may be relevant to chilhuacle chili production, but regardless of that, soil characteristics must also be taken into account for production purposes. For example, for the cultivation of local and commercial varieties of chilies in Yucatán, Mexico, it is recommended to apply 280-200-330 kg ha^{-1} NPK plus 10 t ha^{-1} of chicken manure distributed in three stages as follows: first, at the time of the transplant, all of the chicken manure and half of the N, P, and K; the second application during flowering, applying 25% of N and 50% of the remaining P and K; and finally, the remaining quantity of N after the third harvest [52]. In the production of mirasol chili in loam-textured soils with a mean organic matter content of 1.93%, pH of 7.8, and electrical conductivity of 0.63 dS m^{-1}, in Zacatecas, Mexico, the application of 200-160-100 kg ha^{-1} NPK + calcium has been recommended [53]. In this variety, the application of 210-150-100 kg ha^{-1} NPK increased the total yield and the plants produced better quality dried fruit [54]. Accordingly, the application of 120-60-00 kg ha^{-1} NPK for mirasol variety cultivated in San Luis Potosí (Mexico) highlands has been recommended to be applied in 2 stages: first, before the third irrigation, half of the N and all the P (i.e., 60-60-00 kg ha^{-1} of NPK) are applied, and the remaining 60 kg of N is applied at the onset of flowering. With this fertilization dosage, the mean yield is 1.2 t ha^{-1} of dry chili at planting densities of 30,000 plants per hectare [55].

In general, yields are affected by various factors including genotype, climate, soil fertility, fertilization dose, pest and disease control and management, and harvesting and processing methods. Regional and seasonal variations in the environment, cultural practices, the availability of N, and the absorption efficiency of N by the plant are also considered to be determining factors for yield [56], although neither the total production nor the productivity per plant or per surface area increases synchronously with nutrient supply [57]. In fact, fertilization can have secondary, often unpredictable effects on growth and yield through changes in growth pattern, plant morphology, anatomy, or chemical composition, which can increase or decrease the resistance or tolerance of plants to biotic and abiotic stress factors [58]. Research approaches that target the Capsicum genus that scarcely study genotypes such as the chilhuacle chili are required to determine the precise physical environment and management conditions that should be implemented.

With respect to agronomic management, there are two greenhouse studies of chilhuacle chili. In the first one, different planting densities and the pruning of side stems were evaluated, reporting that the highest yields were obtained with high densities and without pruning, using the Escobar nutrient solution [34, 36]. This solution has been used for the growth of plantings established on substrates such as perlite and rockwool. The solution has the following concentrations (in mEq L^{-1}) of anions, NO_3^- (13.5), $H_2PO_4^-$ (1.5), and SO_4^{2-}

(1.35), and of cations, K^+ (5.5), Ca^{2+} (4.5), and Mg^{2+} (1.5). In the second study, physiotechnical and quality variables of chilhuacle chili fruits from plants cultivated using Escobar [36], Steiner [35], and Urrestarazu [59] nutrient solutions were tested [29]. Plants exhibited a better response to the Steiner and Escobar nutrient solutions as a consequence of the ionic mutual ratio [36], which is defined as a mutual relationship among anions (NO_3^-, $H_2PO_4^-$, and SO_4^{2-}) and a mutual relationship among cations (K^+, Ca^{2+}, and Mg^{2+}). Therefore, if the relationship among them is adequate, the plant can achieve its maximum potential [60]. Although the effects of nutrients on the growth and yield of the chilhuacle chili can be explained in terms of the function of these elements in plant metabolism, in the *Capsicum* genus, the demand for nutrients such as N and K has been observed to increase during the flowering, fruiting, and fruit-filling stages [61]. In varieties of chilies such as jalapeño, habanero, and a number of commercial hybrids, the levels of N and K significantly affect plant growth, stem diameter, number of leaves, fruit yield, pungency, and capsaicin levels [62–64].

5.3. Pests and Diseases. The chilhuacle chili has various problems of a phytosanitary nature, such as the pepper weevil (*Anthonomus eugenii* Cano) and insect vectors of viruses (aphids and whiteflies). The pepper weevil is one of the most important insect pests of this crop [29], especially during the flowering and fruiting stages. The larva of this insect drastically reduces fruit number, causing early fruit drop, premature ripening, and fruit deformation, which together may reduce the harvest by 90% [65, 66]. Although there is no recommended control method described for preventing pepper weevil damage in chilhuacle chili, generally, a chemical control based on chlorpyrifos, oxamyl, and fipronil at doses of 720, 520, and 50 g of active ingredient per hectare, respectively, is recommended [67, 68]. Nevertheless, the best prevention method is to select genotypes displaying resistance against such biotic stress factors. Accordingly, we are currently creating novel genetic diversity using gamma radiation by ^{60}Co and the evaluation of the first-generation of mutant varieties is underway, which may result in a major breakthrough to increase diversity in chilhuacle chili in Mexico.

Though important pests and diseases in chilhuacle chili plants are being documented, only a few insect pests and pathogens have been reported so far. Just recently, we have published the first report of powdery mildew in chilhuacle chili caused by *Leveillula taurica* in southern Mexico [69]. Therefore, more in-depth studies on those phytosanitary problems need to be conducted in order to ensure a sustainable crop production system.

5.4. Fruit Dehydration. The crop is cultivated to produce dried chilies to be sold in local markets in Oaxaca and Puebla, Mexico. Dehydration is performed in areas with slopes greater than 5%, where ripe fruits are placed on stone beds and left under direct sunlight. This procedure has a considerable influence on the color, flavor, texture, and nutritional quality of the dried chilies. This method also prevents any damage associated with excessive moisture [3].

However, the labor is costly and involves a large number of working hours because each day the fruit has to be taken out of the sun, turned over, and stored for the afternoon until a suitable moisture level for sale at the local markets is achieved. Hence, one area for innovation is the postharvest handling procedure, mainly in relation to the dehydration process.

6. Chilhuacle Chili May Be More Than a Spicy Culinary Fruit

In general, products derived from *Capsicum* fruits include fresh, dried, or pickled pepper, ground powders, and processed products such as purees, sauces, and oleoresins. Oleoresins contain a significant amount of esters of capsorubin, capsanthin, cryptoxanthin, zeaxanthin, and other carotenoids [70, 71], used in foodstuff and cosmetics, and serve as a source of the pungent component capsaicin for pharmaceutical products [72, 73] or self-defense weaponry [74]. Bioactive compounds present in *Capsicum* fruits display antioxidant, anticancer, anti-inflammatory, antiulcer, and antiobesity pharmaceutical properties, among others, and they also promise other health benefits [75]. Furthermore, these fruits exhibit a wide range of pharmacological activities, including chemopreventive, analgesic, antilithogenic, antidiarrhoeal, antiallergic, antidiabetic, antihypertensive, hypoglycaemic, antimicrobial, antioxidant, antifungal, and antiviral properties [76].

Compounds known as capsaicinoids cause the pungency of chili pepper fruit. Capsaicinoids include capsaicin, dihydrocapsaicin, nordihydrocapsaicin, homodihydrocapsaicin, and homocapsaicin, with capsaicin being the primary capsaicinoid in chili pepper (i.e., accounting for up to 80% of capsaicinoid content of fruits) [77]. We determined the capsaicinoid concentrations in two groups of mature chilhuacle chili fruits (Table 2). In 78 DAP fresh fruits the concentration of capsaicin (in dry basis) was lower than that observed in dehydrated fruits. Moreover, the concentrations of dihydrocapsaicin and nordihydrocapsaicin, as well as the pungency value, were higher in dehydrated fruits. When comparing these values with other chilies produced in the same region, chilhuacle chili has a higher content of capsaicinoids and pungency value than miahuateco, tecomatlán, and mulato but a lower concentration of capsaicin and less pungency than copi [78]. Our results demonstrate that the concentrations of capsaicin and capsaicinoids and the pungency in dehydrated chilhuacle fruits are higher than those found in other chili varieties including guajillo, ancho, pasilla, and puya. However, values found in fruits of jalapeño, mirasol, morita, serrano, chipotle, de árbol, piquín, and habanero are higher than those found in chilhuacle [79]. Capsaicinoid concentrations in chilies may vary according to the genotype, geographic origin, and climatic conditions where they are produced [80]. Nonetheless, the genotype is the most critical factor [81]. Since capsaicin has therapeutic applications [76–78], chilhuacle chili may deliver diverse health benefits.

Finally, *Capsicum* fruits have an enantiomeric composition of phytochemicals. Many of these are chiral molecules that can exist in plants in different enantiomeric forms, such as the pepper aroma compound linalool [82], which occurs

TABLE 2: Capsaicin and capsaicinoid concentrations and pungency of different chilies native to Mexico.

Chili variety	Capsaicin (mg kg^{-1})	Dihydrocapsaicin (mg kg^{-1})	Nordihydrocapsaicin (mg kg^{-1})	Pungency (SHU)	Reference
Chilhuacle (78 DAP)	161.0	47.0	9.0	3379	—
Chilhuacle (dehydrated)	235.0	142.0	53.0	6420	—
Miahuateco	63.6	45.5	na	1637	[79]
Copi	267.4	167.7	na	6525	[79]
Native Tecomatlán	54.6	35.8	na	1354	[79]
Mulato	29.5	49.5	na	1183	[79]
Guajillo	22.85	36.85	na	961.13	[80]
Ancho	42.82	42.19	na	1368.60	[80]
Pasilla	49.21	68.76	na	1899.34	[80]
Puya	53.88	67.40	na	1952.61	[80]
Jalapeño	373.51	210.40	na	9400.83	[80]
Mirasol	353.77	231.70	na	9426.09	[80]
Morita	338.25	334.94	na	10838.24	[80]
Serrano	627.48	399.77	na	16538.78	[80]
Chipotle	883.04	552.65	na	23114.68	[80]
De Árbol	1293.36	641.74	na	31155.10	[80]
Piquín	2656.74	1031.57	na	59381.77	[80]
Habanero	9097.35	4023.63	na	211247.65	[80]

SHU: Scoville Heat Units. na: not available. DAP: days after pollination.

naturally in one enantiomeric form but experiences racemization during postharvest treatment [75, 83]. In the case of the chilhuacle chili, all these chemical and pharmaceutical properties need to be studied in greater depth in a new research environment improved by the development and validation of new analytical protocols.

7. Conclusions and Perspectives

The chilhuacle chili is recognized as a unique ingredient of Mexican cuisine, especially as the characteristic seasoning of Oaxacan black mole. As an endemic genetic resource of Mexico, it is necessary to implement various innovative strategies throughout the value chain that will ensure its conservation and sustainable use.

For the thematic review of the chilhuacle chili, only 15 publications from Mexico were found, which demonstrates the clear need to further study this crop and its uses. In environmental terms, the effects of global climate change on agriculture in Mexico could reduce national agri-food production by 25.7% by 2080 [84], if relevant strategic measures are not taken to address this global phenomenon. On the other hand, the country's population continues to grow, which also makes it necessary to develop efficient strategies that ensure food and energy security. An advantage in terms of global politics is that Mexico has provided numerous nutritional elements that have enriched international gastronomy because, as a megadiverse country, this nation is the center of origin and genetic diversity of numerous crops such as corn, squash, beans, and chilies. This diversity of food crops has long been valued by international agencies such as UNESCO, which recognized Mexican food as an Intangible Cultural Heritage of Humanity in 2010. Oaxacan black mole and the

chilhuacle chili have high nutritional and gastronomic values that justify any innovation initiative undertaken to improve knowledge of their cultivation and use. Nevertheless, only a few producers are growing this variety to date. In the face of these challenges and opportunities, the lines of research, technological development, and innovation that may be proposed can range from basic studies on the genomic, biochemical, and genetic variability of crop physiology to issues related to its use and possible new applications. Characterization of nutritional and nutraceutical properties, development of new drying and processing strategies for the product, improvements in the organization and training of producers, exploration of new market niches and international markets for chili and Oaxacan black mole, and the general management of the products obtained are all required. The benefits of these initiatives could be obtained by the original producers through, perhaps, the opening of Oaxacan cuisine gourmet restaurants managed by the producers themselves. This last possibility represents one of the major challenges for the development of indigenous peoples and the promotion of their cuisine internationally because, in addition to their high linguistic diversity (considering that each Chinanteco town has its own linguistic variation) and cultural diversity, there are serious limitations in terms of education level, mastery of other languages, and resistance to organization and trade, among others. Therefore, any strategy that aims to improve the use and exploitation of this chili genotype will have to consider the human factor as a determinant of the success of such projects. Our team is currently creating the generation of new mutants, with the aim of increasing the genetic diversity of chilhuacle chili in Mexico, and we have published the first report of powdery mildew in chilhuacle chili caused by Leveillula taurica in southern Mexico. In terms of technological innovations, studies that evaluate the oil content of the seeds are required to support their use as seasoning and flavoring ingredients, while alternative uses of the fruit and seeds in the chemical and pharmaceutical industries await further investigation. The generation of standards for good agricultural practices, fair trade, and the appropriate integration of the links in the value chain will help to better position and protect the product in the marketplace.

Disclosure

The current address of Soledad García-Morales is CONACYT-CIATEJ Plant Biotechnology, 45019 Zapopan, Jalisco, Mexico.

Competing Interests

The authors declare that they have no conflict of interests regarding the publication of this work.

Acknowledgments

The authors are grateful to Mexico's National Science and Technology Council (CONACYT-Mexico), as well as the Plant Nutrition Laboratory and the Sustainable Agri-Food Innovation program of the Colegio de Postgraduados, for the support and facilities granted for this research.

References

[1] T. A. Hill, H. Ashrafi, S. Reyes-Chin-Wo et al., "Characterization of *Capsicum annuum* genetic diversity and population structure based on parallel polymorphism discovery with a 30K unigene pepper GeneChip," *PLoS ONE*, vol. 8, no. 2, Article ID e56200, 2013.

[2] L. Perry and K. V. Flannery, "Precolumbian use of chili peppers in the Valley of Oaxaca, Mexico," *Proceedings of the National Academy of Sciences of the United States of America*, vol. 104, no. 29, pp. 11905–11909, 2007.

[3] V. H. Aguilar-Rincón, T. Corona-Torres, P. López-López et al., *Mexican Chilies and Their Distribution*, SINAREFI, Colegio de Postgraduados, INIFAP, IT-Conkal, UANL, UAN, Texcoco, México, 2010.

[4] C. R. Clement, M. de Cristo-Araújo, G. C. d'Eeckenbrugge, A. A. Pereira, and D. Picanço-Rodrigues, "Origin and domestication of native Amazonian crops," *Diversity*, vol. 2, no. 1, pp. 72–106, 2010.

[5] D. Wang and P. W. Bosland, "The genes of *Capsicum*," *HortScience*, vol. 41, no. 5, pp. 1169–1187, 2006.

[6] M. A. Scaldaferro, A. R. Prina, E. A. Moscone, and J. Kwasniewska, "Effects of ionizing radiation on *Capsicum baccatum* var. *pendulum* (Solanaceae)," *Applied Radiation and Isotopes*, vol. 79, pp. 103–108, 2013.

[7] USDA-ARS, "GRIN species records of *Capsicum*," National Genetic Resources Program, National Germplasm Resources Laboratory, Maryland, Md, USA, 2011.

[8] L. Perry, R. Dickau, S. Zarrillo et al., "Starch fossils and the domestication and dispersal of chili peppers (*Capsicum* spp. L.) in the Americas," *Science*, vol. 315, no. 5814, pp. 986–988, 2007.

[9] SIAP (Servicio de Información Agroalimentaria y Pesquera), "Agri-food and fisheries information service," 2013, http://www.siap.gob.mx/produccion-chile-verde/.

[10] UNESCO, *Traditional Mexican Cuisine: Ancestral, Ongoing Community Culture*, The United Nations for Education, Science and Culture, Mexico City, Mexico, 2010, http://www.unesco.org/culture/ich/index.php?RL=00400.

[11] SICDE, "Sistema de información coyuntural de las delegaciones SAGARPA," SAGARPA Regional Offices' Joint Information System, 2013, http://www.sicde.gob.mx/portal/bin/nota.php?from=470&accion=buscar&subrutina=pagina_1&column=2&busqueda=&orderBy=Notas.FechaNota&order=ASC&fecha=&notaId=80-805372450ff098a2dd3d.

[12] P. López-López, R. Rodríguez-Hernández, and E. Bravo Mosqueda, "Economic Impact of the Huacle Chile (*Capsicum annuum* L.) in the State of Oaxaca," *Revista Mexicana De Agronegocios*, vol. 20, no. 38, pp. 317–328, 2016.

[13] CONAPO, "National population council," in *Indicadores de Demográficos Básicos. Datos de Proyecciones, México*, 2010, http://www.conapo.gob.mx/es/CONAPO/Proyecciones_Datos.

[14] P. López-López and D. Pérez-Bennetts, "Huacle chili (*Capsicum annuum* sp.) in the State of Oaxaca, Mexico," *Agroproductividad*, vol. 8, no. 1, pp. 35–39, 2015.

[15] J. MacVeigh, "Mexican cuise," in *International Cuisine*, J. MacVeigh, Ed., pp. 325–349, Delmar Cengage Learning, New York, NY, USA, 2008.

[16] J. S. Andrews, *Pepper Lady's Pocket Pepper Primer*, American Botanical Council and University of Texas Press, Austin, Tex, USA, 1998.

[17] P. W. Bosland and E. J. Votava, "Taxonomy, pod types and genetic resources," in *Peppers: Vegetable and Spice Capsicums. Crop Production Science in Horticulture*, P. W. Bosland and E. J. Votava, Eds., pp. 14–38, CABI Publishing, New York, NY, USA, 2000.

[18] D. DeWitt and P. W. Bosland, *The Complete Chile Pepper Book: A Gardener's Guide to Choosing, Growing, Preserving, and Cooking*, Timber Press, Portland, Calif, USA, 2009.

[19] É. Castellon-Martínez, J. L. Chávez-Servia, J. C. Carrillo-Rodriguez, and A. M. Vera-Guzman, "Consumption preferences of pepper (*Capsicum annuum* L.) Landraces in the central valleys of oaxaca, Mexico," *Revista Fitotecnia Mexicana*, vol. 35, no. 5, pp. 27–35, 2012.

[20] T. J. Long, "The *Capsicum* through Mexican history," in *El chile, Protagonista de la Independencia y la Revolución*, T. H. Hernández-Pons and S. A. De Miguel, Eds., pp. 7–20, Fundación Herdez, Mexico City, Mexico, 2011.

[21] S. Salles-Filho, C. Gianoni, and P. Jeanne, *Methodological Guide for the Diagnosis of National Agro-Food Innovation Systems in Latin America and the Caribbean*, Innovagro, IICA, San José, Costa Rica, 2012.

[22] A. K. De, *Capsicum: The Genus Capsicum*, CRC Press, London, UK, 2003.

[23] J. Rzedowski, "Provincias florísticas de México," in *Vegetación de México*, Comisión Nacional para el Conocimiento y Uso de la Biodiversidad, pp. 104–121, Editorial Limusa, Mexico City, México, 1st edition, 2006.

[24] P. Dávila, M. C. Arizmendi, A. Valiente-Banuet, J. L. Villaseñor, A. Casas, and R. Lira, "Biological diversity in the Tehuacán-Cuicatlán Valley, Mexico," *Biodiversity & Conservation*, vol. 11, no. 3, pp. 421–442, 2002.

[25] S. Raghavan, "A to Z Spices," in *Handbook of Spices, Seasoning, and Flavorings*, S. Raghavan, Ed., pp. 63–185, CRC Press, Boca Raton, Fla, USA, 2007.

[26] S. Montes-Hernández, "Compilation and analysis of existing information on the species of the genus *Capsicum* grown and cultivated in Mexico," Informe Final, Comisión Nacional para el Conocimiento y Uso de la Biodiversidad (CONABIO), Mexico City, Mexico, 2010.

[27] SIOVM, *Information System for Living Modified Organisms*, Proyecto GEF-CIBIOGEM/CONABIO, Mexico City, Mexico, 2011, http://www.conabio.gob.mx/conocimiento/bioseguridad/doctos/consulta_SIOVM.html.

[28] IPGRI; AVRDC and CATIE, *Descriptors for Capsicum (Capsicum spp.)*, International Plant Genetic Resources Institute, Roma, Italia; Centro Asiático para el Desarrollo y la Investigación Relativos a los Vegetales, Taipei, Taiwán; Centro Agronómico Tropical de Investigación y Enseñanza, Turrialba, Costa Rica, 1995.

[29] R. M. Espinosa, *Response of Huacle chili (Capsicum spp.) to four nutrient solutions in soilless culture under greenhouse conditions [Thesis]*, Instituto Politécnico Nacional, Oaxaca, Mexico, 2011.

[30] E. Castellón-Martínez, J. C. Carrillo-Rodríguez, J. L. Chávez-Servia, and A. M. Vera-Guzmán, "Phenotype variation of chile morphotypes (*Capsicum annuum* L.) native to Oaxaca, Mexico," *Phyton*, vol. 83, no. 2, pp. 225–236, 2014.

[31] K. H. Kraft, J. De Jesús Luna-Ruíz, and P. Gepts, "Different seed selection and conservation practices for fresh market and dried chile farmers in Aguascalientes, Mexico," *Economic Botany*, vol. 64, no. 4, pp. 318–328, 2010.

[32] P. López-López and G. H. Castro, *Chilhuacle chili (Capsicum sp.): A Typical Chili of the Cañada Region of Oaxaca*, Fundación Produce Oaxaca, Oaxaca, Mexico, 2005.

[33] A.-M. Araceli, P. L. Morrell, M. L. Roose, and S.-C. Kim, "Genetic diversity and structure in semiwild and domesticated chiles (*Capsicum annuum*; Solanaceae) from Mexico," *American Journal of Botany*, vol. 96, no. 6, pp. 1190–1202, 2009.

[34] A. L. A. Langlé, *Response of Huacle chili (Capsicum spp.) to different planting densities and prunings under intensive greenhouse management [Thesis]*, Instituto Politécnico Nacional, Oaxaca, Mexico, 2011.

[35] A. A. Steiner, "A universal method for preparing nutrient solutions of a certain desired composition," *Plant and Soil*, vol. 15, no. 2, pp. 134–154, 1961.

[36] J. I. Escobar, "Cultivo del pimiento en sustratos en las condiciones del sudeste español," in *Cultivo sin Suelo: Hortalizas en Clima Mediterráneo*, C. E. Martínez and L. F. García, Eds., pp. 109–113, Ediciones de Horticultura, Barcelona, Spain, 1993.

[37] T. K. Lim, "Fruit," in *Edible Medicinal and Non-Medicinal Plants*, T. K. Lim, Ed., pp. 161–213, Springer, Dordrecht, The Netherlands, 2013.

[38] G. G. Medina, C. B. Cabañas, and L. A. Bravo, "Areas with high potential for chili production," in *Tecnología de Producción de Chile Seco*, L. A. G. Bravo, G. G. Galindo, and R. M. D. Amador, Eds., pp. 177–194, Instituto de Investigaciones Forestales, Agrícolas y Pecuarias, Centro de Investigación Experimental Zacatecas, Zacatecas, Mexico, 2006.

[39] D. Granados-Sánchez, G. M. A. Hernández, and R. G. F. López, "Comprehensive Study of the Tehuacán-Cuicatlán Valley: Genetic Resources of Plants," in *Manejo de la Diversidad de los Cultivos en los Agroecosistemas Tradicionales*, J. L. Chávez-Servia, J. Tuxill, and D. I. Jarvis, Eds., pp. 97–109, Instituto Internacional de Recursos Fitogenéticos, Cali, Colombia, 2005.

[40] R. Vidal-Zepeda, "Las regiones climáticas de México I.2.2," in *Temas Selectos de Geografía de México*, vol. 10, Instituto de Geografía-UNAM, Mexico City, Mexico, 2005.

[41] M. J. J. Reyes, *Technology Components of Impact in Soledad Chili*, Agroproduce, Fundación Produce Oaxaca, Oaxaca, Mexico, 2005.

[42] V. F. Nuez, O. R. Gil, and G. J. Costa, "Non-virotic diseases," in *El Cultivo de Pimientos, Chiles y Ajíes*, V. F. Nuez, O. R. Gil, and G. J. Costa, Eds., pp. 193–247, Mundi-Prensa, Mexico City, Mexico, 2003.

[43] H. M. Reveles, L. A. G. Bravo, and C. B. Cabañas, "Chilli seedling production," in *Tecnología de Producción de Chile Seco*, L. A. G. Bravo, G. G. Galindo, and R. M. D. Amador, Eds., pp. 45–60, Instituto de Investigaciones Forestales, Agrícolas y Pecuarias. Centro de Investigación Experimental Zacatecas, Zacatecas, Mexico, 2006.

[44] T. D. J. De la Cruz, *Nutritional requirements of Poblano chili (Capsicum anuum L.) and its relationship with yield and fruit quality [Thesis]*, Colegio de Postgraduados, Campus Montecillo, Mexico, 2008.

[45] V. M. J. Ayala, *Analysis of growth and seed quality of three types of chili (Capsicum annuum L.) [Thesis]*, Colegio de Postgraduados, Campus Montecillo, Montecillo, Mexico, 2012.

[46] L. D. Ortega-Martínez, J. Sánchez-Olarte, R. Díaz-Ruiz, and J. Ocampo-Mendoza, "Effect of different substrates on tomato seedlings growth (*Lycopersicum esculentum* Mill.)," *Ra Ximhai: Revista Científica de Sociedad, Cultura y Desarrollo Sostenible*, vol. 6, no. 3, pp. 365–375, 2010.

[47] A. A. Steiner, "The universal nutrient solution," in *Proceedings of the 6th International Congress on Soilless Culture. Proceedings International Society for Soilless Culture*, pp. 1–17, Lunteren, The Netherlands, 1984.

[48] P. F. Martínez and D. Roca, "Substrates for soilless culture. Materials, properties and handling," in *Sustrato, Manejo del Clima, Automatización y Control en Sistemas de Cultivo Sin Suelo*, R. V. J. Flórez, Ed., pp. 37–77, Universidad Nacional de Colombia, Bogotá, Colombia, 2011.

[49] H. Pérez-López, F. C. Gómez-Merino, L. I. Trejo-Téllez, S. García-Morales, and L. Y. Rivera-Olivares, "Agricultural ligno-cellulosic waste and volcanic rock combinations differentially affect seed germination and growth of pepper (*Capsicum annuum* L.)," *BioResources*, vol. 9, no. 3, pp. 3977–3992, 2014.

[50] C. Fernández-Bravo, N. Urdaneta, W. Silva, H. Poliszuk, and M. Marín, "Germination of tomato (*Lycopersicon esculentum* Mill.) cv. Río Grande seeds, sown in plug trays, using different substrates," *Revista de la Facultad de Agronomía (LUZ)*, vol. 23, no. 2, pp. 186–193, 2006.

[51] F. I. Salazar-Jara and P. Juárez-López, "Macronutrimental requirement in chili plants (*Capsicum annuum* L.)," *Revista Bio Ciencias*, vol. 2, no. 2, pp. 27–34, 2013.

[52] T. D. J. De la Cruz and B. W. I. Avilés, "Technology for the production of yellow chili," in *Opciones Hortícolas para Suelos Pedregosos*, B. W. I. Avilés, B. F. Santamaría, T. D. J. De la Cruz, and M. L. A. Pérez, Eds., pp. 9–14, Secretaría de Agricultura, Ganadería, Desarrollo Rural, Pesca y Alimentación, Instituto Nacional de Investigaciones Forestales Agrícolas y Pecuarias Centro de Investigación Regional del Sureste, Mexico City, Mexico, 2000.

[53] L. A. Bravo, D. F. Mojarro, C. B. Cabañas, and H. A. Lara, "Influence of drip irrigation and fertigation on dry Mirasol chili production in Zacatecas, Mexico," in *Segunda Convencion*

Mundial del Chile, pp. 210–214, Consejo Nacional de Productores, Chiles de Mexico, Mexico, 2005.

[54] C. B. Cabañas, G. G. Galindo, D. F. Mojarro, L. A. Bravo, and D. J. A. Zegbe, "Fertilización y arreglo topológico en el rendimiento y calidad de fruto del chile Mirasol (*Capsicum annuum* L.) en Zacatecas, México," in *Segunda Convención Mundial del Chile*, pp. 226–230, Consejo Nacional de Productores de Chiles de México, Mexico City, Mexico, 2005.

[55] C. A. Ramiro, *Mirasol or Guajillo Chili in the Plateau of San Luis Potosi*, Folleto para Productores no. 13, Secretaría de Agricultura y Recursos Hidráulicos, Instituto Nacional de Investigaciones Forestales y Agropecuarias, Centro de Investigación Regional Noreste, Campo Experimental Palma de la Cruz San Luis Potosí, San Luis Potosí, Mexico, 1992.

[56] T. K. Hartz, M. LeStrange, and D. M. May, "Nitrogen requirements of drip-irrigated peppers," *HortScience*, vol. 28, no. 11, pp. 1097–1099, 1993.

[57] F. Wiesler, "Nutrition and quality," in *Marschner's Mineral Nutrition of Higher Plants*, P. Marschner, Ed., pp. 271–282, Academic Press, Cambridge, UK, 2012.

[58] H. Huber, V. Römheld, and M. Weinmann, "Relationship between nutrition, plant diseases and pest," in *Marschner's Mineral Nutrition of Higher Plants*, P. Marschner, Ed., pp. 283–298, Academic Press, London, UK, 2012.

[59] G. M. Urrestarazu, "Manual práctico del cultivo sin suelo e hidroponía," in *Practical Handbook of Soilless Culture and Hydroponics*, Mundi-Prensa, Madrid, España; Universidad de Almería, Almería, Spain, 2015.

[60] L. I. Trejo-Téllez and F. C. Gómez-Merino, "Nutrient solutions for hydroponic systems," in *Hydroponics-A Standard Methodology for Plant Biological Researches*, T. Asao, Ed., pp. 1–23, InTech, Rijeka, Croatia, 2012.

[61] D. F. Mojarro, I. A. Bravo, C. B. Cabañas et al., "Fertilization requirements of Mirasol chili, in five different locations, in the state of Zacatecas, Mexico," in *Segunda Convención Mundial del Chile*, pp. 143–148, Consejo Nacional de Productores de Chiles de México, Tampico, Mexico, 2005.

[62] M. A. Báez, L. T. Chávez, P. S. García, L. A. A. Navarro, J. E. Estrada, and A. M. Garza, "Jalapeño pepper production under fertigation as a function of soil-water tension, nitrogen and potassium nutrition," *Terra Latinoamericana*, vol. 20, no. 2, pp. 209–215, 2002.

[63] C. D. Johnson and D. R. Decoteau, "Nitrogen and potassium fertility affects Jalapeno pepper plant growth, pod yield, and pungency," *HortScience*, vol. 31, no. 7, pp. 1119–1123, 1996.

[64] F. Medina-Lara, I. Echevarría-Machado, R. Pacheco-Arjona, N. Ruiz-Lau, A. Guzmán-Antonio, and M. Martinez-Estevez, "Influence of nitrogen and potassium fertilization on fruiting and capsaicin content in habanero pepper (*Capsicum chinense* Jacq.)," *HortScience*, vol. 43, no. 5, pp. 1549–1554, 2008.

[65] G. Bhuvaneswari, R. Sivaranjani, S. Reeth, and K. Ramakrishnan, "Application of nitrogen and potassium efficiency on the growth and yield of chilli *Capsicum annuum* L.," *International Journal of Current Microbiology and Applied Sciences*, vol. 2, no. 12, pp. 329–337, 2013.

[66] D. R. Seal and D. J. Schuster, "Control of pepper weevil, *Anthonomus eugenii*, in west-central and south Florida," *Proceedings of the Florida State Horticultural Society*, vol. 108, pp. 220–225, 1995.

[67] C. J. Mena, "Integrated management strategy against pest insects in chili," in *Tecnología de Producción de Chile Seco*, L. A. G.

Bravo, G. G. Galindo, and R. M. D. Amador, Eds., pp. 97–120, Instituto de Investigaciones Forestales, Agrícolas y Pecuarias, Centro de Investigación Experimental Zacatecas, Zacatecas, Mexico, 2006.

[68] U. E. Garza, *The Pepper Weevil (Anthonomus eugenii) and Its Management on the Huastec Plain*, Instituto Nacional de Investigaciones Forestales, Agrícolas y Pecuarias. Centro de Investigación Regional del Noreste. Campo Experimental Ébano, San Luis Potosí, Mexico, 2001.

[69] V. García-Gaytán, S. García-Morales, H. V. Silva-Rojas, L. I. Trejo-Téllez, and F. C. Gómez-Merino, "First report of powdery mildew in chilhuacle chili (*Capsicum annuum*) caused by *Leveillula taurica* in Southern Mexico," *Plant Disease*, vol. 100, no. 11, p. 2325, 2016.

[70] V. S. Govindarajan and V. S. Salzer, "*Capsicum*—production, technology, chemistry, and quality. Part III. Chemistry of the color, aroma, and pungency stimuli," *C R C Critical Reviews in Food Science and Nutrition*, vol. 24, no. 3, pp. 245–355, 1986.

[71] S. L. Kothari, A. Joshi, S. Kachhwaha, and N. Ochoa-Alejo, "Chilli peppers—a review on tissue culture and transgenesis," *Biotechnology Advances*, vol. 28, no. 1, pp. 35–48, 2010.

[72] J. S. Pruthi, "Advances in post-harvest processing technologies of *Capsicum*," in *Capsicum—The Genus Capsicum*, A. K. De, Ed., pp. 175–213, Taylor & Francis, London, UK, 2003.

[73] T. J. Zachariah and P. Gobinath, "Paprika and chili," in *Chemistry of Spices*, V. A. Parthasarathy, B. Chempakam, and T. J. Zachariah, Eds., pp. 260–286, CABI, Wallingford, UK, 2008.

[74] C. A. Reilly, D. J. Crouch, and G. S. Yost, "Quantitative analysis of capsaicinoids in fresh peppers, oleoresin capsicum and pepper spray products," *Journal of Forensic Sciences*, vol. 46, no. 3, pp. 502–509, 2001.

[75] L. Asnin and S. W. Park, "Isolation and analysis of bioactive compounds in *Capsicum* peppers," *Critical Reviews in Food Science and Nutrition*, vol. 55, no. 2, pp. 254–289, 2015.

[76] F. A. Khan, T. Mahmood, M. Ali, A. Saeed, and A. Maalik, "Pharmacological importance of an ethnobotanical plant: *Capsicum annuum* L.," *Natural Product Research*, vol. 28, no. 16, pp. 1267–1274, 2014.

[77] V. Fattori, M. S. N. Hohmann, A. C. Rossaneis, F. A. Pinho-Ribeiro, and W. A. Verri, "Capsaicin: current understanding of its mechanisms and therapy of pain and other pre-clinical and clinical uses," *Molecules*, vol. 21, no. 7, article 844, 2016.

[78] S. H. Morán-Bañuelos, V. H. Aguilar-Rincón, T. Corona-Torres, F. Castillo-González, R. M. Soto-Hernández, and R. S. Miguel-Chávez, "Capsaicinoids in Chile pepper landraces of Puebla, Mexico," *Agrociencia*, vol. 42, no. 7, pp. 807–816, 2008.

[79] L. Orellana-Escobedo, L. E. Garcia-Amezquita, G. I. Olivas, J. J. Ornelas-Paz, and D. R. Sepulveda, "Capsaicinoids content and proximate composition of Mexican chili peppers (*Capsicum* spp.) cultivated in the State of Chihuahua," *CYTA—Journal of Food*, vol. 11, no. 2, pp. 179–184, 2013.

[80] H. A. Eissa, B. E. Mostafa, and A. S. Hussein, "Capsaicin content and quality characteristics in different local pepper varieties (*Capsicum annum*) and acid-brine pasteurized puree," *Journal of Food Technology*, vol. 5, no. 3, pp. 246–255, 2007.

[81] S. W. Meckelmann, D. W. Riegel, M. van Zonneveld et al., "Capsaicinoids, flavonoids, tocopherols, antioxidant capacity and color attributes in 23 native Peruvian chili peppers (*Capsicum* spp.) grown in three different locations," *European Food Research and Technology*, vol. 240, no. 2, pp. 273–283, 2015.

[82] L. Jiang and K. Kubota, "Differences in the volatile components and their odor characteristics of green and ripe fruits and dried

pericarp of Japanese pepper (*Xanthoxylum piperitum* DC.)," *Journal of Agricultural and Food Chemistry*, vol. 52, no. 13, pp. 4197–4203, 2004.

[83] B. Polak, "Confirmation of chirality of some natural products by the HPLC method," in *High Performance Liquid Chromatography in Phytochemical Analysis*, Chromatographic Science Series, pp. 373–396, CRC Press, 2010.

[84] M. Moyer, "How much is left? a graphical accounting of the limits to what one planet can provide," *Scientific American*, vol. 303, no. 3, pp. 74–81, 2010.

6

Winter Grazing in a Grass-Fed System: Effect of Stocking Density and Sequential Use of Autumn-Stockpiled Grassland on Performance of Yearling Steers

Domingo J. Mata-Padrino, E. E. D. Felton, W. B. Bryan, and D. P. Belesky

West Virginia University, Morgantown, WV 26506, USA

Correspondence should be addressed to Domingo J. Mata-Padrino; djmatapadrino@mail.wvu.edu

Academic Editor: David Clay

Winter grazing can help reduce the need for purchased feeds in livestock production systems, when finishing cattle on pasture. Our objective was to evaluate the influence of stocking density and grazing stockpiled forage on performance of yearling steers during winter. Three grasslands were winter grazed for two years: I, naturalized pastureland, and II and III, sown and managed for hay production during the growing season but grazed in winter. Two stocking densities were used: low 7.41 and high 12.35 steers ha^{-1}. Herbage mass was estimated before and after each grazing event, and disappearance (consumption, weathering, and trampling) was the difference between both. Forage mass and residual differed by stocking density (SD), year (YR), and grazing interval (GI), and disappearance differed by YR and GI. Grass and dead constituents of botanical composition differed by YR and GI. No differences were found for legumes and forbs. CP differed by YR and GI, and NDF and ADF differed only by YR. Steer average daily gain was 0.15 kg d^{-1} in 2011 and 0.68 kg d^{-1} in 2012 and varied by YR and GI. Acceptable gains in 2012 may be a product of environmental conditions that influenced herbage mass and nutritive value during stockpile and animal behavior during winter.

1. Introduction

Sustaining livestock on pasture throughout the year in temperate regions is dictated by weather conditions within and among years. Herbage production during the growing season often is restricted by precipitation deficits even when day length and temperatures are favorable [1]. In autumn and winter, decreasing temperatures and photoperiods contribute to forage plant dormancy and, eventually, the cessation of active growth. In some instances, forage plants adapted to water deficit and high temperatures and forage capable of sustaining some growth at lower temperatures can be used to stabilize variation in herbage production. Since pastures throughout the Appalachian region comprise many different species of grasses, legumes, and forbs, some level of sustained productivity would be more likely when temperature and precipitation trends vary with climate variability [2]. In cool-temperate regions, this might mean that plants adapted to warm-season growing conditions would persist and extend

the interval of active herbage production and thus grazing time in late autumn and winter. However, when pasture growth is less than that required to meet grazing animal needs, stored or purchased feeds must be provided [3]. In cool-temperate regions, winter typically is the time when stored feeds are used to sustain grazing livestock performance on pasture. Winter grazing may contribute to profitability of forage-based production systems by reducing the need for purchased or stored feeds.

Volesky et al. [4] noted that stockpiled forage should be grazed during late autumn and early winter (November and December) in the central US to take advantage of herbage quantity and nutritive value. Delaying herbage use beyond late autumn is likely to result in mass and nutritive value losses [5, 6]. This occurs because of tissue damage caused by exposure to weather [7, 8]. Susceptibility to weather-related damage may be a function of the species stockpiled [9].

Animal performance is influenced by interactions occurring among herbage mass, nutritive value, weather, and grazing behavior, such as selectivity among sward components, grazing time and location, and previous experience [1, 10]. Recommendations for winter grazing often are based upon general knowledge about forage growth developed during the growing season for a given location. However, this approach may neglect changes in grazing animal behavior as herbage growth ceases with less favorable weather-related growing conditions. During winter, livestock may spend less time foraging over a large area and more time on a relatively smaller area of pasture [11]. This typically contrasts with longer periods of foraging that occur during milder weather conditions. Patton et al. [1] noted that grazing influenced the amount of herbage produced and that modest grazing intensity improved production over nongrazed areas. Wade and Carvalho [12] stated that, at an empirical level, animal responses to stocking methods and stocking density are now well known. However, livestock response to the interactions of weather, herbage mass, nutritive value, and grazing behavior is complex and not well understood.

Livestock production systems that include winter grazing as a component of pasture management must take into account season-long influences and interactions of livestock grazing activities with the forage resources available. Considerations include weather, location of winter feeding sites on the pasture, edaphic features, and pasture botanical composition during the stockpiling period. Short-term variations in production and chemical composition could be related to management and weather during the growing season [13], whereas long-term changes could be a product of management and environment interacting with time [1].

Mata-Padrino et al. [14] noted that when grasslands comprising different locally adapted forage species were grazed simultaneously, herbage quality declined more quickly on naturalized permanent grassland, while tall fescue dominant swards maintained herbage quality the longest and orchardgrass dominant swards were intermediate. To validate this observation, we grazed the grasslands in a sequence related to tolerance to late-season weather conditions for this experiment. Our objective was to determine whether average daily gain of steers and total gain per hectare were influenced by stocking density when using stockpiled herbage sequentially.

2. Materials and Methods

2.1. Site Conditions. The experiment was conducted at the West Virginia University Reedsville Experiment Farm, located in Preston County, WV (39°30'N, 79°47'W; 537 m elevation), on the Appalachian High Plateau of the eastern US. Three grasslands grazed during winter from 2004 to 2009 were divided into 24 paddocks of 0.405 ha. The paddocks were grazed by 24 *Bos taurus* steers divided into two groups to create low 7.41 and high 12.35 steers ha^{-1} stocking densities (SD), each replicated three times. We created a sequential grazing scheme based on previously observed resilience of the grassland communities to winter conditions [14]. Each grassland represented one of three grazing intervals (GI) coincident with pasture type (Table 1). In order to maintain

Table 1: Sequence and schedule for winter grazing.

Year		Grazing interval		
		I†	II	III
2011	In	Nov 2	Nov 22	Dec 16
	Out	Nov 22	Dec 16	Jan 4
2012	In	Nov 12	Nov 26	Dec 18
	Out	Nov 26	Dec 18	Jan 9

†I: naturalized permanent grazingland; II: orchardgrass cultivated pastureland; III: tall fescue cultivated pastureland.

the sequence based on grassland endurance to late autumn-winter conditions, the second grazing interval comprised twice the number of paddocks as the initial and final grazing interval.

The experiment on winter grazing of autumn-stockpiled herbage was conducted from August 2011 to January 2013 with herbage stockpiling beginning in August of each year and later grazed from early November until early January. Grasslands were defined as follows and will henceforth be referred to as I, naturalized permanent grazingland, and II and III, sown and managed primarily to hay production during the growing season but grazed during winter. The sequence began with grassland I, which consisted of a mixture of orchardgrass (*Dactylis glomerata* L.), velvet grass (*Holcus lanatus* L.), timothy (*Phleum pratense* L.), other grasses of lesser quality and presence in the sward, white clover (*Trifolium repens* L.), red clover (*T. pratense* L.), and forbs. Forbs included ground ivy (*Glechoma hederacea* L.), dandelion (*Taraxacum officinale* Weber ex F. H. Wigg.), dock (*Rumex* spp. L.), plantain (*Plantago* spp.), and horsenettle (*Solanum carolinense* L.), followed by grassland II that consisted mainly of orchardgrass, timothy, alfalfa (*Medicago sativa* L.), white and red clover, and forbs including Canada thistle (*Cirsium arvense* (L.) Scop.), bedstraw (*Galium aparine* L.), ground ivy, and dock and grassland III that consisted mainly of tall fescue (*Schedonorus arundinaceus* (Schreb.) Dumort.), orchardgrass, and white and red clover along with the forbs bedstraw, plantain, ragweed (*Ambrosia artemisiifolia* L.), and smartweed (*Polygonum pensylvanicum* L.).

Soil was sampled annually (0 to 5 cm depth) at 20 to 25 locations along a predetermined transect in each grassland for estimates of soil nutrient status. Lime (dolomitic) was applied to maintain pH above 5.8 and triple superphosphate and potassium chloride were applied to maintain soil available P and K above 65 kg ha^{-1} and 195 kg ha^{-1}, respectively. Daily maximum and minimum temperatures and precipitation were recorded at the experiment site.

2.2. Plant Measurements

2.2.1. Forage Mass. Herbage mass was determined using the plate meter technique described by E. B. Rayburn and S. B. Rayburn [15] in each paddock within each grassland. Seventy-five sward height measurements were made in each paddock before and after grazing. Sampling sites were located by global satellite positioning at fixed intervals on four transects within each paddock. Means of all sampling sites within a given paddock provided an estimate of the sward height for each. The difference between forage mass determined before grazing and residual herbage mass after each grazing event was assumed to represent herbage disappearance. Disappearance comprises the herbage consumed by grazing animals as well as senesced, damaged, and detached leaves.

Sward height was converted to herbage mass using the following equation:

$$\text{Forage mass}\left(\text{kg ha}^{-1}\right) = 266.15 \cdot \text{sward height (cm)} + 265. \tag{1}$$

This equation was derived from data collected from each of the paddocks in each grassland unit from 2004 through 2009 [14].

2.2.2. Botanical Composition and Nutritive Value. Prior to grazing each paddock, 6 forage samples were collected for botanical composition and nutritive value analyses along the same transects used to estimate herbage mass. Samples were clipped to the soil surface within a $0.07\,\text{m}^2$ quadrat. Samples collected were placed in paper bags and stored at $0°\text{C}$ until processed. Herbage samples were hand-sorted into grasses, legumes, forbs, and dead material. Grasses and forbs were separated further by species. Grasses included orchardgrass, tall fescue, timothy, velvet grass, and other grasses, which represented infrequently occurring sward components. Forbs included bedstraw, Canada thistle, ground ivy, dandelion, dock, horsenettle, smartweed, ragweed, plantain, and other ephemeral and infrequently occurring forbs. Individual species and total sampled area weights were determined and the percentage of each component of the sward was computed. Individual sward components were placed in paper bags and air-dried at $60°\text{C}$ in a forced-draft oven for 48 h. Dry samples were weighed and then recombined by grassland. Two subsamples collected from each grassland were ground using a stainless steel Wiley mill (Model 4, A. H. Thomas, Philadelphia, PA) equipped with a 1 mm particle size screen. All nutritive analyses were performed in duplicate. Dry matter ($105°\text{C}$), ash, and crude protein (CP) (Kjeldahl N $\times$ 6.25) were determined according to AOAC [16] procedures. Neutral detergent fiber (NDF), acid detergent fiber (ADF), and lignin were determined sequentially according to Van Soest et al. [17] using modifications outlined in the ANKOM$^©$ fiber analysis procedure (ANKOM Technology, Macedon, NY).

2.3. Animal Measurements. Steers used during 2011-2012 were born and raised in Virginia, whereas those used in 2012-2013 were born and raised at the Reedsville farm where the experiment was conducted. Steers were assigned each year randomly among paddocks by weight and frame size at the start of the winter grazing. Animals were treated for internal and external parasites at the beginning of the winter grazing. Three grazing intervals were coincident with grazing grasslands I, II, and III in sequence. Animals were weighed at the start of each grazing interval until the conclusion of the experiment, providing a total of four live weight

measurements per animal for each year. A uniform method for weighing livestock is critical when comparing livestock responses to treatment or conditions in both commercial and research settings [18]. We removed grazers from paddocks at 0700 h and then weighed the animals between 1000 and 1100 h for each weighing date. Animals were returned to paddocks at 1200 h on the same day. All animals were handled in the same manner at each weighing event to help minimize differences attributable to grazing behavior. Productivity per hectare was calculated each year, for treatment and for the complete grazing interval.

2.4. Statistical Analysis. Paddock was the experimental unit for grassland and animal measurements. All forage data were analyzed as a completely random design, replicated three times, using the PROC-MIXED procedure of SAS [19]. For grazing management and botanical composition data, the model included year, stocking density grazing interval, and their interactions. Analysis of nutritive value included year, grasslands, and their interaction in the model. In both analyses, a repeated measurement statement and the denominator degrees of freedom using the Kenward-Roger option were included. Average daily gain was calculated for each grazing interval and reported as season-long total gain. The model used to analyze ADG included year, stocking density, grazing interval, and their interactions, and a repeated measurement statement and the denominator degrees of freedom using the Kenward-Roger option were computed. Treatment least squares means were calculated and means compared using LSR when the F test for means was significant ($P < 0.05$). Initial weight each year was compared between stocking densities with one-way ANOVA using the PROC-MIXED procedure of SAS.

3. Results and Discussion

3.1. Weather Conditions. Precipitation and air temperature varied by month and year during the stockpile and winter grazing interval of the experiment years and when compared to location 30-year mean (Figures 1 and 2). Variable weather conditions would contribute to variations observed in herbage accumulation in late summer/early fall and in animal performance during the grazing period. Total precipitation was above normal throughout the duration of the experiment (2011-2012); however, it was less than the 30-year mean in August and November 2012. Temperatures were similar to the 30-year mean and less than the long-term mean only in November 2012.

During stockpile, precipitation and air temperature have a significant influence on forage production and quality and could be more important determinants of sward composition, herbage mass, and nutritive value than management imposed during late summer and autumn [1, 6].

Winter grazing in grass-fed systems is highly dependent on weather [14]. Also, during winter grazing, temperatures and snow cover can affect grazing behavior and animal performance [20–22]. During winter-stocking period of 2011-12, snow covered the grasslands for a few days. Nevertheless,

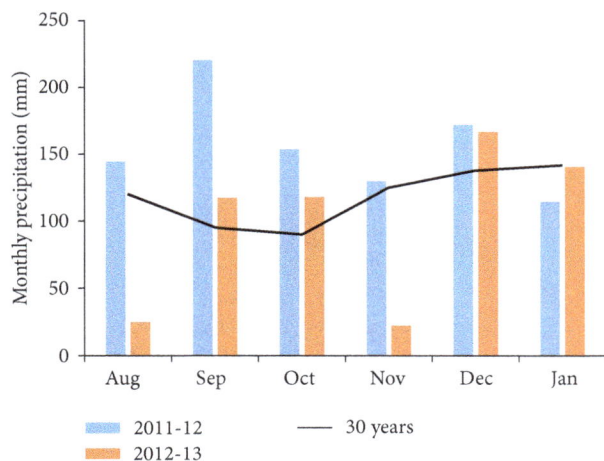

FIGURE 1: Monthly precipitation and 30-year mean values at Reedsville, WV.

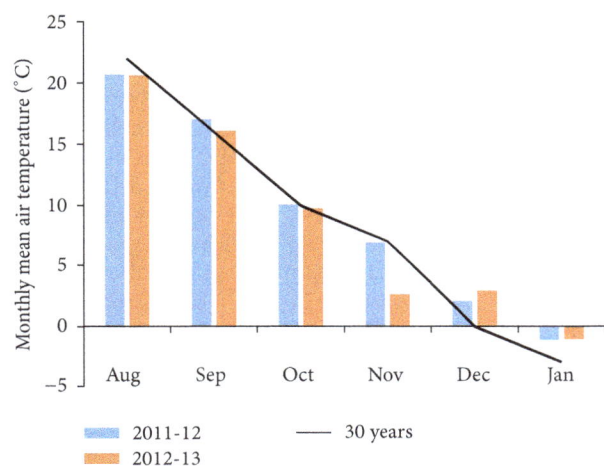

FIGURE 2: Monthly mean air temperature and 30-year mean values at Reedsville, WV.

in December 2012 and January 2013, snow covered the pasture for a longer period, which restricted grazer access to herbage.

3.2. Grassland Characterization.

Riesterer et al. [23] proposed a schedule to use stockpiled grasslands in Wisconsin depending on species endurance. Similarly, our previously reported observations [14] suggested a sequence of use based on resistance to weathering. A thorough understanding of the pregraze pastures was needed to help explain any difference which may occur as a result of stocking density on average daily gain of yearling steers during winter.

3.2.1. Forage Mass.

Stockpiled forage mass, determined prior to each winter grazing interval, averaged 2948 kg ha^{-1} for the two-year experiment (Table 2). Stockpiled forage mass measure at the two stocking densities showed differences ($P = 0.0155$) attributable to weather and soil moisture variability during the stockpile. Year and grazing interval were also different (both $P < 0.0001$), with the difference

TABLE 2: Forage mass, residual forage mass, and forage disappearance of each grassland for two years.

Year	Grazing interval[†]	Forage mass	Residual	Disappearance
			kg ha^{-1}	
2011	I	3900[a]	1701[b]	2199[a]
	II	3164[b]	1517[c]	1647[b]
	III	3257[b]	1480[c]	1777[b]
2012	I	2562[c]	1848[a]	714[d]
	II	2141[d]	1419[d]	722[d]
	III	2664[c]	1592[b]	1072[c]

Means in the same column followed by a different lowercase letter are different at the 0.05 probability level.
[†]I: naturalized permanent grazingland; II: orchardgrass cultivated pastureland; III: tall fescue cultivated pastureland.

probably attributable to precipitation during stockpile and weathering during the winter grazing intervals (Table 4). Interactions were significant for YR * GI ($P < 0.0001$) and SD * YR ($P = 0.0022$). An explanation for these interactions may be the different way cold weather influenced the grasslands during each year and differences associated with pregraze herbage mass and stocking densities. Herbage production was lower for the same grasslands during winter of 2005–2009 when using strip-grazing management [14]. The increase in production obtained in this experiment may be explained by the use of a grazing sequence based on grassland that helped avoid the deterioration of herbage mass and quality resulting from late autumn and winter weather conditions. Prigge et al. [24] and Baker et al. [25] found similar forage production for orchardgrass during autumn at the WVU Reedsville Experiment Farm. Tall fescue mass determined in our experiment was less than that reported in North Carolina [26], but similar to the amounts obtained at Morgantown, West Virginia [6, 7], Wooster, Ohio [27], and Columbia, Missouri [28].

Volesky et al. [4] observed that orchardgrass was more susceptible than tall fescue to effects of snow compression and as a result would present relatively less herbage available for grazing. Collins and Balasko [6] found that herbage mass of stockpiled tall fescue declined from 2500 kg ha^{-1} in December to 2250 kg ha^{-1} by February, demonstrating possible loss attributable to leaf senescence or weathering. Disappearance averaged 2794 kg d^{-1} for the low stocking density and 2954 kg d^{-1} for the higher; performance showed that despite trampling and differences in pregrazing and residual forage mass the steers could remove similar amounts of forage in both stocking densities (Table 3).

Residual forage mass differed between treatments ($P < 0.0312$); however, it is the resultant effect of using the residual herbage on pastures with greater stocking rate as a determinant to move animals to the next grassland. In addition, grazing interval ($P < 0.0001$) showed the effect of weathering as the season advances. Residual herbage was influenced by YR × GI ($P = 0.0011$) and YR × SD ($P = 0.0008$) interactions. Herbage disappearance differed by year and interacted with stocking density (both $P < 0.0001$). Grazing interval

TABLE 3: Forage mass, residual forage mass, and forage disappearance by stocking density for two years.

Year	Stocking density[†]	Forage mass	Residual	Disappearance
			kg ha^{-1}	
2011	Low SD	3234[a]	1661[a]	1737[a]
	High SD	3509[b]	1447[b]	1847[a]
2012	Low SD	2355[c]	1565[c]	781[b]
	High SD	2399[c]	1574[c]	834[b]

Means in the same column followed by a different lowercase letter are different at the 0.05 probability level.
[†]SD: stocking density. Low SD = 7.41 steers ha^{-1} and high SD = 12.35 steers ha^{-1}.

TABLE 4: Analysis of variance of forage mass, residual herbage, forage disappearance, and forage use as a function of stocking density, grazing interval, year, and their interactions.

Source of variation	Forage mass	Residual	Disappearance
		P value	
Stocking density (SD)	0.0155	0.0312	NS
Year (YR)	<0.0001	0.0623	<0.0001
Grazing interval (GI)	<0.0001	<0.0001	0.0009
SD * YR	0.0022	0.0008	<0.0001
YR * GI	<0.0001	0.0011	NS
SD * GI	NS	NS	NS
SD * YR * GI	0.0010	0.0175	0.0427

[†]SD: stocking density. Low SD = 7.41 steers ha^{-1} and high SD = 12.35 steers ha^{-1}.

showed differences for herbage disappearance but did not differ between stocking densities (Table 4). Forage mass in this experiment was determined by canopy height; those measurements would be influenced by snow compression and trampling. Furthermore, the effects of weathering and animal impact would depend on stocking densities and sward components. Despite the influence of limiting factors, analyzing forage mass in winter grazing situations may help explain the balance between herbage consumption, sward decomposition, and subsequently animal performance.

3.2.2. Botanical Composition. Grass and forbs were influenced by year and grazing interval regardless of stocking density. All principal botanical components were influenced by year interacting with grazing interval (Table 5). Differences in endurance of grass species in each grazing interval and precipitation and temperature patterns during the two years that the experiment lasted may have contributed to this response, even with the different pattern of snow precipitation and time that snow was covering grassland both years. In 2011, grazing intervals II and III had much more dead material than grazing interval I and more than any interval in 2012 (Figure 3). Grass was the predominant component in 2011 in grazing interval I, a relatively wet and warmer year, and primarily was represented by "other grasses." Conversely, during 2012, cultivated cool-season grasses represented a greater proportion in the grass component (Figure 4), but again the botanical composition of naturalized grasslands

TABLE 5: Analysis of variance for main components of grassland botanical composition as a function of stocking density, grazing interval, year, and their interactions.

Source of variation	Grass	Legume	Forb	Dead
		P value		
Stocking density (SD)	NS	NS	NS	NS
Year (YR)	0.0124	NS	NS	<0.0001
Grazing interval (GI)	0.0425	NS	NS	<0.0001
SD * YR	NS	NS	NS	NS
YR * GI	<0.0001	0.0348	0.0089	<0.0001
SD * GI	NS	NS	NS	NS
SD * YR * GI	NS	NS	NS	NS

grazed in interval I performed differently, with a decreasing grass proportion. Tall fescue was the most reduced cultivated cool-season grass during 2011. Proportion of other grasses and forbs along with higher air temperatures during stockpile and lower air temperatures in December 2011 may explain that reduction in tall fescue proportion and the patterns of senesced herbage observed during winter grazing in 2011-2012. Legume did not have a considerable contribution to the forage available for winter grazing. Legume level increased in 2012 compared with 2011, grazing interval II with the greatest amount of legume. Conversely, the legume proportion in grazing interval III, corresponding with tall fescue pasture, was less in 2012 compared with 2011. The interaction YR * GI for legume component could be explained by changes in the legume component. Hitz et al. (2000) reported also a reduction in legume proportion when associated with fescue in winter grazing. The greatest percentages of forbs occurred during December 2011 during grazing interval II, an orchardgrass dominant pasture, with bedstraw the dominant forb (Figure 5). Hobbs et al. [29] found that orchardgrass pastures had an encroachment of weeds estimated in the range of 12–20% that was similar to forb proportion in orchardgrass dominant pasture. In both years, tall fescue dominant pastures compared to other pastures had fewer forbs, which is consistent with published results for tall fescue pastures [29, 30]. Bedstraw was the most prevalent forb noted during winter grazing in grazing intervals II and III and thrived during dry and relatively cooler conditions occurring during the stockpiling interval of late summer and autumn. Ground ivy was dominant in grazing interval I but was also more abundant in all pastures when warmer temperatures occurred during stockpiling in 2011-2012. Dandelion seemed to be present in greater amounts during the relatively colder stockpiling interval of the 2012-2013 season (Figure 5), regardless of grazing interval.

3.2.3. Chemical Composition. Crude protein (CP) concentration differed during the two years of the experiment ($P = 0.0003$) and also differed by grazing interval (Tables 6 and 7). In 2011, CP concentration differed little between grazing intervals, but in 2012 the CP during grazing interval I was less than the CP of subsequent intervals. Variations

TABLE 6: Crude protein (CP), neutral detergent fiber (NDF), acid detergent fiber (ADF), and lignin of forage collected from each grassland during a period of two years.

Year	Grazing interval associated with grassland[†]	CP	NDF	ADF	Lignin
			$g \cdot kg^{-1}$		
2011	I	138[c]	572[b]	300[a]	47
	II	146[c]	545[bc]	255[b]	45
	III	145[bc]	619[cd]	281[c]	62
2012	I	138[c]	575[b]	296[a]	46
	II	190[a]	510[d]	243[c]	58
	III	152[a]	613[d]	284[c]	64

Means in the same column followed by a different lowercase letter are different at $P < 0.05$.

[†]I: naturalized permanent grazingland; II: orchardgrass cultivated pastureland; III: tall fescue cultivated pastureland.

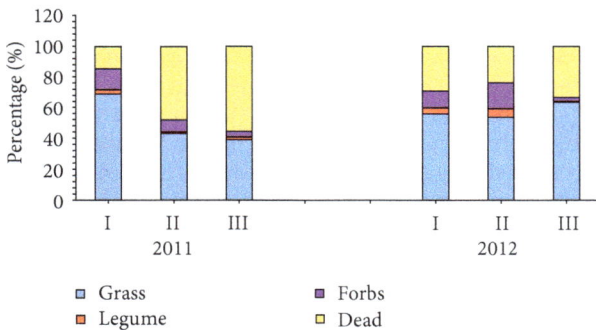

FIGURE 3: Botanical composition of grasslands used for winter grazing in three consecutive grazing intervals (I, II, and III) at Reedsville, WV. Grazing Interval: I = naturalized permanent grazingland, II = orchardgrass cultivated pastureland, and III = tall fescue cultivated pastureland.

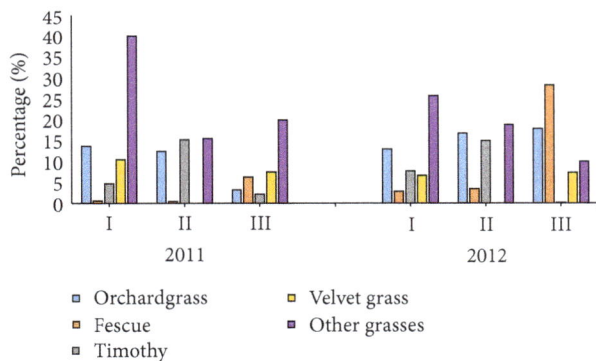

FIGURE 4: Proportion of dominant species in the grass component of grasslands used for winter grazing in three consecutive grazing intervals (I, II, and III) at Reedsville, WV. Grazing Interval: I = naturalized permanent grazingland, II = orchardgrass cultivated pastureland, and III = tall fescue cultivated pastureland.

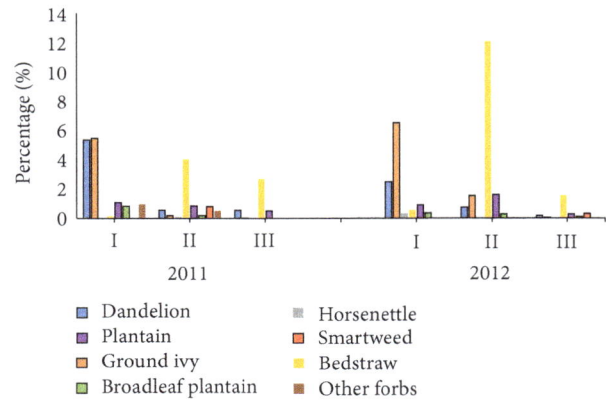

FIGURE 5: Proportion of dominant species in the forb component of grasslands used for winter grazing in three consecutive grazing intervals (I, II, and III) at Reedsville, WV. Grazing interval: I = naturalized permanent grazingland, II = orchardgrass cultivated pastureland, and III = tall fescue cultivated pastureland.

in weather, N fertilization, and botanical composition contributed to these differences. Naturalized grassland had the lowest concentration of CP and the greatest proportion of other grasses. Concentrations of CP were the greatest in the cultivated grasslands, being greater in orchardgrass than in tall fescue dominant grasslands. This is consistent with other published work from the region [14, 25, 31]. Declining CP through fall and winter has been reported in some studies with fescue, orchardgrass, and other cool-season grasses, but fescue was consistently reported to be more resistant to the effects of winter weather conditions [4, 9, 14, 32, 33]. Collins and Balasko [6, 7] reported that, between mid-December and February, in vitro dry matter digestibility of tall fescue decreased significantly. Forage CP concentration in this experiment was greater than previously published data from this location [14, 25, 32]. Results suggest that sequential grazing of forage from most to least susceptible to deterioration related to winter weather conditions may have contributed to sustained herbage nutritive value. Time of sampling may also be a factor that contributes to differences in CP concentration between grazing intervals. Pérez-Prieto et al. [34] found that ryegrass grasslands performed similarly in winter and they related this to leafy conditions of the sward. Herbage ADF and NDF concentrations differed with year (Table 7), being higher in 2011. Grazing interval I maintained similar values both years but higher concentrations of dead material in the cultivated grasslands could be associated with higher proportions of dead material in 2011 [35]. They also

TABLE 7: Analysis of variance for crude protein (CP), neutral detergent fiber (NDF), acid detergent fiber (ADF), and lignin as a function of grazing interval and year and their interaction.

Source of variation	CP	NDF	ADF	Lignin
		P value		
Grazing interval (GI)	0.0010	NS	NS	NS
Year (YR)	0.0003	<0.0001	0.0004	NS
YR * GI	0.0007	NS	NS	NS

TABLE 9: Analysis of variance for average daily gain by stocking density, grazing interval, year, and their interactions.

Source of variation	*P* value
Stocking density (SD)	0.0662
Year (YR)	<0.0001
Grazing interval (GI)	<0.0001
SD * YR	NS
YR * GI	<0.0001
SD * GI	NS
SD * YR * GI	NS

TABLE 8: Average daily gain by grazing interval and treatment over two years.

Year	Grazing interval[†]	ADG (kg d^{-1})	SEM[⁕]	Stocking density[‡]	ADG (kg d^{-1})	SEM[⁕]
2011	I	0.28	0.085	Low SD	0.21	0.065
	II	0.44				
	III	−0.26		High SD	0.09	0.056
	Total	0.15	0.043			
2012	I	2.00	0.085	Low SD	0.73	0.065
	II	0.79				
	III	−0.05		High SD	0.62	0.056
	Total	0.68	0.043			

[†]I: naturalized permanent grazingland; II: orchardgrass cultivated pastureland; III: tall fescue cultivated pastureland.
[‡]SD: stocking density. Low SD = 7.41 steers ha^{-1} and high SD = 12.35 steers ha^{-1}.
[⁕]SEM: standard error.

found that orchardgrass fiber concentration was less than that of tall fescue. Volesky et al. [4] reported for orchardgrass NDF concentrations between 562 and 629 g kg^{-1}, from December through January with slightly greater NDF for tall fescue during the same period. Meyer et al. [28] reported NDF and ADF concentrations similar to those presented here. Cultivated grasslands during the winter of 2012-2013 had chemical composition that was comparable to high quality forage in spring (e.g., high CP and low fiber). Lignin concentration was similar across all grasslands.

3.3. Animal Performance. Initial body weight of steers was not different between stocking densities and averaged 249 kg for both years: 241 kg d^{-1} in 2011 and 256 kg d^{-1} in 2012. Steer performance differed with year ($P < 0.0001$) and tended to differ with stocking density ($P = 0.0662$). The group managed at 7.41 steers ha^{-1} (low density) consistently gained more steers on the higher density (7.41 steers ha^{-1}) and there was no stocking density by year interaction (Tables 8 and 9). Paddocks stocked at the lower density for the same length of time led consistently to greater residual herbage mass; however, no differences for forage use were observed between the two stocking densities ($P = 0.6048$). Differences attributable to year may arise from the influence of weather on animal behavior and forage botanical and chemical composition. Winter 2011-2012 was slightly warmer with minimal snow cover of short duration (one day in December 2011

and two days in January 2012) during grazing interval III while tall fescue was grazed. Snow covered grasslands for short intervals during December 2012 during grazing interval II. Pasture was virtually continuously snow-covered during grazing interval III in late December and early January of 2013. Dunn et al. [21] suggested that grazing experience during winter influences foraging behavior. We observed that steers tended to graze more intensively in a limited area when temperatures were low or snow cover occurred, which was also observed by Krysl and Hess [22]. Wade and Carvalho [12] concluded that an inverse relationship between herbage intake and animal performance is a function of stocking density, where higher densities contribute to decreased available herbage per animal. In winter, that relationship may have a stronger impact as the animal grazing behavior changes. Botanical composition of pastures to be winter-grazed is usually determined at the end of the stockpile; however, changes in the short term may influence herbage quality and animal winter grazing behavior could affect overall performance [20, 21].

Animal productivity (kg ha^{-1}) was much higher for winter grazing in 2012-2013 than in 2011-2012 (Figure 6). In addition, the lower stocking rate led to more gain per hectare in 2011-2012; however, in 2012-2013, it was the higher stocking rate that produced the most animal gain per hectare. This response might be explained by the influence of weather on forage quality and on animal behavior. In 2011-2012, forage mass, disappearance, and forage use were higher than in 2012-2013. However, forage quality was compromised in 2011-2012 by greater amounts of dead material, especially for grasslands used during grazing intervals II and III.

The results for productivity show that the effect of stocking density was the opposite in 2011-2012 of what it was in 2012-2013. In spite of lower forage mass and utilization, steers foraged more effectively in 2012-2013 under drier and cooler conditions, with more snow cover.

4. Conclusions

Our results support the conclusion that performance targets for ADG during winter grazing must consider environmental conditions during the grazing periods and the influence of weather during the late summer-autumn stockpile interval on forage quantity and quality available for winter grazing. However, it would be required to consider as well the influence of

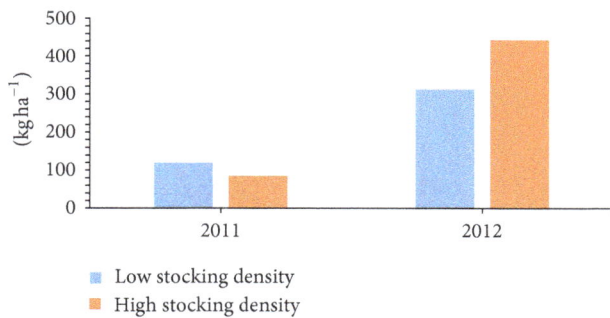

FIGURE 6: Productivity ($kg\,ha^{-1}$) by stocking density, during winter grazing over two years.

SD on steer ADG. Our outcomes revealed a favorable trend by decreasing SD during winter grazing. The proportion of senesced herbage as a fraction of sward botanical composition during winter is related to weather conditions during the stockpile interval. Weather also influenced animal grazing behavior during winter grazing and subsequently influenced animal performance and appeared to supersede aspects of management, such as grassland composition and stocking rate. Grassland responses were not influenced by stocking densities which implies that productivity per unit of land area during winter depended upon forage quality and the source of grazing animals.

Competing Interests

The authors declare that they have no competing interests.

Acknowledgments

The authors thank Rodney Kiser and the WVU Reedsville farm personnel for their help with site management, livestock handling, and field data collection and Eric Nestor and Kara Haught for conducting herbage nutritive value chemical analyses. Funding for this research was provided by the "Economic Pasture-Based Beef Systems for Appalachia" USDA-ARS (1932-21630-002-00D), Beaver, WV, specific cooperative agreement partnership with West Virginia University, Virginia Tech, and Clemson University.

References

[1] B. D. Patton, X. Dong, P. E. Nyren, and A. Nyren, "Effects of grazing intensity, precipitation, and temperature on forage production," *Rangeland Ecology & Management*, vol. 60, no. 6, pp. 656–665, 2007.

[2] D. P. Belesky and D. P. Malinowski, "Grassland communities in the USA and expected trends associated with climate change," *Acta Agrobotanica*, vol. 69, no. 2, article 1673, 2016.

[3] J. P. Flores and B. Tracy, "Impacts of winter hay feeding on pasture soils and plants," *Agriculture, Ecosystems and Environment*, vol. 149, pp. 30–36, 2012.

[4] J. D. Volesky, B. E. Anderson, and M. C. Stockton, "Species and stockpile initiation date effects on yield and nutritive value of irrigated cool-season grasses," *Agronomy Journal*, vol. 100, no. 4, pp. 931–937, 2008.

[5] E. B. Rayburn, R. E. Blaser, and D. D. Wolf, "Winter tall fescue yield and quality with different accumulation periods and n rates," *Agronomy Journal*, vol. 71, no. 6, pp. 959–963, 1979.

[6] M. Collins and J. A. Balasko, "Effects of N fertilization and cutting schedules on stockpiled tall fescue. I. Forage yield," *Agronomy Journal*, vol. 73, no. 5, pp. 803–807, 1981.

[7] M. Collins and J. A. Balasko, "Effects of N fertilization and cutting schedules on stockpiled tall fescue. II. Forage quality," *Agronomy Journal*, vol. 73, no. 5, pp. 821–826, 1981.

[8] J. C. Burns and D. S. Chamblee, "Summer accumulation of tall fescue at low elevations in the humid Piedmont: II. Fall and winter changes in nutritive value," *Agronomy Journal*, vol. 92, no. 2, pp. 217–224, 2000.

[9] V. S. Baron, A. C. Dick, M. Bjorge, and G. Lastiwka, "Stockpiling potential of perennial forage species adapted to the Canadian western Prairie Parkland," *Agronomy Journal*, vol. 96, no. 6, pp. 1545–1552, 2004.

[10] J. Hodgson, J. M. Rodriguez Capriles, and J. S. Fenlon, "The influence of sward characteristics on the herbage intake of grazing calves," *The Journal of Agricultural Science*, vol. 89, no. 3, pp. 743–750, 1977.

[11] J. C. Malechek and B. M. Smith, "Behavior of range cows in response to winter weather," *Journal of Range Management*, vol. 29, no. 1, pp. 9–12, 1976.

[12] M. H. Wade and P. C. F. Carvalho, "Defoliation patterns and herbage intake on pastures," in *Grassland Ecophysiology and Grazing Ecology*, G. Lemaire, J. Hodgson, A. Moraes, C. Nabinger, and P. C. F. Carvalho, Eds., pp. 233–248, CAB International, New York, NY, USA, 2000.

[13] M. H. Poore and M. E. Drewnoski, "Review: utilization of stockpiled tall fescue in winter grazing systems for beef cattle," *Professional Animal Scientist*, vol. 26, no. 2, pp. 142–149, 2010.

[14] D. Mata-Padrino, E. Felton, and W. B. Bryan, "Winter management of yearling steers in a grass-fed beef production system," *Agronomy Journal*, vol. 107, no. 3, pp. 1048–1054, 2015.

[15] E. B. Rayburn and S. B. Rayburn, "A standardized plate meter for estimating pasture mass in on-farm research trials," *Agronomy Journal*, vol. 90, no. 2, pp. 238–241, 1998.

[16] Association of Official Analytical Chemists, *Official Methods of Analysis*, Association of Official Analytical Chemists, Arlington, Va, USA, 17th edition, 2002.

[17] P. J. Van Soest, J. B. Robertson, and B. A. Lewis, "Methods for dietary fiber, neutral detergent fiber, and nonstarch polysaccharides in relation to animal nutrition," *Journal of Dairy Science*, vol. 74, no. 10, pp. 3583–3597, 1991.

[18] A. K. Watson, B. L. Nuttelman, T. J. Klopfenstein, L. W. Lomas, and G. E. Erickson, "Impacts of a limit-feeding procedure on variation and accuracy of cattle weights," *Journal of Animal Science*, vol. 91, no. 11, pp. 5507–5517, 2013.

[19] R. C. Littell, G. A. Milliken, W. W. Stroup, and R. D. Wolfinger, *SAS Systems for Mixed Models*, SAS Institute Inc., Cary, NC, USA, 1996.

[20] S. K. Beverlin, K. M. Havstad, E. L. Ayers, and M. K. Petersen, "Forage intake responses to winter cold exposure of free-ranging beef cows," *Applied Animal Behaviour Science*, vol. 23, no. 1-2, pp. 75–85, 1989.

[21] R. W. Dunn, K. M. Havstad, and E. L. Ayers, "Grazing behavior responses of rangeland beef cows to winter ambient temperatures and age," *Applied Animal Behaviour Science*, vol. 21, no. 3, pp. 201–207, 1988.

[22] L. J. Krysl and B. W. Hess, "Influence of supplementation on behavior of grazing cattle," *Journal of Animal Science*, vol. 71, no. 9, pp. 2546–2555, 1993.

[23] J. L. Riesterer, D. J. Undersander, M. D. Casler, and D. K. Combs, "Forage yield of stockpiled perennial grasses in the upper midwest USA," *Agronomy Journal*, vol. 92, no. 4, pp. 740–747, 2000.

[24] E. C. Prigge, W. B. Bryan, and E. S. Goldman-Innis, "Early- and late-season grazing of orchardgrass and fescue hayfields overseeded with red clover," *Agronomy Journal*, vol. 91, no. 4, pp. 690–696, 1999.

[25] M. J. Baker, E. C. Prigge, and W. B. Bryan, "Herbage production from hay fields grazed by cattle in fall and spring," *Journal of Production Agriculture*, vol. 1, no. 3, pp. 275–279, 1988.

[26] M. H. Poore, M. E. Scott, and J. T. Green Jr., "Performance of beef heifers grazing stockpiled fescue as influenced by supplemental whole cottonseed," *Journal of Animal Science*, vol. 84, no. 6, pp. 1613–1625, 2006.

[27] J. P. Schoonmaker, S. C. Loerch, J. E. Rossi, and M. L. Borger, "Stockpiled forage or limit-fed corn as alternatives to hay for gestating and lactating beef cows," *Journal of Animal Science*, vol. 81, no. 5, pp. 1099–1105, 2003.

[28] A. M. Meyer, M. S. Kerley, R. L. Kallenbach, and T. L. Perkins, "Comparison of grazing stockpiled tall fescue versus feeding hay with or without supplementation for gestating and lactating beef cows during winter," *Professional Animal Scientist*, vol. 25, no. 4, pp. 449–458, 2009.

[29] C. S. Hobbs, T. W. Jr. High, and I. Jr. Dyer, "Orchardgrass and fescue pastures for producing yearling slaughter steers," in *University of Tennessee Agricultural Experiment Station Bulletin*, University of Tennessee Agricultural Experiment Station, Oak Ridge, Tenn, USA, 1965.

[30] K. Saikkonen, K. Ruokolainen, O. Huitu et al., "Fungal endophytes help prevent weed invasions," *Agriculture, Ecosystems and Environment*, vol. 165, pp. 1–5, 2013.

[31] K. A. Archer and A. M. Decker, "Autumn-accumulated tall fescue and orchardgrass. I. growth and quality as influenced by nitrogen and soil temperature," *Agronomy Journal*, vol. 69, no. 4, pp. 601–605, 1977.

[32] J. A. Balasko, "Effects of N, P, and K fertilization on yield and quality of tall fescue forage in winter," *Agronomy Journal*, vol. 69, no. 3, pp. 425–428, 1977.

[33] W. R. Ocumpaugh and A. G. Matches, "Autumn-winter yield and quality of tall fescue," *Agronomy Journal*, vol. 69, no. 4, pp. 639–643, 1977.

[34] L. A. Pérez-Prieto, J. L. Peyraud, and R. Delagarde, "Pasture intake, milk production and grazing behaviour of dairy cows grazing low-mass pastures at three daily allowances in winter," *Livestock Science*, vol. 137, no. 1-3, pp. 151–160, 2011.

[35] K. A. Archer and A. M. Decker, "Autumn-accumulated tall fescue and orchardgrass. ii. effects of leaf death on fiber components and quality parameters," *Agronomy Journal*, vol. 69, no. 4, pp. 605–609, 1977.

7

Phytotoxicity and Benzoxazinone Concentration in Field Grown Cereal Rye (*Secale cereale* L.)

C. La Hovary,[1] D. A. Danehower,[2] G. Ma,[3] C. Reberg-Horton,[4]
J. D. Williamson,[3] S. R. Baerson,[5] and J. D. Burton[3]

[1]*Department of Plant and Microbial Biology, North Carolina State University, Raleigh, NC 27695, USA*
[2]*Avoca, Inc., P.O. Box 129, 841 Avoca Farm Road, Merry Hill, NC 27957, USA*
[3]*Department of Horticultural Science, North Carolina State University, Raleigh, NC 27695, USA*
[4]*Department of Crop Science, North Carolina State University, Raleigh, NC 27695, USA*
[5]*USDA-ARS, Natural Products Utilization Research Unit, University, MS 38677, USA*

Correspondence should be addressed to J. D. Burton; jim_burton@ncsu.edu

Academic Editor: Othmane Merah

Winter rye (*Secale cereale* L.) is used as a cover crop because of the weed suppression potential of its mulch. To gain insight into the more effective use of rye as a cover crop we assessed changes in benzoxazinone (BX) levels in rye shoot tissue over the growing season. Four rye varieties were planted in the fall and samples harvested at intervals the following spring. Two different measures of phytotoxic compound content were taken. Seed germination bioassays were used as an estimate of total phytotoxic potential. Dilutions of shoot extracts were tested using two indicator species to compare the relative toxicity of tissue. In addition, BX (DIBOA, DIBOA-glycoside, and BOA) levels were directly determined using gas chromatography. Results showed that rye tissue harvested in March was the most toxic to indicator species, with toxicity decreasing thereafter. Likewise the BX concentration in rye shoot tissue increased early in the season and then decreased over time. Thus, phytotoxicity measured by bioassay and BX levels measured by GC have a similar but not identical temporal profile. The observed decrease in phytotoxic potential and plant BX levels in rye later in the season appears to correlate with the transition from vegetative to reproductive growth.

1. Introduction

The agricultural use of cover crops has many potential benefits including reduced soil erosion, decreased nutrient runoff, increased soil organic matter, improved soil tilth, the ability to scavenge residual nitrogen, and the suppression of weed growth [1, 2]. Suppression of weed growth is of particular importance to organic growers, due to restrictions in herbicide use [3, 4]. In general, cover crops suppress weed growth by both physical competition and chemical interference. Chemical interference involves the production and release of allelopathic chemicals, primary and secondary metabolites that are phytotoxic or have growth regulating properties. Allelopathy refers to a broad range of biochemical interactions between organisms [5–8]. Many examples of allelopathic chemicals from plants have been described, including sorgoleone from *Sorghum* Moench spp. [9], juglone

from walnut [10], artemisinin from *Artemisia* L. spp. [6], and the benzoxazinones (BXs) and their metabolites from corn, wheat, and rye [11–15].

Cereal rye (*Secale cereale* L.) is widely used as a winter cover crop because of its winter hardiness, high biomass productivity, low cost, and effectiveness in controlling weeds in multiple crops that are planted into the killed cover crop in the spring [16–19]. Several allelopathic chemicals have been identified in rye foliage, including the benzoxazinones (BXs), or cyclic hydroxamic acids, such as DIBOA (2,4-dihydroxy-2*H*-1,4-benzoxazin-3(4*H*)-one) [20, 21] and its metabolites, and organic acids (β-phenyllactic acid and β-hydroxybutyric acid) [21, 22]. DIBOA and other BXs have been the focus of extensive research because they have broad biological activity, including antibacterial, antifungal, and antifeedant properties [8, 21, 23]. Further emphasizing the versatility of this chemistry, BXs have also been under investigation for use as pharmaceuticals [24–27]. DIBOA and its metabolites are major allelochemicals in rye, as is demonstrated by their inhibition of seedling growth in bioassays [28, 29]. Additional evidence for the activity of BXs *in vivo* was provided by studies with wheat, where they accounted for 69% of the variation in root exudate phytotoxicity [30]. Analysis of the decomposition of rye shoot tissue in soil produced a complex BX profile over time [31], and microbiological studies have demonstrated that BOA (benzoxazolin-2(3*H*)-one), the predominant BX degradation product [32], is transformed by soil borne bacteria into 2-amino-phenoxazin-3-one (APO) that is even more phytotoxic in bioassays [33]. All BOA present in an extract is typically assumed to be a degradation product of DIBOA, itself a degradation product of the DIBOA-glycoside that was stored in the vacuole [32].

The use of allelochemicals to manage weeds can theoretically be achieved by one of two approaches. One method involves the use of crops that possess and deliver allelochemicals by root exudation during growth in the field [8]. Alternatively, allelopathic cover crops can be utilized. These are killed prior to planting the main crop, and both the allelochemicals leached from the living cover crop and those released from the resulting mulch act to suppress weeds. The time of kill varies depending on soil and temperature conditions, as well as susceptibility of the spring crop to the physical/chemical properties of the cover crop, that is, its ability to emerge through a layer of residues or withstand allelopathic effects [19]. Uptake of allelochemicals from rye residue may be from direct contact with tissue fragments or through the soil. This study focuses on foliage leachates rather than root exudates because the residues, especially in no-tillage cropping systems, tend to be concentrated near the seed germination zone where they have a greater potential to chemically interfere with weed seedling growth [28].

The allelochemical concentration in rye tissue is not constant during the growing season. The BX concentration ($mg \cdot g^{-1}$ dry weight of tissue) in fall planted rye reaches a maximum soon after planting or in early spring, depending on weather conditions. In Maryland, USA, Rice et al. [15] observed a constant decrease in BX concentration in "Abruzzi" rye from November to June, whereas, in North Carolina, BX concentration in "Abruzzi" and other varieties decreased from March through May [14]. Not only does the BX/DIBOA concentration change, but the phytotoxicity of the rye aqueous extract (as determined by bioassay) also changes over the season in a manner that largely parallels the BX content [14, 15, 34]. Unfortunately, accumulation profiles of potentially allelopathic secondary metabolites do not always coincide with the requirements of effective early season weed control at the time the main crop is planted [35]. Thus, the observed decrease in the allelochemical concentration in maturing rye combined with variations in the maturation rate due to annual variation in environmental factors could result in differences in the allelochemical content from year to year. These differences likely account for the inconsistent weed control sometimes reported for rye cover crops [36]. Teasdale et al. [37] have reported that although residues of fall planted rye incorporated in the soil the following spring inhibited indicator species growth,

only low levels of most toxic BXs were detected in the soil. This suggested that other compounds may have been responsible for the observed phytotoxicity. At the time of sampling, however, measured tissue BX concentrations in that study were much lower than the values encountered at a comparable time in this study, as well as potential "field rates" (BX concentration × biomass/area). Although phytotoxic compounds other than BXs need to be investigated, the sheer amount of BXs synthesized by rye in North Carolina justifies that they are the focus of this study.

Weed control utilizing allelopathic cover crops is an attractive weed management option for both conventional and organic growers. The variation in weed suppressive efficacy of cover crop mulches indicates the need for better understanding of the chemical ecology of these systems. A clearer picture of the effects of the specific developmental, environmental, physiological, and genetic factors that regulate allelochemical production is needed to make such crops useful and reliable agronomic tools [38]. The goals of this research were to more fully characterize temporal changes in BX content of fall planted, field grown rye in order to obtain a better understanding of rye's natural allelopathic potential. In a previous field study in North Carolina [14], three time points, from March 1 to April 26, had been selected for tissue sampling, whereas the present study includes seven from February 6 to May 1 (only showing data for five since the last two showed no significant difference). Also, serial dilutions were made to determine the D_{50} of rye extracts for each time point, instead of a single dilution used to determine a percent inhibition rate. A different analytical method for tissue BX content was used: GC-FID instead of HPLC-Vis. Finally, a second year of trials was included in the study. This should, in turn, help optimize management or a variety of choices so allelochemical content in foliar tissues of rye cover crops provides better, more consistent weed control in the field.

2. Materials and Methods

2.1. Plant Material. Field experiments were conducted over a two-year period. In year one (2002-2003) rye was planted on November 15th at the Cherry Research Station near Goldsboro, NC, in a clay-loam soil that had previously been in pasture. Each variety was planted into 1.90 × 18.3 m plots with a grain drill set for 18 cm row spacing and a depth of 4 cm. Seeding rate was $100 \, kg \cdot ha^{-1}$. Prior to planting, plots were fertilized with $90 \, kg \, N \cdot ha^{-1}$ of nitrogen applied as ammonium nitrate (34-0-0). Rye varieties used are commonly planted in North Carolina by growers who chose to include cover crops in their rotation (Reberg-Horton, personal communication); they were the winter varieties "Wheeler" and "Aroostook," which require vernalization, as well as the facultative varieties "Wrens Abruzzi" and "Bonel," which do not require vernalization for initiation of flowering. Plots were distributed in a randomized complete block design with five replications. Shoot tissue was harvested on seven dates (Julian date (JD) in parentheses): February 6 (37), 15 (46), and 25 (56), March 6 (65) and 25 (84), April 12 (102), and May 1 (121) and processed as described below.

TABLE 1: Soils analysis results for research plots at Cherry Research Station near Goldsboro, NC (year 1), and the Lake Wheeler Field Laboratory near Raleigh, NC (year 2). Analysis provided by the Environmental and Agricultural Testing Service at North Carolina State University.

	Total C	Total N	P	K	Ca	Mg	S	Cu	Mn	Zn
	%					$mg \cdot kg^{-1}$				
Year 1	0.71	0.06	468.58	243.90	562.57	220.89	80.68	1.78	18.20	7.05
Year 2	0.53	0.05	450.28	225.03	387.06	237.96	74.81	1.52	36.95	6.06

The second-year experiment (2005-2006) was conducted to better characterize the decline in BX content and allelopathic potential of rye tissue that occurs in March in North Carolina. Rye was planted on December 3 at the Lake Wheeler Field Laboratory near Raleigh, NC, also in a clay-loam soil that had been in pasture. Plot size, fertilizer, seeding rate, row spacing, and planting depth were as described above. Rye varieties used were "Aroostook," "Wheeler" (winter types), and "Wrens Abruzzi" (facultative type). Plots were distributed in a randomized complete block design with five replications. Shoot tissue was harvested on five dates (Julian date in parentheses): March 2 (61), 11 (70), 17 (76), 24 (83), and 31 (90) and processed as described below. Soil analysis results for both test sites are presented in Table 1.

2.2. Tissue Harvest, Handling, and Storage.

Tissue was harvested from a $0.5 \, m^2$ area by cutting the stems just above the soil surface. Tissue was then oven dried at 60°C for 3 days, ground to a powder in a Wiley mill with a 1 mm screen, and stored at −20°C pending analysis. Samples were used both for determining BX concentration and in bioassays. The harvest and storage methods used here provided a reliable, reproducible source of tissue for both bioassay and chemical analyses of the relatively large quantities of field grown plants [14, 15, 34]. Preliminary assessment found minimal variation in BX levels when comparing tissues that were flash-frozen and then freeze dried to those that were oven dried as described above (not shown). In addition, dried, ground tissue showed no differences over time for either bioassays or BX analyses when stored at −20°C. Further, although DIBOA and other BXs have been reported to be unstable in aqueous solution [32], they appear to be much more stable in plant tissue. In fact, Yenish et al. [35] reported that only 50% of the BX content in field rye tissue had disappeared 10 to 12 days after being cut and let in fiberglass mesh bags in the field.

2.3. Bioassay of Rye Allelopathic Potential.

In preliminary trials, minimal differences in phytotoxicity were observed for tissue extracts from the last three sampling dates of the first year. Thus, only tissues from the first five harvests were used for bioassays. To assess changes in the allelopathic potential of rye foliage over time, the D_{50} (dilution of extract that inhibits indicator species root elongation by 50% compared to a water only control) was calculated from serial dilutions of the extract. The D_{50} value was derived from the equation assigned to the curve best fitting the relationship between root elongation (percent of control) and the dilution ($mL \cdot g^{-1}$). All pairwise comparisons were adjusted with the Tukey-Kramer method. One gram of ground tissue was added to

50.0 mL of deionized water and extracted by agitation on an orbital shaker for 1 h. The extract was then vacuum-filtered through Whatman number 1 filter paper and its pH was measured. It was then stored for up to 12 h at 4°C or at −20°C for longer periods prior to bioassay. Serial dilutions of the filtered extract ranged from 1 : 50 to 1 : 1600. Aliquots (5.0 mL) of the diluted extract were added to 90 × 15 mm round Petri dishes containing two Whatman number 1 filter papers, and each dilution was run in duplicate. Ten seeds of redroot pigweed (Amaranthus retroflexus) or perennial ryegrass (Lolium perenne) were placed on the filter papers and the dishes were sealed with Parafilm and incubated in a germination chamber at 28°C in the dark. Root elongation of each seedling was measured to the nearest mm after 72 h.

2.4. Determination of BX Content.

The benzoxazinone content of tissues was analyzed by GC [38]. Dried rye shoot tissue (0.5 g, see above) was extracted in 10.0 mL of 50% ethanol in a 75 mL SPE (solid phase extraction) reservoir (Alltech Associates, Deerfield, IL) fitted with a bottom frit (20 μ porosity). Samples were vortexed for 1 min and the resulting extract was then filtered through the reservoir using a vacuum manifold (VWR Scientific, West Chester, PA). The extracted tissue was then washed with 5.0 mL of deionized (DI) water and filtered under vacuum until the plant tissue was dry. The ethanol filtrate was transferred to a volumetric flask and brought to 25.0 mL with DI water. A 10.0 mL aliquot of the filtrate was partitioned three times against equal volumes (5.0 mL) of ethyl acetate (EtOAc). The EtOAc extracts were combined and dried overnight with anhydrous sodium sulfate. A 10.0 mL aliquot of the resulting anhydrous EtOAc extract was mixed with 750 μL of a 200 μg/mL octadecanol internal standard dissolved in toluene and the resulting mixture reduced to dryness under a N_2 stream at 40°C. The dried residue was dissolved in 400 μL of 1 : 1 MSTFA–DMF (derivatization agent N-methyl-n-trimethylsilyltrifluoroacetamide and dimethylformamide solvent), capped under N_2, and heated at 75°C for 30 min. The resulting derivatized solutions were then transferred to autosampler vials and a septum cap was attached. Compounds in the plant extracts were then analyzed and quantified using an Agilent 6890N gas chromatograph (Agilent Technologies, Santa Clara, CA) equipped with a 30 m DB-5 column (0.32 mm diameter, 1.5 μm film thickness, J&W Scientific, Agilent Technologies, Santa Clara, CA) and an Agilent 7683 autosampler. Derivatized samples (0.5 μL) were introduced using splitless injection with an injector temperature of 250°C and separated using a helium (UHP) carrier gas linear velocity of $43.0 \, cm \cdot s^{-1}$. The initial oven temperature was 100°C, followed by a 5°C/min increase to a final temperature of 300°C. The oven temperature increase began upon injection, and the final temperature was maintained for 30 min. Flame ionization detection was utilized with a detector temperature of 325°C, and data were collected using a PerkinElmer TotalChrom 6.2 data system (PerkinElmer, Waltham, MA). Peaks were quantified by comparison to internal standards using appropriate multilevel calibration curves for BOA (benzoxazolin-2($3H$)-one), DIBOA (2,4-dihydroxy-$2H$-1,4-benzoxazin-3($4H$)-one), and DIBOA-Glu (2(2,4-dihydroxy-$2H$-1,4-benzoxazin-3($4H$)-one)-β-D-glucopyranoside). Limits of detection were 173.8 pmol/injection for BOA, 16.13 pmol/injection for DIBOA, and 72.38 pmol/injection for DIBOA-Glu [40]. The DIBOA and DIBOA-Glu standards were prepared in Dr. W. S. Chilton's and Dr. D. A. Danehower's Laboratories, respectively, at NC State University [40]. The BOA standard was purchased from Sigma-Aldrich (St. Louis, MO).

2.5. Statistical Analysis. Data were analyzed over time using a PROC MIXED repeated measures ANOVA model using SAS Software Version 9.1.3 (SAS Institute Inc., 2000–2004) with the Tukey adjustment method to test for significance of differences among varieties and sampling dates. Normality of data was verified using the univariate procedure and homogeneity of variance was confirmed using Levene's test.

3. Results

3.1. Phytotoxicity of Rye Extracts. Previous analyses of rye tissue phytotoxicity were based on single dilution bioassays that did not anticipate the dramatic changes in phytotoxicity that were observed over the experimental season [14]. To make more accurate comparisons over these large changes in toxicity, both pigweed and perennial ryegrass seeds were set to germinate on serial dilutions of crude aqueous tissue extracts. The inhibition of germination and growth of the seedlings were compared to a water control, and the dilution that inhibited growth by 50% (D_{50}) was calculated. Germination rates in the controls were 92.5% and 97.5% for pigweed and ryegrass, respectively. Ungerminated seeds, which showed no root growth, were included in the growth measurement pool (i.e., their root elongation was equal to zero). pH of extracts was measured for the initial dilution and proved to be similar across sampling dates (data not shown). Preliminary bioassays were conducted to estimate a suitable range of dilutions, and replicate bioassays were then conducted at dilutions that would bracket the D_{50}. The second-year field experiment focused on the latter part of the season (in March, JDs 60 and 90) to obtain a more precise measurement of the change in rye leaf tissue phytotoxicity and benzoxazinone content during their decline phase.

3.2. Bioassay Using Pigweed as an Indicator. For the first year's pigweed bioassays (2002-2003), seed germination and root elongation were sensitive to aqueous extracts from all rye varieties tested (Figure 1). The phytotoxicity of these extracts decreased as the season progressed, with the apparent rate

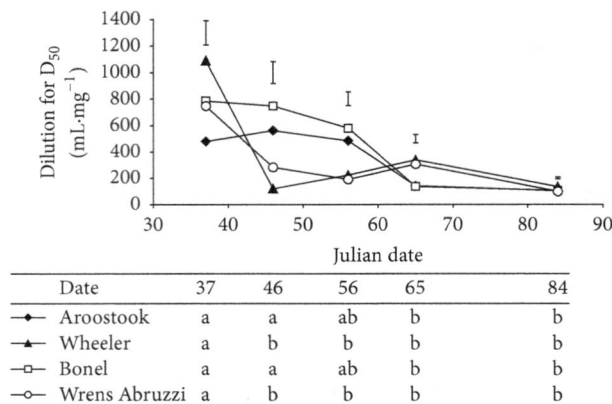

Date	37	46	56	65	84
Aroostook	a	a	ab	b	b
Wheeler	a	b	b	b	b
Bonel	a	a	ab	b	b
Wrens Abruzzi	a	b	b	b	b

FIGURE 1: Inhibition of pigweed germination and growth by extracts of four field grown rye varieties, harvested at five different dates (Julian dates) during the spring of the first year of experiments. D_{50} is the concentration required to cause 50% growth inhibition. Significant differences over time within each variety according to the Tukey-Kramer method ($p < 0.05$) are indicated by different letters. Vertical bars represent standard errors at each sampling date.

of decrease varying among varieties. As noted above, preliminary trials showed minimal differences in phytotoxicity for tissue extracts from the last two sampling dates (April 12 and May 1; JD 102 and 121, resp.); therefore only tissues from the first five harvests were fully assessed. For "Aroostook" and "Bonel," extracts from the first three harvest dates (February 6, 15, and 25; JDs 37, 46, and 56, resp.) were the most phytotoxic. The phytotoxicity of the extracts from these varieties then decreased at the fourth harvest on March 6 (JD 65) and remained at that level through the harvest on March 25 (JD 84). For "Wheeler" and "Wrens Abruzzi," extracts from the first harvest date were the most phytotoxic, with toxicity decreasing by the second harvest and remaining low until the last harvest. For each individual sampling date, there was no difference in phytotoxicity between varieties except at the first and second harvest dates. During the second year (2005-2006), pigweed bioassays between March 2 and 31 (JDs 61 through 90) showed a lesser decline in phytotoxicity over time (Figure 2) and showed no significant differences between varieties. Phytotoxicity was the same on March 6 and 11 (JDs 61 and 70) and then decreased on March 17 (JD 76) to a minimum that remained stable until the last sampling on March 31 (JD 90). Phytotoxicity values in the second year were similar to those measured in the first year during equivalent sampling periods, except for the first sampling on March 2 (JD 61) when Bonel and Aroostook had much lower toxicity than at the same time the first year.

3.3. Bioassay Using Ryegrass as an Indicator. In the first-year ryegrass bioassays, seed germination and root elongation were also sensitive to aqueous extracts of all rye varieties (Figure 3), with varying sensitivities according to the rye variety. For "Aroostook" extracts from the first two harvests (February 6 and 15; JDs 37 and 46) were the most phytotoxic. Phytotoxicity then decreased at the third harvest February 25 (JD 56) and remained low until the last harvest on

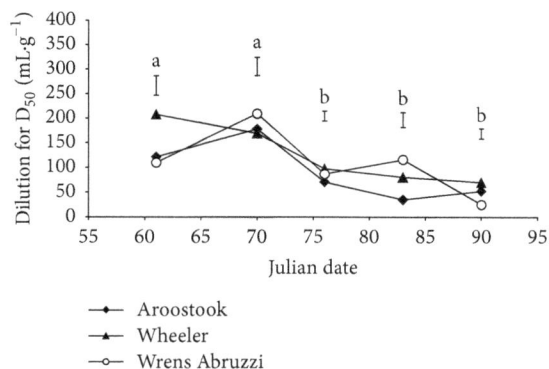

FIGURE 2: Inhibition of pigweed germination and growth by extracts of three field grown rye varieties, harvested at five different dates (Julian dates) during the spring of the second year of experiments. D_{50} is the concentration required to cause 50% growth inhibition. Values associated with the same letter are not significantly different according to the Tukey-Kramer method ($p < 0.05$). Vertical bars represent standard errors at each sampling date.

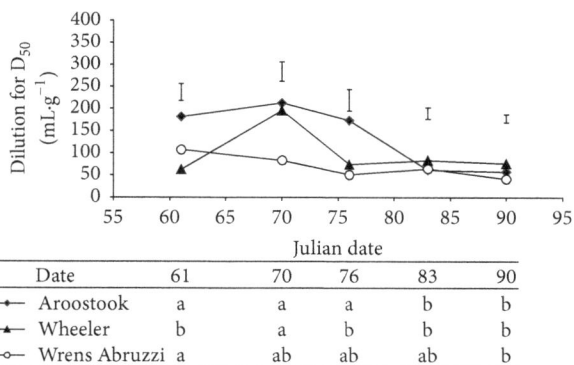

Date	61	70	76	83	90
Aroostook	a	a	a	b	b
Wheeler	b	a	b	b	b
Wrens Abruzzi	a	ab	ab	ab	b

FIGURE 4: Inhibition of ryegrass germination and growth by extracts of three field grown rye varieties, harvested at five different dates (Julian dates) during the spring of the second year of experiments. D_{50} is the concentration required to cause 50% growth inhibition. Values associated with the same letter are not significantly different according to the Tukey-Kramer method ($p < 0.05$). Vertical bars represent standard errors at each sampling date.

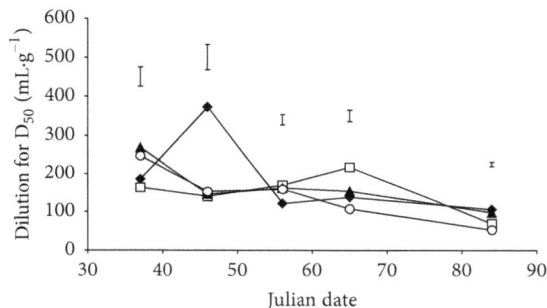

Date	37	46	56	65	84
Aroostook	ab	a	b	b	b
Wheeler	a	a	a	a	a
Bonel	a	a	a	a	a
Wrens Abruzzi	a	a	a	b	b

FIGURE 3: Inhibition of ryegrass germination and growth by extracts of four field grown rye varieties, harvested at different dates (Julian dates) during the spring of the first year of experiments. D_{50} is the concentration required to cause 50% growth inhibition. Significant differences over time within each variety according to the Tukey-Kramer method ($p < 0.05$) are indicated by different letters. Vertical bars represent standard errors at each sampling date.

March 25 (JD 84). For "Wrens Abruzzi," extracts from the first three harvests were the most phytotoxic. Phytotoxicity then decreased at the fourth harvest (March 6; JD 65) and remained stable until the last sampling date. There were no significant variations in phytotoxicity of "Bonel" or "Wheeler" over the duration of the experiment. For each individual sampling date, there was generally no difference in phytotoxicity between varieties except at the second harvest (March 11; JD 70), when "Aroostook" extracts were significantly more phytotoxic than those of all other varieties ($p \leq 0.01$).

The relatively constant, low levels of ryegrass phytotoxicity observed between March 6 and 25 (JDs 65 and 84) in year 1 of the study were essentially reproduced by

the second-year results (Figure 4), but at later sampling dates. In contrast, however, although extracts from all rye varieties were inhibitory to ryegrass seed germination and root elongation during the second year, significant differences between varieties were observed ($p = 0.05$). Phytotoxicity of "Aroostook" extracts was maximal and equivalent on March 2 through March 17 (JDs 61, 70, and 76), then decreased on March 24 (JD 83), and remained stable until the last sampling on March 31 (JD 90). Phytotoxicity of "Wheeler" extracts increased, for reasons that are unclear, between March 2 and 11 (JDs 61 and 70) then returned to its initial level March 17 (JD 76) to remain stable until the last sampling date. Finally, phytotoxicity of "Wrens Abruzzi" extracts gradually decreased between JDs 61 and 90, with no significant difference between March 2 and 24 (JDs 61 to 83).

3.4. Benzoxazinone Content. In the first year of this study, BX levels in extracts of rye shoot tissue varied during the season (Figure 5). BX concentrations were similar among varieties at all dates except for Wrens Abruzzi, for which higher values than other varieties were measured at the first two harvest dates and lower values were measured at the third harvest date. Maximum concentrations of BX were found in all varieties by the first week in March (JD 65), with levels decreasing to a minimum by the next sampling date (March 25; JD 84).

In the second year, BX concentrations decreased over the month of March (JDs 61 and 90) (Figure 6), with no significant difference observed between varieties ($p = 0.79$). BX concentrations were similar from March 2 to 17 (JDs 61 and 76), then decreased by March 24 (JD 83), and remained the same thereafter.

As with the bioassays, BX profiles varied between years one and two of the study. The average BX concentration for the initial harvest in year two (925 $\mu g \cdot g^{-1}$ dry wt. at JD 61; Figure 6) was much lower than at a similar date in year one (1902 $\mu g \cdot g^{-1}$ dry wt. at JD 65; Figure 5). By March 25 (JD 84),

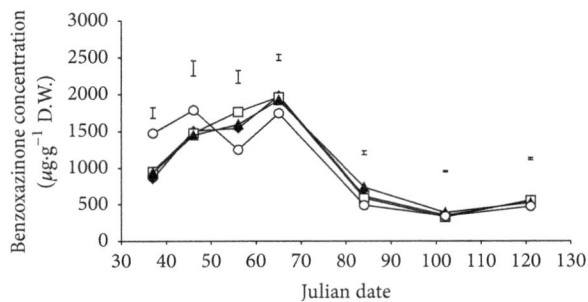

FIGURE 5: Variation of total benzoxazinone concentration (sum of DIBOA-glucoside, DIBOA, and BOA) in shoot tissue of four rye cultivars between dates 37 and 121 (Julian dates) during the first year of experiments. Significant differences over time within each variety according to the Tukey-Kramer method ($p < 0.05$) are indicated by different letters. Vertical bars represent standard errors at each sampling date.

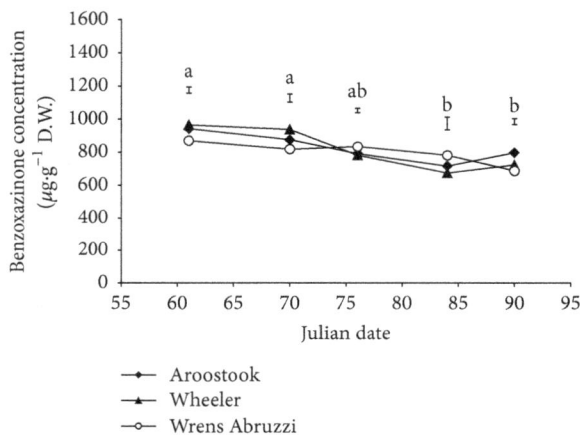

FIGURE 6: Variation of total benzoxazinone content (sum of DIBOA-glucoside, DIBOA, and BOA) in shoot tissue of three rye cultivars between dates 61 and 90 (Julian dates) of the second year of experiments. Values associated with the same letter are not significantly different according to the Tukey-Kramer method ($p < 0.05$). Vertical bars represent standard errors at each sampling date.

however, average BX content was similar (year 2: 723 $\mu g \cdot g^{-1}$ dry wt. on JD 83 (Figure 6); year 1: 605 $\mu g \cdot g^{-1}$ dry wt. at JD 84 (Figure 5)). The differences between year one and year two are likely due, at least in part, to later planting in the second year caused by weather constraints. Consequently, the initial harvest in year two was delayed (Figure 7) to obtain an adequate sample size. In addition, temperatures in year two were higher than in year one (Figures 8 and 9), resulting in a significant increase in growing degree days (GDDs; Figure 7). As a result, at corresponding dates after planting, plants in year two had accumulated from 27 to 48% more GDDs than in year one.

The highest BX concentrations observed here were in year one and ranged from 1744 to 1981 $\mu g \cdot g^{-1}$ dry tissue (Figure 5). In year two, the highest BX concentrations ranged from 870 to 965 $\mu g \cdot g^{-1}$ dry tissue (Figure 6).

3.5. Total Field Accumulation of BX. The total amount of BX in a field at any time depends on both the concentration in the plant and the total plant biomass. Total dry matter accumulation for different rye varieties throughout the growing season was recorded. In year one (Figure 10), "Wheeler" accumulated dry matter more slowly than "Bonel" ($p = 0.015$) or "Wrens Abruzzi" ($p = 0.001$). In contrast, the other winter variety "Aroostook" showed no significant difference in biomass accumulation from other varieties.

Although the tissue concentration of BX reached a maximum in early March (JD 65) and then decreased over the remainder of the season (Figure 5), the continued accumulation of biomass resulted in two peaks of total field BX (Figure 11). These two peaks occurred in early March (JD 65) and early May (JD 121) with BX "rate" ranging from 3.0 to 5.0 kg·ha^{-1} at each of these points in time. The first peak in early March (JD 65) corresponded to the maximum concentration of BX in rye leaf tissue and occurred during the transition from vegetative to reproductive growth. The second peak (JD 121) was on the last sampling date, when maximum biomass was recorded. Although the BX concentration in rye tissue at the end of the growing season had decreased to about 1/4 of its early season maximum, total BX field accumulation was at its maximum due to the large accumulation of above ground biomass. All varieties accumulated equivalent total BX per hectare throughout the experiment, with the exception of "Wheeler," which had accumulated less dry matter by the last sampling date (Figure 10).

In year two total BX per hectare increased steadily over the sampling period (Figure 12), with no statistically significant differences between varieties for either total field BX or dry matter accumulation (Figure 13). Total BX per ha was low through early March (JDs 61 to 70). In mid-March (JD 76) total BX for "Aroostook" and "Wrens Abruzzi" began to increase and continued to increase through the end of March (JD 90). Increase in total field BX for "Wheeler" increased after mid-March (JD 76).

4. Discussion

4.1. Bioassays Show Maximum Toxicity prior to Reproductive Growth. In year one, phytotoxicity of extracts from field grown rye tissue varied depending on the variety, harvest date, and indicator species used (Figures 1 and 3). In general, not only were extracts from tissues harvested earlier in the season more phytotoxic than those harvested later, but greater variation was seen between varieties. In addition, extracts from the earliest harvests inhibited pigweed root elongation more than ryegrass root elongation. This corresponds with previous reports indicating that dicot species are typically more sensitive than monocots to rye allelochemicals [20, 41]. However, phytotoxicity of extracts of all varieties at

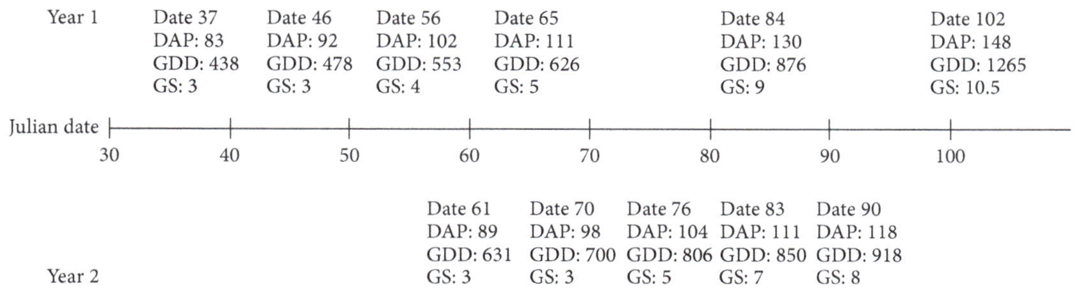

FIGURE 7: Accumulated days after planting (DAP), growing degree days (GDD), and observed Feekes growth stage (GS, for variety Aroostook) at successive sampling dates (Julian dates) for year one and year two of the field experiment.

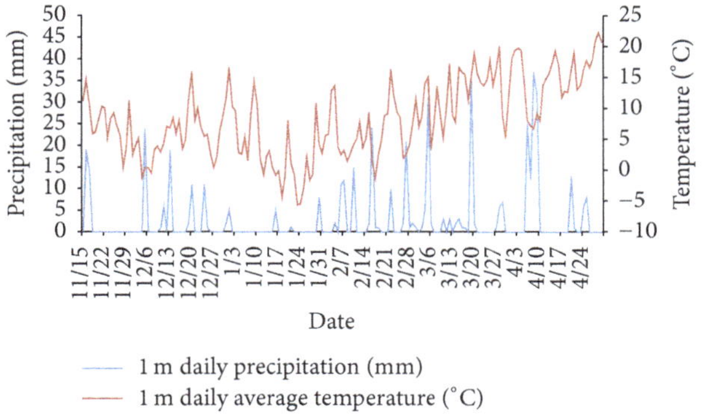

FIGURE 8: Average daily precipitation and temperature values recorded at the Cherry Research Station near Goldsboro, NC, during year 1.

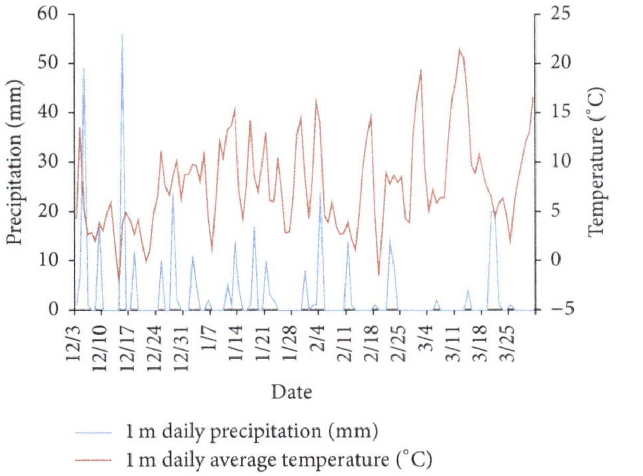

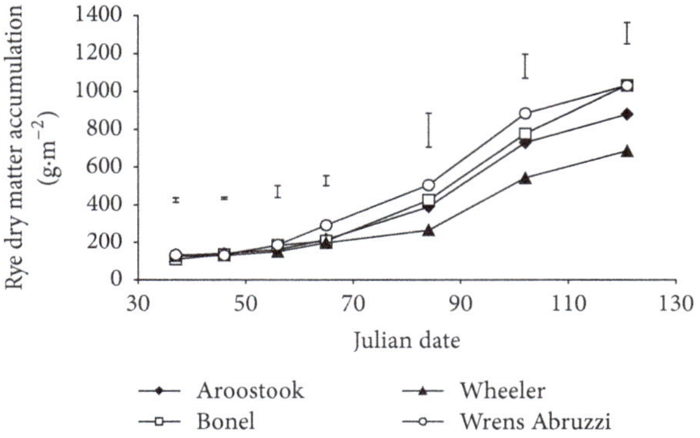

FIGURE 10: Accumulation of above ground dry matter in four rye varieties over time (Julian dates) during the first year of experiments. Vertical bars represent standard errors at each sampling date.

FIGURE 9: Average daily precipitation and temperature values recorded at the Lake Wheeler Field Laboratory near Raleigh, NC, during year 2.

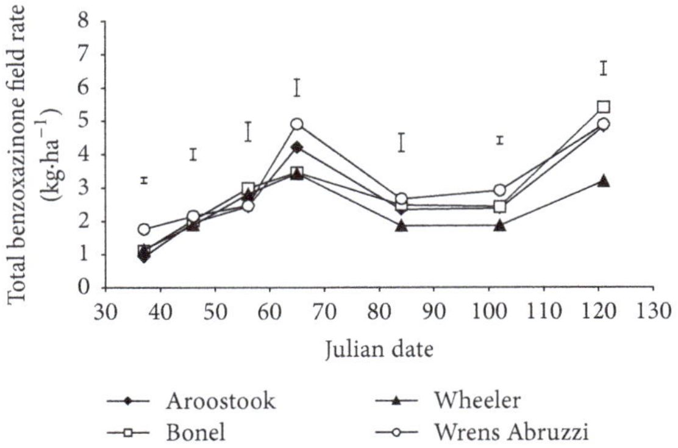

FIGURE 11: Benzoxazinone rate (concentration × biomass) at successive dates in rye foliage of four rye varieties during the first year of experiments. Vertical bars represent standard errors at each sampling date.

the third, fourth, and fifth harvests was much lower and was essentially the same for both pigweed and ryegrass bioassays. Extracts from tissues harvested in the second year between the beginning and end of March showed low and generally decreasing levels of phytotoxicity, similar to those seen over the comparable period in year 1 and consistent with the results of previous abbreviated analyses [14, 15]. The greater detail provided by the present study revealed that the maximum phytotoxicity of rye plant extracts occurred when plants were at stage 3 or stage 4 of the Feekes scale, just prior to the transition from vegetative to reproductive growth [39] (Figures 1 and 2; Table 2).

4.2. Benzoxazinone in Tissues Is Also Maximal prior to Initiation of Reproductive Growth. The range of BX concentrations measured in this study is commensurate with findings by other groups: In studies by Burgos et al. [34] field grown shoot

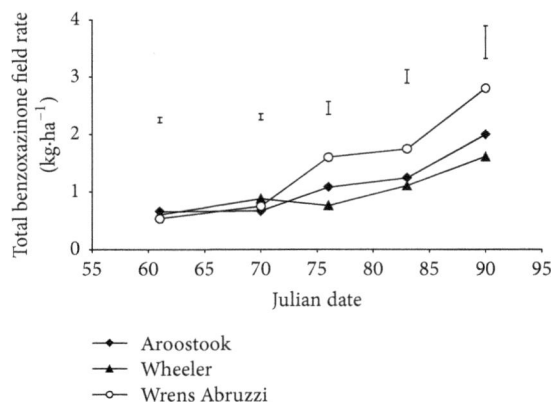

FIGURE 12: Benzoxazinone field rate (concentration × biomass) at successive dates in rye foliage of four rye varieties during the second year of experiments. Vertical bars represent standard errors at each sampling date.

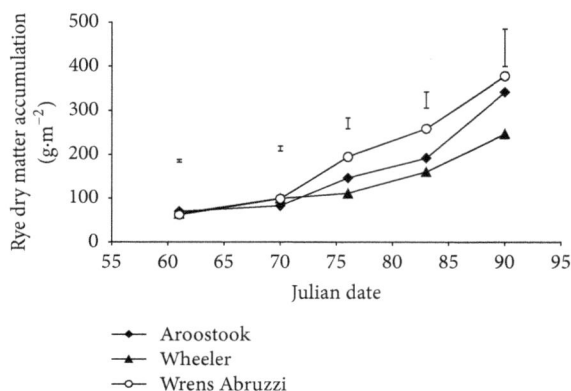

FIGURE 13: Accumulation of dry matter for three rye varieties over time (Julian dates) during the second year of experiments. Vertical bars represent standard errors at each sampling date.

TABLE 2: Average growth stage (Feekes scale [39]) of different rye varieties observed at successive sampling dates during the first and second year of experiments ("Bonel" not used in the second year).

| | Date | Average growth stage (Feekes scale) | | | |
		Aroostook	Bonel	Wheeler	Wrens Abruzzi
	37	3	3	3	3
	46	3	3	3	4
	56	4	4	3	4
Year 1	65	5	5	4	6
	84	9	9	8	10.1
	102	10.5	10.5	10.4	10.5
	121	10.54	10.54	10.53	10.54
	61	3	—	3	3
	70	3	—	3	4
Year 2	76	5	—	3	7
	83	7	—	4	8
	90	8	—	5	9

tissue of various varieties (planted on October 22 in Arkansas, harvested on April 30, at Feekes 10) had 137 to 1240 $\mu g \cdot g^{-1}$ dry tissue total BX. "Abruzzi" rye planted in mid-September in Maryland contained a maximum total BX of 404 $\mu g \cdot g^{-1}$ dry tissue in early November [15]. In North Carolina, four rye varieties planted in mid-November had 1000 to 1800 $\mu g \cdot g^{-1}$ dry tissue DIBOA [14] and, in Denmark, thirteen spring-harvested (Feekes 7 to 8) varieties had an average total BX of 1944 $\mu g \cdot g^{-1}$ dry tissue [40].

In our analyses, the primary BX found in tissues was DIBOA-glycoside. Some DIBOA aglycone and BOA were also present (data not shown), but the proportion of each varied only slightly over the sampling period. This variation was likely not entirely due to phenological differences, as sample handling and processing can also independently affect BX stability.

In fact, the observed composition of the tissue BX confirms the relative stability of BXs under the isolation and assay conditions used here.

In year 1, BX concentrations in all varieties increased from the first harvest in early February (JD 37) to their highest levels at the end of the first week in March (JD 65). Although this generally mirrors the higher levels of phytotoxicity seen in bioassays of tissues harvested earlier in the season, there is not an exact correlation. For instance, phytotoxicity is already at its highest level in the assays from the first harvest in early February while BX levels are still increasing. Additionally, although BX profiles of the various varieties were quite similar, "Aroostook" and "Bonel" remained phytotoxic through the end of February, while toxicity of "Wrens" and "Wheeler" begins to decrease much earlier. Given the similarities in BX concentration profiles of the various varieties, this seems unlikely to be caused solely by differences in the decline of BX accumulation and suggests that compounds other than BXs are also involved in extract toxicity. Although not measured in this study, reports do indicate that rye contains other phytotoxic compounds including phenolics and organic acids [22, 37, 40]; in a study reported by Teasdale et al. [37], BXs contributed very little to toxicity measured in assays where rye was incorporated to field soil. The plants in that study, however, had much lower BX concentrations at kill time (145 to 161 $\mu g \cdot g^{-1}$ for the first year and 12 to 55 $\mu g \cdot g^{-1}$ for the second year) than what was measured in this study at the end of the sampling season (472 to 554 $\mu g \cdot g^{-1}$ for the first year and 687 to 795 $\mu g \cdot g^{-1}$ for the second year, albeit at an earlier developmental stage) and much lower field rates of BX (0.9 kg·ha^{-1} and 0.08 kg·ha^{-1} for the first and second years, resp.) than calculated in this study (3.21 to 5.41 kg·ha^{-1} and 1.6 to 2.79 kg·ha^{-1} for the first and second years, resp.).

The change from vegetative to reproductive growth correlated even more clearly with the large decrease in BX concentration in all varieties seen between March 6 (JD 65; Feekes stages 4 to 6) and March 25 (JD 84), at which time all varieties were into reproductive growth (Feekes of 8 and 10.1; Table 2). In year two, the decline was not as dramatic, but BX levels did decrease as reproductive growth progressed. The change from vegetative to reproductive growth is known to initiate reallocation of assimilated C and N from older leaves to developing reproductive tissues [42]. Thus the observed decrease in BX content coinciding with the onset of reproductive growth might reflect such metabolic reallocation to the reproductive tissue.

The specific roles of growth stage and leaf senescence in secondary metabolite and allelochemical content in plants have not been extensively explored. However, it is well known that when cereal grasses and other plants shift from vegetative to reproductive growth, they begin a patterned reallocation of accumulated C and N products from older leaves to the flowering tissue. For instance, the programmed catabolism of chlorophyll and photosynthetic proteins and the remobilization of the recycled C and N to reproductive tissues are characteristic of this transition [43–46]. Thus, one might expect a similar reallocation or directed export of accumulated defense compounds and a resultant decrease in levels of these compounds in the senescing tissue. Interestingly, current data on secondary metabolites in senescing leaves in other plants indicate that the content of at least some of these compounds increases [43, 46, 47]. For example, higher levels of flavonoid and anthocyanin secondary metabolites and/or increased specific expression of their biosynthetic genes have been observed in senescing leaves of *Arabidopsis* and poplar [42, 45]. This is hypothesized to be a response to an increased need for defense compounds during senescence, because senescing tissues are more susceptible to pathogens. Though there is no data for rye, senescence in the flag leaf of wheat is characterized by an increase in the expression of specific secondary metabolite genes, including BX2 and BX4 (P450 monooxygenases), BX6 (dioxygenase), and BX9 (UDP-glucosyltransferase), which are involved in BX biosynthesis [47]. In rice, a similar increase in serotonin levels and tryptophan biosynthetic pathway genes was observed in senescing flag leaves [46]. Although these results might lead us to expect that levels of BX or defense related secondary metabolites would increase during senescence and flowering, one actually sees a decrease in BX content at the whole plant level in rye and wheat [44] as flowering progresses. It is likely that comparisons of secondary metabolite levels in specific tissues (e.g., flag leaves) versus entire plants (which represent a range of tissue types at various stages of development) would clarify the issue.

4.3. Total Field BX.

For effective weed suppression using rye mulches, there must be both enough biomass for physical suppression and enough allelochemical in the field to provide chemical competition at the time the main crop is to be planted. Total field BX, of course, depends on both BX concentration in the plant and total plant biomass. The extended, more detailed sampling conducted here allowed us to produce a better picture of these BX field "rates" throughout the season. Unfortunately the highest BX concentrations in rye are early in the season, presumably a time when rye uses BX to protect itself from biological competition. BX concentrations decrease thereafter and in previous studies (e.g., [14]) this was mirrored by a decrease in total field BX during later part of the season for all varieties. In contrast, field BX accumulation in the present study showed an increase at the end of the season due to an increase in biomass large enough offset lower shoot BX concentrations.

The variation observed between studies as well as from year to year within studies highlights some of the complexities inherent in using allelopathic cover crops for weed suppression. In addition to differences in tissue BX concentrations between two growing years, differences in biomass accumulation due to the potential impact of environmental variation, soils, and management practices impact BX content and stability in rye. Factors that affect both biomass accumulation and BX concentration can include planting date, rainfall, soil type, and drainage, as well as fertilization and soil microbiology. For example, the soil type in the Reberg-Horton et al. study [14] was a sandy-loam, and some early desiccation of the older leaves was observed. This, in turn, likely decreased BX concentration in older, vegetative leaf tissue due to stress-induced desiccation and/or leaf abscission, while simultaneously decreasing total biomass. In contrast, the soils in the present study had a greater water holding capacity, and early desiccation was not observed (data not shown). Management practices, such as delaying rye control, allow for more biomass accumulation, therefore increasing ground cover and potentially reducing weed competition or increasing the interference with the emerging new crop [48, 49]. The method of control, chemical or mechanical, may influence rye regrowth and also affect weed competition and new crop emergence [19]. Apart from variety selection, the amount of rye biomass present depends greatly on growing conditions throughout the fall and winter [50]. This enhances of course the unpredictable nature of a cover crop's efficiency to control weeds.

This and other studies underscore the need for management practices that optimize both biomass production and allelochemical production if we are to obtain consistent performance using rye as a cover crop. In fact, these studies suggest several potential management strategies for rye cover crops that could result in elevated levels of allelochemicals and enhanced weed suppression. One strategy would involve the late winter or early spring planting of a late maturing rye that has a high vernalization requirement. These conditions would delay or prevent flowering, and the resulting increase in the duration of vegetative growth should result in greater accumulation of BXs near the time of the spring planting dates. This approach, however, might reduce accumulated biomass and thus reduce the cover crop's ability to smother weeds via competition for light resources. Thus, the use of typically high biomass producers such as "Wheeler" might result in higher field BX levels late in the season. Although "Wheeler" did not show appreciably higher biomass or BX content in the present study, this appeared to be due to a delay in initial growth. This is perhaps because "Wheeler," a winter variety, is the latest maturing of the varieties tested (Table 2). "Wheeler," however, is in fact a high biomass producer [14] and would likely continue to accumulate biomass, reaching a maximum later than the other varieties.

Another possible management strategy would involve spring mowing of fall planted rye. The ensuing regrowth would result in the presence of younger tissue later in the season. This might also circumvent biomass issues. De Bruin et al. [48] reported that mowing early during vegetative growth has minimal impact on the regrowth potential of rye. An additional approach not previously available in cover crop allelopathy studies is the potential effect of "stay-green" genes, which delay leaf senescence [51]. Such a delay might also result in maintenance of allelochemical levels in vegetative tissues. This germplasm could be utilized alone or with either of the above management strategies.

In summary, this study not only confirms earlier reports of variations in allelochemical content in rye shoot tissue during the growing season, but also reveals a previously unobserved relationship between rye growth stage and allelochemical level. In addition, the use of tissue extract dose-response bioassays combined with GC analysis of the same extracts established a parallel between allelochemical content and allelopathic potential. Although these results in themselves contribute to characterize allelopathic patterns in rye, the further implication that developmental and senescence processes have an impact on synthesis of BX in rye suggests exciting new avenues of exploration.

Abbreviations

BX: Benzoxazinone
DIBOA: 2,4-Dihydroxy-2H-1,4-benzoxazin-3(4H)-one
BOA: Benzoxazolin-2(3H)-one
JD: Julian date.

Conflict of Interests

The authors declare that there is no conflict of interests regarding the publication of this paper.

Acknowledgments

The authors thank Matthew Finney and Ken Fager for their technical assistance as well as Dr. D. Boos, Department of Statistics, NCSU, for help with the statistical analysis.

References

[1] L. A. Weston, "Utilization of allelopathy for weed management in agroecosystems," *Agronomy Journal*, vol. 88, no. 6, pp. 860–866, 1996.

[2] M. E. Foley, "Genetic approach to the development of cover crops for weed management," *Journal of Crop Production*, vol. 2, no. 1, pp. 77–93, 1999.

[3] G. G. Nagabhushana, A. D. Worsham, and J. P. Yenish, "Allelopathic cover crops to reduce herbicide use in sustainable agricultural systems," *Allelopathy Journal*, vol. 8, no. 2, pp. 133–146, 2001.

[4] National Organic Program, *Code of Federal Regulations, Title 7: Agriculture, Part 205*, National Organic Program, Washington, DC, USA, 2009.

[5] Inderjit and S. O. Duke, "Ecophysiological aspects of allelopathy," *Planta*, vol. 217, no. 4, pp. 529–539, 2003.

[6] L. A. Weston and S. O. Duke, "Weed and crop allelopathy," *Critical Reviews in Plant Sciences*, vol. 22, no. 3-4, pp. 367–389, 2003.

[7] T. L. Weir, S.-W. Park, and J. M. Vivanco, "Biochemical and physiological mechanisms mediated by allelochemicals," *Current Opinion in Plant Biology*, vol. 7, no. 4, pp. 472–479, 2004.

[8] R. G. Belz, "Allelopathy in crop/weed interactions—an update," *Pest Management Science*, vol. 63, no. 4, pp. 308–326, 2007.

[9] M. A. Czarnota, R. N. Paul, F. E. Dayan, C. I. Nimbal, and L. A. Weston, "Mode of action, localization of production, chemical nature, and activity of sorgoleone: a potent PSII inhibitor in *Sorghum* spp. root exudates," *Weed Technology*, vol. 15, no. 4, pp. 813–825, 2001.

[10] A. M. Hejl and K. L. Koster, "Juglone disrupts root plasma membrane H$^+$-ATPase activity and impairs water uptake, root respiration, and growth in soybean (*Glycine max*) and corn (*Zea mays*)," *Journal of Chemical Ecology*, vol. 30, no. 2, pp. 453–471, 2004.

[11] A. Gierl and M. Frey, "Evolution of benzoxazinone biosynthesis and indole production in maize," *Planta*, vol. 213, no. 4, pp. 493–498, 2001.

[12] A. Friebe, "Role of benzoxazinones in cereals," *Journal of Crop Production*, vol. 4, no. 2, pp. 379–400, 2001.

[13] Z. Huang, T. Haig, H. Wu, M. An, and J. Pratley, "Correlation between phytotoxicity on annual ryegrass (*Lolium rigidum*) and production dynamics of allelochemicals within root exudates of an allelopathic wheat," *Journal of Chemical Ecology*, vol. 29, no. 10, pp. 2263–2279, 2003.

[14] S. C. Reberg-Horton, J. D. Burton, D. A. Danehower et al., "Effect of time on the allelochemical content of ten cultivars of rye (*Secale cereale* L.)," *Journal of Chemical Ecology*, vol. 31, no. 1, pp. 179–192, 2005.

[15] C. P. Rice, B. P. Yong, F. Adam, A. A. Abdul-Baki, and J. R. Teasdale, "Hydroxamic acid content and toxicity of rye at selected growth stages," *Journal of Chemical Ecology*, vol. 31, no. 8, pp. 1887–1905, 2005.

[16] A. D. Worsham and U. Blum, "Allelopathic cover crops to reduce herbicide inputs in cropping systems," in *Proceedings of the 1st International Weed Control Congress*, vol. 2, pp. 577–579, Melbourne, Australia, February 1992.

[17] J. P. Barnes and A. R. Putnam, "Rye residues contribute to weed suppression in no-tillage cropping systems," *Journal of Chemical Ecology*, vol. 9, no. 8, pp. 1045–1057, 1983.

[18] K. N. Reddy, "Impact of rye cover crop and herbicides on weeds, yield, and net return in narrow-row transgenic and conventional soybean (glycine max)," *Weed Technology*, vol. 17, no. 1, pp. 28–35, 2003.

[19] L. R. Westgate, J. W. Singer, and K. A. Kohler, "Method and timing of rye control affects soybean development and resource utilization," *Agronomy Journal*, vol. 97, no. 3, pp. 806–816, 2005.

[20] J. P. Barnes and A. R. Putnam, "Evidence for allelopathy by residues and aqueous extracts of rye (*Secale cereale*)," *Weed Science*, vol. 34, no. 3, pp. 384–390, 1986.

[21] H. M. Niemeyer, "Hydroxamic acids derived from 2-hydroxy-2H-1,4-benzoxazin-3(4H)-one: fey defense chemicals of cereals," *The Journal of Agricultural and Food Chemistry*, vol. 57, no. 5, pp. 1677–1695, 2009.

[22] D. G. Shilling, L. A. Jones, A. D. Worsham, C. E. Parker, and R. F. Wilson, "Isolation and identification of some phytotoxic compounds from aqueous extracts of rye (*Secale cereale* L.)," *Journal of Agricultural and Food Chemistry*, vol. 34, no. 4, pp. 633–638, 1986.

[23] A. Gierl, S. Gruen, U. Genschel, R. Huettl, and M. Frey, "Evolution of indole and benzoxazinone biosynthesis in *Zea mays*," in *Recent Advances in Phytochemistry*, vol. 38, chapter 4, pp. 69–83, Elsevier, 2004.

[24] M. Patel, R. J. McHugh Jr., B. C. Cordova et al., "Synthesis and evaluation of benzoxazinones as HIV-1 reverse transcriptase inhibitors. Analogs of Efavirenz (Sustiva)," *Bioorganic and Medicinal Chemistry Letters*, vol. 9, no. 22, pp. 3221–3224, 1999.

[25] P. J. Rybczynski, R. E. Zeck, D. W. Combs et al., "Benzoxazinones as PPARγ agonists. Part 1: SAR of three aromatic regions," *Bioorganic & Medicinal Chemistry Letters*, vol. 13, no. 14, pp. 2359–2362, 2003.

[26] W. Huang, P. Zhang, J. F. Zuckett et al., "Design, synthesis and structure-activity relationships of benzoxazinone-based factor Xa inhibitors," *Bioorganic and Medicinal Chemistry Letters*, vol. 13, no. 3, pp. 561–566, 2003.

[27] T. R. Belliotti, D. J. Wustrow, W. A. Brink et al., "A series of 6- and 7-piperazinyl- and -piperidinylmethylbenzoxazinones with dopamine D4 antagonist activity: discovery of a potential atypical antipsychotic agent," *Journal of Medicinal Chemistry*, vol. 42, no. 25, pp. 5181–5187, 1999.

[28] J. P. Barnes and A. R. Putnam, "Role of benzoxazinones in allelopathy by rye (*Secale cereale* L.)," *Journal of Chemical Ecology*, vol. 13, no. 4, pp. 889–906, 1987.

[29] J. P. Barnes, A. R. Putnam, B. A. Burke, and A. J. Aasen, "Isolation and characterization of allelochemicals in rye herbage," *Phytochemistry*, vol. 26, no. 5, pp. 1385–1390, 1987.

[30] R. G. Belz and K. Hurle, "Differential exudation of two benzoxazinoids—one of the determining factors for seedling allelopathy of *Triticeae* species," *Journal of Agricultural and Food Chemistry*, vol. 53, no. 2, pp. 250–261, 2005.

[31] S. S. Krogh, S. J. M. Mensz, S. T. Nielsen, A. G. Mortensen, C. Christophersen, and I. S. Fomsgaard, "Fate of benzoxazinone allelochemicals in soil after incorporation of wheat and rye sprouts," *Journal of Agricultural and Food Chemistry*, vol. 54, no. 4, pp. 1064–1074, 2006.

[32] F. A. Macías, A. Oliveros-Bastidas, D. Marín, D. Castellano, A. M. Simonet, and J. M. G. Molinillo, "Degradation studies on benzoxazinoids: soil degradation dynamics of (2R)-2-O-β-D-glucopyranosyl-4-hydroxy-(2H)-1,4-benzoxazin-3(4H)-one (DIBOA-Glc) and its degradation products; phytotoxic allelochemicals from gramineae," *Journal of Agricultural and Food Chemistry*, vol. 53, no. 3, pp. 554–561, 2005.

[33] I. S. Fomsgaard, A. G. Mortensen, and S. C. K. Carlsen, "Microbial transformation products of benzoxazolinone and benzoxazinone allelochemicals—a review," *Chemosphere*, vol. 54, no. 8, pp. 1025–1038, 2004.

[34] N. R. Burgos, R. E. Talbert, and J. D. Mattice, "Cultivar and age differences in the production of allelochemicals by *Secale cereale*," *Weed Science*, vol. 47, no. 5, pp. 481–485, 1999.

[35] J. P. Yenish, A. D. Worsham, and W. S. Chilton, "Disappearance of DIBOA-glucoside, DIBOA, and BOA from rye (*Secale cereale* L.) cover crop residue," *Weed Science*, vol. 43, no. 1, pp. 18–20, 1995.

[36] N. G. Creamer, M. A. Bennett, B. R. Stinner, J. Cardina, and E. E. Regnier, "Mechanisms of weed suppression in cover crop-based production systems," *HortScience*, vol. 31, no. 3, pp. 410–413, 1996.

[37] J. R. Teasdale, C. P. Rice, G. Cai, and R. W. Mangum, "Expression of allelopathy in the soil environment: soil concentration and activity of benzoxazinoid compounds released by rye cover crop residue," *Plant Ecology*, vol. 213, no. 12, pp. 1893–1905, 2012.

[38] M. Schulz, A. Marocco, V. Tabaglio, F. A. Macias, and J. M. G. Molinillo, "Benzoxazinoids in rye allelopathy—from discovery to application in sustainable weed control and organic farming," *Journal of Chemical Ecology*, vol. 39, no. 2, pp. 154–174, 2013.

[39] E. C. Large, "Growth stages in cereals, illustrations of the Feekes scale," *Plant Pathology*, vol. 3, no. 4, pp. 128–129, 1954.

[40] M. M. Finney, D. A. Danehower, and J. D. Burton, "Gas chromatographic method for the analysis of allelopathic natural products in rye (*Secale cereale* L.)," *Journal of Chromatography A*, vol. 1066, no. 1-2, pp. 249–253, 2005.

[41] S. C. K. Carlsen, P. Kudsk, B. Laursen, S. K. Mathiassen, A. G. Mortensen, and I. S. Fomsgaard, "Allelochemicals in rye (*Secale cereale* L.): cultivar and tissue differences in the production of benzoxazinoids and phenolic acids," *Natural Product Communications*, vol. 4, no. 2, pp. 199–208, 2009.

[42] P. O. Lim, H. J. Kim, and H. G. Nam, "Leaf senescence," *Annual Review of Plant Biology*, vol. 58, pp. 115–136, 2007.

[43] B. B. Mogensen, T. Krongaard, S. K. Mathiassen, and P. Kudsk, "Quantification of benzoxazinone derivatives in wheat (*Triticum aestivum*) varieties grown under contrasting conditions in Denmark," *Journal of Agricultural and Food Chemistry*, vol. 54, no. 4, pp. 1023–1030, 2006.

[44] M. Kantar, C. Sheaffer, P. Porter, E. Krueger, and T. E. Ochsner, "Growth stage influences forage yield and quality of winter rye," *Forage & Grazinglands*, vol. 9, no. 1, 2011.

[45] H. J. Ougham, P. Morris, and H. Thomas, "The colors of autumn leaves as symptoms of cellular recycling and defenses against environmental stresses," *Current Topics in Developmental Biology*, vol. 66, pp. 135–160, 2005.

[46] K. Kang, Y.-S. Kim, S. Park, and K. Back, "Senescence-induced serotonin biosynthesis and its role in delaying senescence in rice leaves," *Plant Physiology*, vol. 150, no. 3, pp. 1380–1393, 2009.

[47] P. L. Gregersen and P. B. Holm, "Transcriptome analysis of senescence in the flag leaf of wheat (*Triticum aestivum* L.)," *Plant Biotechnology Journal*, vol. 5, no. 1, pp. 192–206, 2007.

[48] J. L. De Bruin, P. M. Porter, and N. R. Jordan, "Use of a rye cover crop following corn in rotation with soybean in the upper Midwest," *Agronomy Journal*, vol. 97, no. 2, pp. 587–598, 2005.

[49] R. A. Mischler, W. S. Curran, S. W. Duiker, and J. A. Hyde, "Use of a rolled-rye cover crop for weed suppression in no-till soybeans," *Weed Technology*, vol. 24, no. 3, pp. 253–261, 2010.

[50] S. W. Duiker and W. S. Curran, "Rye cover crop for corn production in the northern Mid-Atlantic region," *Agronomy Journal*, vol. 97, no. 5, pp. 1413–1418, 2005.

[51] S. Hörtensteiner, "Stay-green regulates chlorophyll and chlorophyll-binding protein degradation during senescence," *Trends in Plant Science*, vol. 14, no. 3, pp. 155–162, 2009.

8

Prediction of Canola Residue Characteristics Using Near-Infrared Spectroscopy

Tami L. Stubbs[1] and Ann C. Kennedy[2]

[1]*Department of Crop and Soil Sciences, Washington State University, 115 Johnson Hall, Pullman, WA 99164-6420, USA*
[2]*Northwest Sustainable Agroecosystems Research Unit, USDA-ARS, 215 Johnson Hall, Pullman, WA 99164-6421, USA*

Correspondence should be addressed to Tami L. Stubbs; tlstubbs@wsu.edu

Academic Editor: Manuel Tejada

Little work has been done to characterize and quantify the residue traits affecting decomposition of winter and spring canola (*Brassica napus* L.) residue in dryland farming systems of the Pacific Northwest United States. Traditional methods of characterizing residue fiber and nutrients are time-consuming and expensive and require large quantities of chemical reagents. The goal of this research was to determine whether near-infrared spectroscopy (NIRS) could accurately predict neutral detergent fiber (NDF), acid detergent fiber (ADF), acid detergent lignin (ADL), carbon (C), and nitrogen (N) of canola stems, litter, and roots and decomposition of canola stems. Canola residue varied in decomposition, fiber, and nutrients by year, location, and type. NIRS predictions were successful for NDF and ADF in 2011 (standard error of prediction (SEP) < 2.67; R^2 > 0.95) and NDF, ADF, and N in 2012 (SEP < 2.38; R^2 > 0.91). Other predictions for residue fiber and nutrient characteristics were considered moderately successful. Prediction of canola residue decomposition with NIRS was useful for screening purposes. Near-infrared spectroscopy shows promise for rapidly and reproducibly predicting some canola residue fiber and nutrient traits and may be useful for estimating residue decomposition potential in dryland conservation cropping systems.

1. Introduction

Production of winter and spring canola (*Brassica napus* L.) is increasing in the Pacific Northwest (PNW) United States as marketing opportunities and profitability for growers surge upward. Worldwide, canola production has increased dramatically, and demand for canola oil and meal continues to rise [1]. In the United States in 2008-2009, domestic consumption of canola oil was more than 2.5 times domestic production [1]. Cultivars of winter and spring canola perform well in conventional and conservation farming systems, and higher yielding cultivars and herbicide resistant varieties have made these crops more economically feasible. A review by Johnston et al. [2] notes that canola is well-suited in conservation tillage systems, including no-till, and can result in increased wheat yields in rotation compared to continuous wheat crops.

In the dryland farming areas of the inland PNW, average annual rainfall varies from 150 mm to 500 mm. Soils in the eastern portion of the region with higher rainfall and steeper slopes are cropped annually and are susceptible to erosion by water. The traditional cropping system in the western segment is a winter wheat-summer fallow system with conventional, high disturbance tillage, which leaves soils prone to wind erosion. Growers throughout the region are working to implement conservation farming practices to curb erosion and enhance soil quality; however, residue management is of concern to growers who wish to employ management practices such as no-till farming. There is limited information available on the decomposition of oilseed crop residue in both conventional and no-till systems [3, 4], and little is known about the residue characteristics of canola cultivars that are currently grown or in development for production in dryland and irrigated cropping systems of the PNW.

A combination of factors must be considered in determining decomposition of crop residues. Among those characteristics are fiber [5] and nutrient content of the residue [6, 7]. Neutral detergent fiber (NDF) of residue includes hemicellulose, cellulose, and lignin, which are the insoluble cell wall components. Acid detergent fiber (ADF) consists of the cellulose and lignin portions, with hemicellulose removed; and acid detergent lignin (ADL) is the portion remaining after cellulose is removed. Along with fiber components of residue, nutrient content, especially the C/N of residue, is critical in determining potential of residue to decompose under field conditions. Traditional wet chemistry and combustion methods of fiber, carbon, and nitrogen analysis are time-consuming, expensive, and destructive to the sample. Near-infrared spectroscopy (NIRS) is a potential method for estimating decomposability, fiber, and nutrient content of crop residues that is reproducible, inexpensive, rapid, and nondestructive to the sample [8]. Previous research has shown NIRS to be an acceptable method to predict fiber components [9–11] and nitrogen [12, 13] of crop residue.

The objectives of this research were to develop calibration equations for winter and spring canola residue NDF, ADF, ADL, N, C, C/N, and decomposition and to determine the feasibility of using NIRS to predict those characteristics in canola residue from various cultivars produced in different years in distinct growing locations. Our goal is to develop NIRS as a rapid screening method for predicting residue composition and decomposition potential in dryland oilseed crops. As growers seek diversification in their wheat-based cropping systems and marketing opportunities for oilseed crops advance, information on canola residue decomposition will be useful to growers who wish to incorporate winter and spring canola into their rotations. Analysis of canola residue nutrient content can be used in determining nitrogen fertilizer efficiency. Additionally, canola residue may be managed for the greatest economic success and soil quality benefits in both conventional and conservation farming systems.

2. Materials and Methods

2.1. Residue Collection. Canola residue was collected after harvest from University of Idaho Brassica Breeding and Research Program canola variety trials in 2011 and 2012. Winter canola residue was collected from trials at Odessa, WA, Moscow, ID, and Genesee, ID, and spring canola residue was sampled from trials located at Davenport, WA, Colfax, WA, and Moscow, ID [14]. Shoot residue from seven cultivars and four replications was sampled at each site, and root residue was sampled to a depth of 30 cm from select locations. Canola litter (leaves, pods, and small stems) was collected from each site. Of the seven spring canola cultivars sampled in 2011, only three were included in the 2012 trials, and so four new spring canola cultivars were substituted in the 2012 sampling. The sampling locations were selected to represent varying amounts of average annual precipitation in eastern Washington and northern Idaho. Winter and spring canola was produced under rainfed conditions at all sites except the irrigated winter canola site at Odessa, WA. Each cultivar was grown under the same management (tillage, planting, seeding rate, fertilization, and weed control) as the grower/cooperator. The trial design at each location was a randomized complete block with four replications. The

experimental design was a completely randomized split split-plot, where year was the main plot and location and cultivar were the subplots. Stems and roots were separated, and roots were washed with water and dried. For fiber, nutrient, and NIRS analyses, residue samples were ground to pass a one mm sieve.

2.2. Laboratory Reference Data. Ground canola residue was enclosed in filter bags (ANKOM Technology Corp., Fairport, NY) for determination of neutral detergent fiber (NDF), acid detergent fiber (ADF), and acid detergent lignin (ADL) using a modification of the Van Soest et al. [15] procedure. Neutral detergent fiber and acid detergent fiber were determined sequentially following processing with an ANKOM 200 Fiber Analyzer (ANKOM Technology Corp., Fairport, NY). Acid detergent lignin was determined following digestion in 72% H_2SO_4. Ground residue samples were analyzed by dry combustion using a LECO TruSpec Analyzer (St. Joseph, MI) to determine total carbon (C) and nitrogen (N) and calculate C/N ratios. For consistency, we chose to analyze litter (leaves, pods, and small stems), stem, and root residues separately, and only stem residue was used in the decomposition study.

Decomposition of canola stem residue was determined in a laboratory incubation study using a procedure similar to Summerell and Burgess [16]. Canola stem residue was cut into 5 cm lengths and oven-dried. A layer of Ritzville silt loam soil (oven-dried and brought to field capacity) was placed in a one-pint glass canning jar, followed by approximately 2.0 g of oven-dry canola stem residue, and covered with a second layer of soil to simulate buried residue. Another 2.0 g of stem residue was placed on the soil surface to simulate surface residue. Residue was dried and weighed prior to placing in soil. Jars were sealed with parafilm to allow air exchange and incubated at 22°C. Residue from 2012 was used for the study and included the seven cultivars each of winter and spring canola from each site. Each cultivar was replicated three times. After an 11-week incubation period, canola residue was removed from soil, washed, dried, and reweighed to determine mass of residue lost during incubation.

2.3. Near-Infrared Spectroscopy. Near-infrared spectroscopy (NIRS) is a rapid, low cost, nondestructive secondary method that was used to predict fiber and nutrient content of canola root and shoot residue. Calibration and prediction equations were developed for canola residue enclosed in stationary metal ring cups (36 mm inside diameter) and scanned with a FOSS XDS Rapid Content Analyzer (Foss NIRSystems, Laurel, MD) using ISIscan software, version 3.10 (Infrasoft International, State College, PA). Samples were scanned twice using the wavelength range 400–2498 nm at 2 nm intervals. The mean of the two scans was used for data analysis. Canola stem, root, and litter data for the two years of the study were analyzed separately and each set was randomly divided into two parts for developing calibration equations (Table 1; 2011, $n = 314$; 2012, $n = 215$) and for validation of equations (Table 2; 2011, $n = 314$; 2012, $n = 215$). Separate calibration equations and validation of equations were performed for the canola residue decomposition study (Table 3). For both the calibration and validation sets, $n = 118$ buried residue

TABLE 1: Calibration statistics for prediction of NDF, ADF, ADL, N, C, and C/N in 2011 and 2012 winter and spring canola residue using NIRS. Residue consisted of stems, litter (small stems, leaves, and pods), and roots.

(a) 2011 winter and spring canola residue

Component	Math treatment[a]	n	Mean[b]	SD	SEC	R^2	SECV	1 − VR	SD/SECV
NDF	2, 10, 10, 1	297	71.57	10.21	2.10	0.96	2.24	0.95	4.56
ADF	2, 10, 10, 1	299	55.90	8.95	1.42	0.98	1.61	0.97	5.55
ADL	2, 10, 10, 1	304	13.12	2.39	0.75	0.90	0.83	0.88	2.72
N	2, 10, 10, 1	305	0.80	0.49	0.08	0.98	0.09	0.97	5.41
C	2, 4, 4, 1	299	43.80	2.99	0.61	0.96	0.74	0.94	4.05
C/N	2, 4, 4, 1	301	73.12	36.82	8.08	0.95	9.93	0.93	3.71

(b) 2012 winter and spring canola residue

Component	Math treatment[a]	n	Mean[b]	SD	SEC	R^2	SECV	1 − VR	SD/SECV
NDF	2, 4, 4, 1	205	73.20	7.08	1.29	0.97	1.59	0.95	4.44
ADF	2, 4, 4, 1	206	59.02	5.96	1.19	0.96	1.40	0.94	4.25
ADL	2, 10, 10, 1	205	13.34	2.01	0.75	0.86	0.83	0.83	2.41
N	3, 5, 5, 1	209	0.815	0.51	0.06	0.99	0.09	0.97	5.41
C	1, 4, 4, 1	207	42.34	2.09	0.62	0.91	0.69	0.89	3.03
C/N	2, 10, 10, 1	207	73.12	45.39	10.88	0.94	12.99	0.92	3.49

NDF: neutral detergent fiber; ADF: acid detergent fiber; ADL: acid detergent lignin; N: nitrogen; C: carbon; C/N: carbon to nitrogen ratio; NIRS: near-infrared spectroscopy; SD: standard deviation; SEC: standard error of calibration; R^2: coefficient of determination; SECV: standard error of cross validation; 1 − VR: 1 minus variance ratio.
[a]The scatter correction SNV and detrend was used. Math treatment: derivative number, gap (nm), smooth (number of smoothing points), and second smooth.
[b]Mean percent NDF, ADF, ADL, C, and N.

TABLE 2: Reference measurements and validation results for the prediction of 2011 and 2012 winter and spring canola residue NDF, ADF, ADL, N, C, and C/N using NIRS.

(a) 2011 winter and spring canola residue

| Component | n | Reference measurements | | | | Validation results | | | | |
		Range	Measured mean[a]	Measured SD	NIRS predicted mean	Bias	R^2	SEP	Slope	RPD
NDF	308	38.10–87.91	69.6	11.56	69.86	−0.262	0.95	2.67	0.99	4.33
ADF	308	29.43–68.59	54.66	9.61	54.79	−0.131	0.96	1.85	1.00	5.19
ADL	308	6.86–18.45	12.89	2.46	12.74	0.148	0.89	0.87	0.92	2.83
N	308	0.247–3.062	0.828	0.53	0.836	−0.008	0.89	0.18	0.97	2.94
C	308	30.85–49.80	43.68	3.22	43.64	0.038	0.83	1.34	0.97	2.40
C/N	308	11.95–193.94	73.56	38.12	72.60	0.958	0.87	14.08	0.94	2.71

(b) 2012 winter and spring canola residue

| Component | n | Reference measurements | | | | Validation results | | | | |
		Range	Measured mean[a]	Measured SD	NIRS predicted mean	Bias	R^2	SEP	Slope	RPD
NDF	214	48.69–84.36	72.31	7.68	72.24	0.07	0.91	2.38	1.05	3.23
ADF	214	39.89–69.25	58.31	6.30	58.16	0.149	0.91	1.91	1.05	3.30
ADL	214	7.59–44.43	13.15	1.93	13.08	0.067	0.75	0.99	0.87	1.95
N	214	0.181–3.446	0.877	0.55	0.864	0.013	0.92	0.158	1.05	3.48
C	214	33.41–46.01	42.04	2.38	42.14	−0.095	0.83	0.99	1.08	2.40
C/N	214	9.69–245.68	72.71	48.26	71.28	1.438	0.89	16.5	1.07	2.92

NDF: neutral detergent fiber; ADF: acid detergent fiber; ADL: acid detergent lignin; N: nitrogen; C: carbon; C/N: carbon to nitrogen ratio; NIRS: near-infrared spectroscopy; SD: standard deviation; R^2: coefficient of determination; SEP: standard error of prediction; RPD: ratio of prediction to deviation (SD/SEP).
[a]Mean percent NDF, ADF, ADL, C, and N.

TABLE 3: Calibration statistics (a) and reference measurements and validation results (b) for prediction of 2012 canola stem residue decomposition in a laboratory incubation study using NIRS.

(a) Calibration statistics

Component	Math treatment[a]	n	Mean[b]	SD	SEC	R^2	SECV	$1 - VR$	SD/SECV
Buried residue	3, 5, 5, 1	116	49.03	10.79	3.15	0.92	5.21	0.77	2.07
Surface-placed residue	3, 5, 5, 1	117	43.98	10.68	5.06	0.78	5.65	0.72	1.89

(b) Reference measurements and validation results

		Reference measurements			Validation results				
	n	Range	Measured mean	Measured SD	NIRS predicted mean	Bias	R^2	SEP	Slope
Buried residue	118	27.35–73.86	49.03	11.58	49.01	0.028	0.63	7.12	0.88
Surface-placed residue	118	23.93–63.82	43.49	11.41	44.59	−1.093	0.67	6.67	0.97

NIRS: near-infrared spectroscopy; SD: standard deviation; SEC: standard error of calibration; R^2: coefficient of determination; SECV: standard error of cross validation; $1 - VR$: 1 minus variance ratio; SEP: standard error of prediction.
[a]The scatter correction SNV and detrend was used. Math treatment: derivative number, gap (nm), smooth (number of smoothing points), and second smooth.
[b]Mean percent of stem mass lost.

samples and $n = 118$ surface-placed residue samples. Reference analysis and NIRS analysis were performed on each of the samples in both sets. Random selection of sample sets and calibration and validation statistics were completed using WinISI software, version 4.0 (Infrasoft International, State College, PA). Calibration equations were derived using modified partial least-squares (MPLS) and cross-validation techniques. The scatter correction of standard normal variant and detrend (SNV-D) was applied, along with several different math treatments for derivative order number, gap, and first smoothing. The second smoothing was set at 1 to indicate no second smoothing. Principal component analysis was used to identify and remove spectral outliers. Samples having spectra with Mahalanobis distance (H) values greater than 3.0 were considered outliers and were removed from the file. The appropriate calibration equation for each component was determined by selecting the one with the lowest standard error of cross-validation (SECV) and the 1 − variance ratio (1 − VR) closest to 1 [8]. The ratio of standard deviation (SD)/SECV was calculated [17] and used to determine calibration equations that were acceptable for quantitative prediction of fiber characteristics (Tables 1 and 3). Correlations between NIRS-predicted values and reference values (fiber wet chemistry, LECO TruSpec, or residue decomposition) were determined using Pearson correlation coefficients [18] with JMP software [19].

3. Results and Discussion

3.1. Calibration Results. In the present study, the residue of commonly grown cultivars of winter and spring canola produced at multiple locations were analyzed for neutral detergent fiber (NDF), acid detergent fiber (ADF), acid detergent lignin (ADL), carbon (C), and nitrogen (N) contents and decomposition in lab incubations. Traditional methods to quantify fiber and nutrient content of plant residues are laborious; destroy the plant sample; and are unfriendly to the environment. Near-infrared spectroscopy (NIRS) represents a secondary method to estimate fiber and nutrient content which is of lower cost and nondestructive to the crop

sample. Near-infrared spectroscopy is reliable and accurate and requires very little sample preparation [8], while allowing for rapid analysis of multiple properties at one time [20]. Sample characteristics are quantitatively predicted by NIRS using the near-infrared absorbance of a sample to measure organic functional groups. In the present study, stem, root, and litter residue from winter and spring canola cultivars grown at multiple locations over two years were scanned with a FOSS XDS Rapid Content Analyzer, and those results were compared to laboratory reference data from fiber, nutrient, and decomposition experiments to develop calibration equations for prediction of those characteristics. Predictions of canola shoot, root, and litter residue characteristics, as well as canola stem residue decomposition, using NIRS were conducted as part of the present study.

Calibration statistics for prediction equations of winter and spring canola residue traits are listed in Table 1. Separate calibrations were done for each year of the study, with Table 1(a) showing results for residue from 2011 and Table 1(b) for residue from 2012. The math treatments which resulted in most optimal calibration statistics were used for the traits of NDF, ADF, ADL, N, C, and C/N. Using equations developed with specific math treatments improves the predictability for individual data sets [21]. Equations that yielded the 1 − VR values closest to 1 and the lowest SECV values were chosen for predictions. For 2011 winter and spring canola residue, NDF, ADF, and ADL, 1 − VR values were 0.95, 0.97, and 0.88, and SECV values were 2.24, 1.61, and 0.83. Nitrogen, C, and C/N 1 − VR and SECV values for 2011 were 0.97, 0.94, and 0.93 and 0.09, 0.74, and 9.93, respectively. The NDF, ADF, and ADL 1 − VR and SECV values for 2012 canola residue were 0.95, 0.94, and 0.83 and 1.59, 1.40, and 0.83. The 2012 1 − VR and SECV for N, C, and C/N were 0.97, 0.89, and 0.92 and 0.09, 0.69, and 12.99. Deaville et al. [17] developed guidelines for the predictive ability of calibration equations using SD/SECV values. Under their protocol, calibrations with SD/SECV ratios >3.0 are acceptable for prediction of characteristics using NIRS. Ratios >2.5 but <3.0 indicate calibrations that would be useful for screening, and equations with SD/SECV <2.5 are not useful for making predictions.

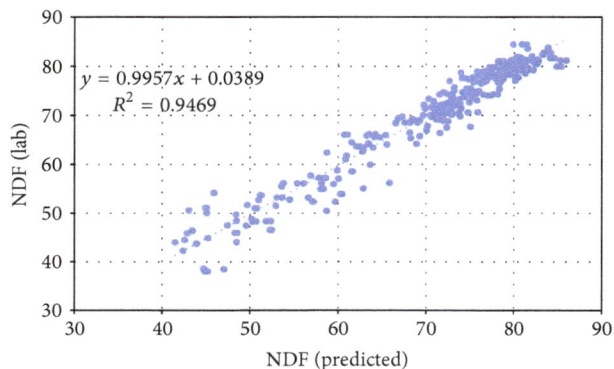

FIGURE 1: Linear regression relationship between laboratory reference methods and NIRS-predicted values for NDF (R^2 = 0.95) of winter and spring canola from 2011 on a dry weight basis.

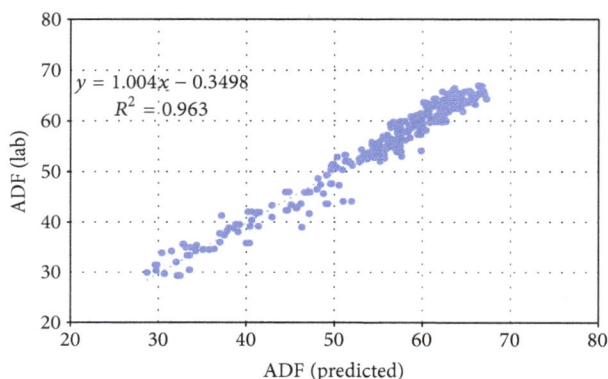

FIGURE 2: Linear regression relationship between laboratory reference methods and NIRS-predicted values for ADF (R^2 = 0.96) of winter and spring canola from 2011 on a dry weight basis.

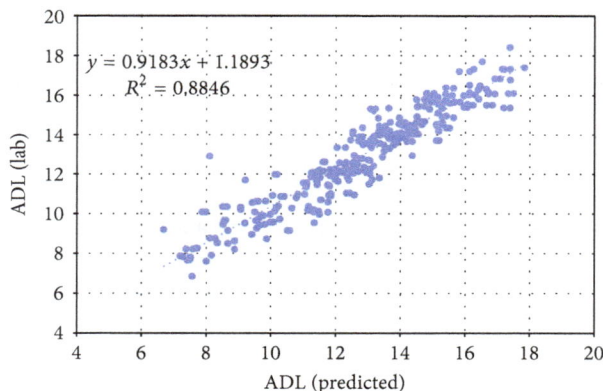

FIGURE 3: Linear regression relationship between laboratory reference methods and NIRS-predicted values for ADL (R^2 = 0.89) of winter and spring canola from 2011 on a dry weight basis.

In this study, all SD/SECV values for 2011 were greater than 3.0 except ADL (2.72; Table 1), indicating that successful calibrations were developed for the traits of NDF, ADF, N, C, and C/N (SD/SECV ranging from 3.71 to 5.55). The same was true for 2012, with SD/SECV values ranging from 3.03 to 5.41 for all traits but ADL (2.41; Table 1). In 2011, the SD/SECV of 2.72 for ADL fell within the range that would be acceptable for screening of traits; however, in 2012 the SD/SECV for ADL was only 2.41, which is outside the range considered acceptable for a screening prediction.

3.2. Validation Results. Reference measurements and validation results are shown in Table 2(a) for 2011 winter and spring canola residue and Table 2(b) for 2012 canola residue samples. In both years, acceptable predictions were attained for each of the residue traits. With 2011 residue, R^2 values ranged from 0.83 to 0.96, and R^2 values for 2012 residue ranged from 0.75 to 0.92. For all traits in both years, slope values were near 1, with the lowest slope = 0.87 for ADL in 2012. Comparisons of laboratory reference values for each trait as compared to NIRS-predicted values are shown in Figures 1–6. Because similar results were obtained for the two years of the study, only results from 2011 are shown. In determining whether NIRS is an acceptable method for prediction of residue traits, Mathison et al. [22] used an additional calculation, ratio of prediction to deviation (RPD), which is the measured standard deviation (SD) divided by the standard error of prediction (SEP) of the validation results. When RPD values are >2.5, the predictions are considered to be acceptable, while RPD values of 10 are excellent [22]. In this study, all RPD values indicated acceptable predictions except for ADL in 2012 (RPD = 1.95) and C in both years of the study (2011 RPD = 2.40; 2012 RPD = 2.40). The highest RPD values were NDF (4.33) and ADF (5.20) in 2011. Malley et al. [23] include a separate calculation, RER, in determining suitable predictions using NIRS. The RER is equal to the range of measured reference values divided by the SEP with RER values of >20 considered to be excellent, RER 15–20 successful, RER 10–15 moderately successful, and RER 8–10 moderately useful [23]. In this study, we calculated RER

values ranging from 12.73 to 37.21 (data not shown); however, these values gave conflicting results when compared to the guidelines using R^2 and RPD values in determining success of predictions. Roggo et al. [24] suggested that RPD values of >3 and RER values >10 may have use for screening characteristics in their study of sugar beet (*Beta vulgaris*) quality. In our study, only NDF and ADF consistently met that level for RPD, while the RER values for each characteristic would indicate acceptable predictions for screening fiber and nutrient traits of canola residue.

3.3. Prediction of Canola Residue Decomposition. Canola stem residue decomposition was measured in a laboratory incubation study [14], and the percent decomposition was compared to NIRS-predicted values. The calibration statistics and validation results are shown in Table 3. The SD/SECV values were 2.07 for buried residue, as in a conventionally tilled farming system, and 1.89 for residue that was surface-placed, as in no-till farming. Using the guidelines proposed by Deaville et al. [17], these calibration equations would be unacceptable for predicting canola residue decomposition using NIRS. Validation results yielded R^2 values of 0.63 for

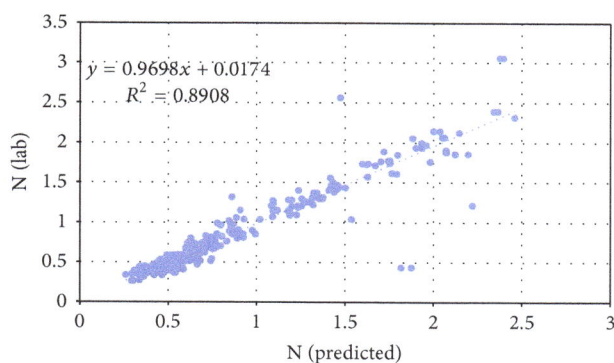

FIGURE 4: Linear regression relationship between laboratory reference methods and NIRS-predicted values for N ($R^2 = 0.89$) of winter and spring canola from 2011 on a dry weight basis.

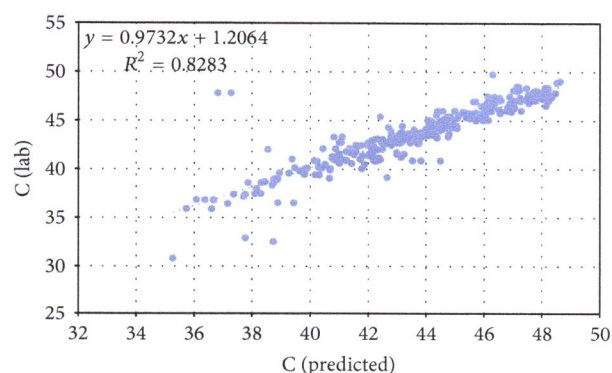

FIGURE 6: Linear regression relationship between laboratory reference methods and NIRS-predicted values for C/N ($R^2 = 0.87$) of winter and spring canola from 2011 on a dry weight basis.

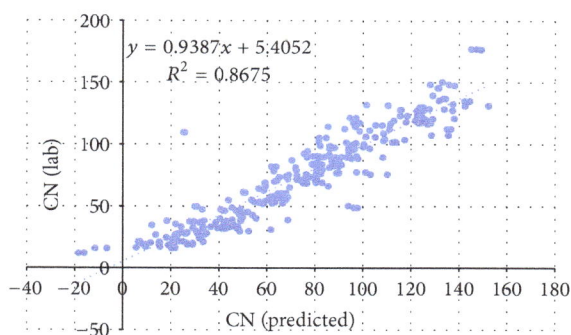

FIGURE 5: Linear regression relationship between laboratory reference methods and NIRS-predicted values for C ($R^2 = 0.83$) of winter and spring canola from 2011 on a dry weight basis.

buried residue and 0.67 for surface-placed residue. The RPD values were 1.63 for buried canola stems and 1.71 for surface-placed residue, and both values are considered unsuitable for screening; however, Pearson correlation coefficients were all >0.77 when comparing laboratory canola residue decomposition with NIRS-predicted values ($P > 0.05$; data not shown). Difficulties in cleaning soil from decomposing residue may have led to variability in the reference measurements [16]. Shepherd et al. [25] found that NIRS has the potential for accurately determining decomposition of organic residues. Additional studies with a larger number of canola residue samples may be needed to develop stronger calibrations in order to use NIRS for predicting canola residue decomposition in soil.

3.4. Correlation between Laboratory Reference Measurements and NIRS-Predicted Measurements. NIRS-predicted values were significantly correlated ($P > 0.05$) with measured reference values for each of the residue fiber and nutrient characteristics tested when calculated across the two years, six study locations, two crop types, and all plant components (Table 4). The six traits of NDF, ADF, ADL, N, C, and C/N were all well-correlated with one another, with correlation coefficients

≥0.92 (level of significance <0.0001). Nitrogen was negatively correlated with each of the other characteristics. Correlations between the fiber fractions (NDF, ADF, and ADL) were higher than those with the nutrient contents (N, C, and C/N). We previously found that NIRS was a successful method for predicting NDF, ADF, and ADL in cereal crop residue, but not for C and N [26]. Velasco and Mollers [12] successfully used NIRS to accurately predict N concentration in *Brassica napus* L. tissues. Additional work has shown the usefulness of NIRS in predicting features of canola seed [27] and/or meal [28]. Long et al. [29] found that NIRS could measure the oil content of canola seed prior to crushing. Wittkop et al. [30] used NIRS to predict NDF, ADF, and ADL in intact seed of various genotypes of oilseed rape (*Brassica napus* L.). They found good R^2 values for ADL and adequate predictions for NDF and ADF when looking for low fiber genotypes. Others have studied properties of canola residue for decomposition [4, 31], biofuel production [32], and pulp or paper production [33]; however, we are unaware of attempts to develop NIRS calibrations for prediction of canola residue properties for those purposes.

Like other crop plants [34, 35], various components of the canola postharvest residue have differing nutrient and fiber contents [14] and have varying rates of residue decomposition [31]. We found that NIRS-predicted values for fiber and nutrient content of canola stems, litter, and roots were well-correlated when compared to laboratory reference methods (Table 5). The highest correlation coefficient for stem residue was for ADF (0.97) and the lowest was for ADL (0.90). Canola litter, which included leaves, pods, and very small stems, had the highest correlation for ADF as well (0.92) and the lowest for C (0.62). The highest Pearson correlation coefficient for canola roots was with N (0.98), and the lowest was with ADL (0.80).

Crop residues, including canola, vary in fiber and nutrient content among years [3, 36], crop types [3, 37], and growing locations [36, 38]. There was significant correlation between laboratory reference measurements and NIRS predictions for all traits in both years, both crop types, and the six field experiment locations (Table 6; $P > 0.05$; level of significance < 0.0001 for all comparisons). The highest correlations were found for samples when separated by year, with the highest

TABLE 4: Pearson correlation coefficients for canola residue traits measured using laboratory reference methods and compared to NIRS predicted values across all years, locations, plant components, and canola types. For each pair, the level of significance was <0.0001 at $P < 0.05$ ($n = 522$).

	NDF	ADF	ADL	N	C	C/N
NDF	0.9683	0.9511	0.8773	−0.5546	0.6322	0.6198
ADF		0.9760	0.8396	−0.5636	0.6204	0.6516
ADL			0.9187	−0.4931	0.6217	0.5182
N				0.9496	−0.7810	−0.8484
C					0.9162	0.7867
C/N						0.9349

NIRS: near-infrared spectroscopy; NDF: neutral detergent fiber; ADF: acid detergent fiber; ADL: acid detergent lignin; N: nitrogen; C: carbon; C/N: carbon to nitrogen ratio.

TABLE 5: Pearson correlation coefficients for residue traits of canola plant components measured using laboratory reference methods and compared to NIRS predicted values across all years, locations, and canola types. For each pair, the level of significance was <0.0001 at $P < 0.05$.

	n	NDF	ADF	ADL	N	C	C/N
Stems	346	0.9661	0.9722	0.8974	0.9702	0.9624	0.9442
Litter	65	0.8062	0.9226	0.7539	0.7791	0.6239	0.722
Roots	110	0.9013	0.8883	0.8025	0.975	0.8242	0.9116

NIRS: near-infrared spectroscopy; NDF: neutral detergent fiber; ADF: acid detergent fiber; ADL: acid detergent lignin; N: nitrogen; C: carbon; C/N: carbon to nitrogen ratio.

TABLE 6: Pearson correlation coefficients for spring and winter canola residue traits measured using laboratory reference methods and compared to NIRS predicted values over two years and three locations for each crop type. For each pair, the level of significance was <0.0001 at $P < 0.05$.

	n	NDF	ADF	ADL	N	C	C/N
Year							
2011	158	0.9255	0.936	0.9148	0.9798	0.9607	0.9366
2012	153	0.9302	0.9298	0.8278	0.9647	0.9473	0.9506
Crop type							
Spring	161	0.753	0.7825	0.8507	0.8964	0.9319	0.8201
Winter	150	0.8261	0.8539	0.7886	0.9514	0.911	0.8959
Location							
Colfax, spring	55	0.8051	0.7311	0.8354	0.9052	0.9566	0.6923
Davenport, spring	51	0.804	0.8421	0.817	0.9088	0.902	0.8976
Moscow, spring	55	0.6666	0.7285	0.8925	0.7321	0.929	0.7346
Odessa, winter	58	0.8139	0.8502	0.8002	0.9225	0.6885	0.8059
Genesee, winter	42	0.6347	0.7001	0.5523	0.9744	0.8863	0.879
Moscow, winter	50	0.8614	0.8583	0.8398	0.9353	0.9429	0.8685

NIRS: near-infrared spectroscopy; NDF: neutral detergent fiber; ADF: acid detergent fiber; ADL: acid detergent lignin; N: nitrogen; C: carbon; C/N: carbon to nitrogen ratio.

correlation for N in 2011 (0.98) and the lowest for ADL in 2011 (0.83). When samples were separated by crop type, the highest correlation was found for N in winter canola (0.95) and the lowest for NDF in spring canola (0.75). For the spring canola locations, the correlation coefficient was highest for C from the Colfax, WA, location (0.96) and lowest for NDF from Moscow, ID, spring canola (0.67). In winter canola, the highest correlation coefficient was for N from Genesee, ID, residue (0.97), and the lowest coefficient was also from the Genesee location for ADL (0.55). Bruun et al. [39] were able to distinguish differences in ash content and degradability among cultivars of wheat straw grown at two sites in Denmark using NIRS predictions.

In previous work, we found that there were no clear differences in fiber or nutrient characteristics among residue samples from various cultivars of winter and spring canola grown in eastern Washington and northern Idaho [14]. In the present study, n values were too low to develop calibration equations with adequate sets of validation data for individual winter and spring canola cultivars (data not shown). Redaelli and Berardo [11] found that NIRS could accurately predict fiber components of oat (*Avena sativa* L.) hulls grown in Italy and that fiber components varied with cultivar, but location played a lesser role in that variation. Oilseed rape genotypes showed variability in ADL with NIRS analysis according to Wittkop et al. [30]. Future work in this area might focus on obtaining a greater number of canola residue samples over more years or experimental locations to determine whether NIRS predictions are feasible for defining differences in residue traits among cultivars, establishing usefulness of NIRS for prediction of fiber and nutrient content of individual plant components, and verifying fertilizer uptake and nutrient efficiency in *Brassica* crops.

4. Conclusions

NIRS was successful in predicting the characteristics of NDF, ADF, N, and C/N of winter and spring canola residue from two years of field studies. NIRS-predicted values were less consistent for ADL, where the prediction was successful for only one of the two years of the study. In this work, we found that NIRS was useful only for screening canola residue C and decomposition in the laboratory incubation studies, not for prediction of these two traits. NIRS predictions show promise for rapidly discerning differences in certain fiber and nutrient characteristics among canola residue samples while using smaller quantities of chemicals and performing less costly analyses. Ultimately, the rapid screening of crop residues for nutrient and fiber characteristics will provide information to crop producers and scientists for designing conservation cropping systems that make efficient use of fertilizers, increase soil organic matter, prevent soil erosion, and lead to soils with greater productivity.

Conflicts of Interest

The authors declare that they have no conflicts of interest.

Acknowledgments

The authors gratefully acknowledge the funding received from the Pacific Northwest Regional Canola Research Program and the assistance from Jim Davis, Jack Brown, and Megan Wingerson of the University of Idaho Brassica Breeding and Research Program. They are grateful to Jeremy Hansen, Brianna Kemp, and Mackenzie Owen for excellent technical assistance.

References

[1] United States Department of Agriculture—Economic Research Service, "Soybeans & Oil Crops-Canola," 2012, http://www.ers.usda.gov/topics/crops/soybeans-oil-crops/canola.aspx%23trade#.UrCXUeKmbdU.

[2] A. M. Johnston, D. L. Tanaka, P. R. Miller et al., "Oilseed crops for semiarid cropping systems in the northern Great Plains," *Agronomy Journal*, vol. 94, no. 2, pp. 231–240, 2002.

[3] Y. K. Soon and M. A. Arshad, "Comparison of the decomposition and N and P mineralization of canola, pea and wheat residues," *Biology and Fertility of Soils*, vol. 36, no. 1, pp. 10–17, 2002.

[4] A. J. Franzluebbers, M. A. Arshad, and J. A. Ripmeester, "Alterations in canola residue composition during decomposition," *Soil Biology and Biochemistry*, vol. 28, no. 10-11, pp. 1289–1295, 1996.

[5] D. C. Wolf and G. H. Wagner, "Carbon transformations and soil organic matter formation," in *Principles and Applications of Soil Microbiology*, D. M. Sylvia, J. J. Fuhrman, R. G. Hartel, and D. A. Zuberer, Eds., pp. 285–332, Prentice-Hall, Upper Saddle River, NJ, USA, 2005.

[6] J. H. Smith and C. L. Douglas, "Wheat straw decomposition in the field," *Soil Science Society of America Journal*, vol. 35, no. 2, pp. 269–272, 1971.

[7] C. A. Palm and A. P. Rowland, "A minimum dataset for characterization of plant quality for decomposition," in *Driven by Nature: Plant Litter Quality and Decomposition*, G. Cadisch and K. E. Giller, Eds., pp. 379–392, CAB International, Wallingford, UK, 1997.

[8] Foss North America, *ISIscan & WinISI Software Training Class*, 2008.

[9] M. K. D. Rambo, E. P. Amorim, and M. M. C. Ferreira, "Potential of visible-near infrared spectroscopy combined with chemometrics for analysis of some constituents of coffee and banana residues," *Analytica Chimica Acta*, vol. 775, pp. 41–49, 2013.

[10] Z. Nie, G. F. Tremblay, G. Bélanger et al., "Near-infrared reflectance spectroscopy prediction of neutral detergent-soluble carbohydrates in timothy and alfalfa," *Journal of Dairy Science*, vol. 92, no. 4, pp. 1702–1711, 2009.

[11] R. Redaelli and N. Berardo, "Prediction of fibre components in oat hulls by near infrared reflectance spectroscopy," *Journal of the Science of Food and Agriculture*, vol. 87, no. 4, pp. 580–585, 2007.

[12] L. Velasco and C. Mollers, "Use of near-infrared reflectance spectroscopy to assess nitrogen concentration in different plant tissues of rapeseed," *Communications in Soil Science and Plant Analysis*, vol. 31, no. 19-20, pp. 2987–2995, 2000.

[13] G.-C. Zhang, Z. Li, X.-M. Yan et al., "Rapid analysis of apple leaf nitrogen using near infrared spectroscopy and multiple linear regression," *Communications in Soil Science and Plant Analysis*, vol. 43, no. 13, pp. 1768–1772, 2012.

[14] T. L. Stubbs and A. C. Kennedy, "Characterization and decomposition of residue from winter and spring canola cultivars," *Agronomy*, In press.

[15] P. J. Van Soest, J. B. Robertson, and B. A. Lewis, "Methods for dietary fiber, neutral detergent fiber, and nonstarch polysaccharides in relation to animal nutrition," *Journal of Dairy Science*, vol. 74, no. 10, pp. 3583–3597, 1991.

[16] B. A. Summerell and L. W. Burgess, "Decomposition and chemical composition of cereal straw," *Soil Biology and Biochemistry*, vol. 21, no. 4, pp. 551–559, 1989.

[17] E. R. Deaville, D. J. Humphries, and D. I. Givens, "Whole crop cereals. 2. Prediction of apparent digestibility and energy value from in vitro digestion techniques and near infrared reflectance spectroscopy and of chemical composition by near infrared reflectance spectroscopy," *Animal Feed Science and Technology*, vol. 149, no. 1-2, pp. 114–124, 2009.

[18] R. G. D. Steel, J. H. Torrie, and D. A. Dickey, *Principles and Procedures of Statistics*, McGraw-Hill, New York, NY, USA, 1997.

[19] SAS Institute Inc, *Using JMP 11*, SAS Institute Inc, Cary, NC, USA, 2013.

[20] J. Stuth, A. Jama, and D. Tolleson, "Direct and indirect means of predicting forage quality through near infrared reflectance

spectroscopy," *Field Crops Research*, vol. 84, no. 1-2, pp. 45–56, 2003.

[21] A. Ruano-Ramos, A. García-Ciudad, and B. García-Criado, "Determination of nitrogen and ash contents in total herbage and botanical components of grassland systems with near infra-red spectroscopy," *Journal of the Science of Food and Agriculture*, vol. 79, no. 1, pp. 137–143, 1999.

[22] G. W. Mathison, H. Hsu, R. Soofi-Siawash et al., "Prediction of composition and ruminai degradability characteristics of barley straw by near infrared reflectance spectroscopy," *Canadian Journal of Animal Science*, vol. 79, no. 4, pp. 519–523, 1999.

[23] D. F. Malley, C. McClure, P. D. Martin, K. Buckley, and W. P. McCaughey, "Compositional analysis of cattle manure during composting using a field-portable near-infrared spectrometer," *Communications in Soil Science and Plant Analysis*, vol. 36, no. 4–6, pp. 455–475, 2005.

[24] Y. Roggo, L. Duponchel, and J.-P. Huvenne, "Quality evaluation of sugar beet (*Beta vulgaris*) by near-infrared spectroscopy," *Journal of Agricultural and Food Chemistry*, vol. 52, no. 5, pp. 1055–1061, 2004.

[25] K. D. Shepherd, B. Vanlauwe, C. N. Gachengo, and C. A. Palm, "Decomposition and mineralization of organic residues predicted using near infrared spectroscopy," *Plant and Soil*, vol. 277, no. 1-2, pp. 315–333, 2005.

[26] T. L. Stubbs, A. C. Kennedy, and A.-M. Fortuna, "Using NIRS to predict fiber and nutrient content of dryland cereal cultivars," *Journal of Agricultural and Food Chemistry*, vol. 58, no. 1, pp. 398–403, 2010.

[27] N. H. Hom, H. C. Becker, and C. Möllers, "Non-destructive analysis of rapeseed quality by NIRS of small seed samples and single seeds," *Euphytica*, vol. 153, no. 1-2, pp. 27–34, 2007.

[28] P. Si, R. J. Mailer, N. Galwey, and D. W. Turner, "Influence of genotype and environment on oil and protein concentrations of canola (*Brassica napus* L.) grown across southern Australia," *Australian Journal of Agricultural Research*, vol. 54, no. 4, pp. 397–407, 2003.

[29] D. S. Long, J. D. McCallum, F. L. Young, and A. W. Lenssen, "In-stream measurement of canola (*Brassica napus* L.) seed oil concentration using in-line near infrared reflectance spectroscopy," *Journal of Near Infrared Spectroscopy*, vol. 20, no. 3, pp. 387–395, 2012.

[30] B. Wittkop, R. J. Snowdon, and W. Friedt, "New NIRS calibrations for fiber fractions reveal broad genetic variation in *Brassica napus* seed quality," *Journal of Agricultural and Food Chemistry*, vol. 60, no. 9, pp. 2248–2256, 2012.

[31] P. V. Blenis, P. S. Chow, and G. R. Stringam, "Effects of burial, stem portion and cultivar on the decomposition of canola straw," *Canadian Journal of Plant Science*, vol. 79, no. 1, pp. 97–100, 1999.

[32] N. George, Y. Yang, Z. Wang, R. Sharma-Shivappa, and K. Tungate, "Suitability of canola residue for cellulosic ethanol production," *Energy and Fuels*, vol. 24, no. 8, pp. 4454–4458, 2010.

[33] R. Hosseinpour, P. Fatehi, A. J. Latibari, Y. Ni, and S. Javad Sepiddehdam, "Canola straw chemimechanical pulping for pulp and paper production," *Bioresource Technology*, vol. 101, no. 11, pp. 4193–4197, 2010.

[34] H. P. Collins, L. F. Elliott, R. W. Rickman, D. F. Bezdicek, and R. I. Papendick, "Decomposition and interactions among wheat residue components," *Soil Science Society of America Journal*, vol. 54, no. 3, pp. 780–785, 1990.

[35] M. Quemada and M. L. Cabrera, "Carbon and nitrogen mineralized from leaves and stems of four cover crops," *Soil Science Society of America Journal*, vol. 59, no. 2, pp. 471–477, 1995.

[36] S. C. Rao, "Regional environment and cultivar effects on the quality of wheat straw," *Agronomy Journal*, vol. 81, no. 6, pp. 939–943, 1989.

[37] N. Z. Lupwayi, G. W. Clayton, J. T. O'Donovan, K. N. Harker, T. K. Turkington, and W. A. Rice, "Decomposition of crop residues under conventional and zero tillage," *Canadian Journal of Soil Science*, vol. 84, no. 4, pp. 403–410, 2004.

[38] T. L. Stubbs, A. C. Kennedy, P. E. Reisenauer, and J. W. Burns, "Chemical composition of residue from cereal crops and cultivars in dryland ecosystems," *Agronomy Journal*, vol. 101, no. 3, pp. 538–545, 2009.

[39] S. Bruun, J. W. Jensen, J. Magid, J. Lindedam, and S. B. Engelsen, "Prediction of the degradability and ash content of wheat straw from different cultivars using near infrared spectroscopy," *Industrial Crops and Products*, vol. 31, no. 2, pp. 321–326, 2010.

9

Relative Efficacy of Liquid Nitrogen Fertilizers in Dryland Spring Wheat

Olga S. Walsh[1] **and Robin J. Christiaens**[2]

[1]*Department of Plant, Soil, and Entomological Sciences, Southwest Research and Extension Center, University of Idaho, 29603 U of I Lane, Parma, ID 83660, USA*
[2]*Private Enterprise, University of Idaho, 29603 U of I Lane, Parma, ID 83660, USA*

Correspondence should be addressed to Olga S. Walsh; owalsh@uidaho.edu

Academic Editor: David Clay

The study was conducted in 2012 and 2013 at three locations in North Central and Western Montana (total of 6 site-years) to evaluate the relative efficacy of three liquid nitrogen (N) fertilizer sources, urea ammonium nitrate (UAN, 32-0-0), liquid urea (LU, 21-0-0), and High NRGN (HNRGN, 27-0-0-1S), in spring wheat (*Triticum aestivum* L.). In addition to at-seeding urea application at 90 kg N ha^{-1} to all treatments (except for the unfertilized check plot), the liquid fertilizers were applied utilizing an all-terrain vehicle- (ATV-) mounted stream-bar equipped sprayer at a rate of 45 kg N ha^{-1} at Feekes 5 growth stage (early tillering). Three dilution ratios of fertilizer to water were accessed: 100/0 (undiluted), 66/33, and 33/66. The effects of N source and the dilution ratio (fertilizer/water) on N uptake (NUp), N use efficiency (NUE), spring wheat grain yield (GY), grain protein (GP) content, and protein yield (PY) were assessed. The dilution ratios had no effect on GY, GP, PY, NUp, and NUE at any of the site-years in this study. Taking into account agronomic and economic factors, LU can be recommended as the most suitable liquid N fertilizer source for spring wheat cropping systems of the Northern Great Plains.

1. Introduction

Wheat is the main food grain produced in the United States [1]. Wheat accounts for approximately 20% of the total food calories consumed worldwide. Overall, approximately 35% of the world's population regularly depends on wheat for their nourishment. In the US, the consumption of wheat per capita exceeds that of any other food staple. Besides supplying carbohydrates, wheat also contains valuable proteins, minerals, and vitamins and essential amino acids like lysine [2]. Currently, the United States exports an average of 26.0 million metric tons of all wheat classes annually and leads in hard red winter and soft red winter wheat exports [3]. While N is considered the most common nutrient limiting yield of wheat and other cereal crops [4], N use efficiency (NUE) is currently between 40 and 50% for most cereal crop production systems [5]. A notable increase from the late 1990s estimates for NUE being 33% [6] is largely due to continuous advances in fertilizer management strategies and novel fertilizer technologies.

For many years, fertilization was driven by maximizing and sustaining crop yields as the main goal [7]. With the harmful effects of inefficient nutrient management practices resulting in soil, water, and air environments becoming a major concern, increasing fertilizer use efficiency has surfaced as a newly defined goal for crop producers. The most sensible and ethical solution to meet crops' nutrient demand is developing of more efficient crop fertilizer practices [8]. Establishing effective N management systems, updating N application guidelines, and improving NUE are the key challenges that must be addressed to sustain and enhance the sustainability of wheat production. Sustaining global food security and minimizing the negative impact of agriculture intensification on environmental quality are the most challenging issues the researchers and crop growers are facing today [9]. One of the key ways the producer can conserve fertilizer energy is utilizing fertilizer more efficiently, which entails optimizing crop yield with a minimum amount of fertilizer [10]. At least 50% of food produced in the world today is only possible due to commercial N, phosphorus (P), and potassium (K) fertilizer application to crops [11]. Commercial fertilizers are available in different forms, grades, and formulations; they can be solid (dry granular), liquids (fluid products), or gaseous (usually stored in a liquid form and transforming to gas when applied). In 2014 alone, the US crop farms expenditures related to fertilizers (including lime and soil conditioners) were $23.2 billion, surpassed only by land rent and labor costs [12].

According to the Ohio State University's Extension [13], "are liquid fertilizers equal to or better than dry fertilizers?" and "are liquid fertilizers more available than granular fertilizers?" are among the top 20 most asked agronomic questions. Inconsistent results in comparing liquid and dry N sources in wheat have been reported in literature. Many studies support the conclusion that there are no differences in the efficiency between the liquid and dry fertilizers [13, 14]. Some researchers concluded that significant ammonia loss occurs from liquid N fertilizers, which in fact decreases NUE [15]. Other research results suggest that liquid products may be superior in regard to crop yield and quality as well as being more environmentally friendly, due to superior plant availability and more efficient uptake [16]. Fluid fertilizers have been shown to have increased fertilizer use efficiently in several studies [17, 18].

In a long-term experiment in Oklahoma, comprising 8 growing seasons and 10 locations, liquid N fertilizer has resulted in a 19% advantage in NUE compared to dry granular N fertilizer in winter wheat. They also found the liquid N to be more profitable, even taking into account the per-unit cost advantage of the dry product over the liquid [19]. A combined application of compatible liquid N fertilizers and chemicals, such as herbicides and pesticides, could result in substantial monetary, time, and labor savings. Liquid fertilizers are easily transported, stored, and calibrated for precise application [20]. Compared to a mix created by combining several dry fertilizers, blending of liquid products results in a much more homogeneous mixture, where each drop has the uniform analysis [21]. The higher production cost of fluid fertilizers due to higher energy requirements may be balanced by higher efficiency resulting from a more consistent and uniform application [13]. The analysis of US fertilizer market share has shown that the utilization of liquid fertilizers is on the increase compared to dry fertilizer sources [21, 22]. The success of liquid fertilizers in corn production suggests that they will be of benefit in small grain cereals as well [23].

Application of liquid fertilizers, especially to crop canopy, has been recognized as the least recommended option for N application by some researchers [24]. Application of liquid N products at high concentrations often results in leaf burn as water evaporates and the fertilizer salts remain behind. Early in the growing season, foliar application may cause leaf burn; furthermore, mid- to late-season application can cause foliar diseases and reduce grain yields due to burn injury. The

documented yield reductions due to sprayed liquid N vary by application conditions and N rates; 400 to 800 kg ha^{-1} yield losses have been frequently reported. Some growers spray liquid N to wheat using flat fan or flood-jet nozzles which often can be a cause of significant leaf injury, even at early wheat growth stages, and may reduce early-season plant health critical for the grain formation [25]. Others note that leaf burn is often generally cosmetic and rarely causes yield reduction [26]. Edwards et al. [27] observed no leaf burn with application of liquid N products to wheat canopy, even at high temperatures of 25–30°C. As noted by Arnall et al. [28], liquid N fertilizers like UAN can cause leaf burning which can be considerable at higher rates, but, normally, the burning does not cause serious leaf injury and often does not impact yields, unless the product was sprayed on already significantly stressed crop. Stream nozzles and stream bars enable placement of liquid fertilizers in a concentrated band on the soil surface; this minimizes the opportunity for immobilization by soil microbial organisms. Some research has shown that streaming liquid products can lead to more efficient N use [26]. Streaming, applying the fluid fertilizers in narrow bands in either large drops or small streams, results in a concentration of the material in very small areas, which minimizes the potential for N loss. As the large drops get in contact with the plant material, there is less potential for injury, because the drops tend to roll off the plant to the soil surface. Arnall et al. [28] noted that streaming using stream bars is a preferred application method for fluid fertilizers.

Diluting fluid N fertilizers with water prior to application is one of the ways often recommended to reduce crop damage due to leaf burn. Diluting UAN 50%-50% with water reduced leaf burn in 2 of 3 years of the study; wheat recovered within 3 weeks and grain yields were not reduced [29]. The South Dakota State University's Extension Service recommends diluting liquid N to be diluted 1:1 with water to reduce leaf burn [30]. Similarly, the North Dakota State University's Extension Service advises growers to dilute UAN with water (1:1) to minimize the potential for leaf burn [31]. Furthermore, it is suggested not to apply liquid fertilizers to wheat at the rate exceeding 68 kg N ha^{-1} [32] and to corn at the rate exceeding 35 kg N ha^{-1} [33]. Gregoire [34] recorded a significant loss in yield when UAN was applied at the 45 kg N ha^{-1} rate. On the other hand, many growers are reluctant to dilute N fertilizers with substantial amounts of water because of the need to refill the tanks more often, which slows down the application time and increases application cost.

Several liquid N fertilizers varying in analysis are currently available on the market. These products include N or a blend of N and other macro- and micronutrients. Some of N foliar fertilizers include UAN, LU, and HNRGN. Urea ammonium nitrate is the most commonly used fluid N fertilizer. Urea ammonium nitrate (28-0-0 or 32-0-0) is a nonpressurized solution that can be used in a variety of agricultural crops. The versatile liquid mix of urea and ammonium nitrate has been available to growers for a long time. It offers fast acting and long lasting plant nutrient supply in a combination of three forms of nitrogen. Nitrate-N provides quick response and ammonic-N a longer lasting response and continuous nutrition from the water soluble organic N in urea [35]. Liquid urea is a water-based urea solution (20-0-0). The noted benefits of LU include slower uptake by the plant, which helps to maintain N levels within the soil-plant system. Liquid urea is suggested for application during the warm periods in the growing season to quickly correct N deficiency [36]. The primary advantage of LU compared to UAN is that it is less corrosive and, thus, poses a lesser risk of leaf burn [37]. On the other hand, the percentage of N in LU is lower and the transportation costs are usually greater per unit of N [38]. As the manufacturer of LU indicates on the product label, the ratio of LU to water should not exceed 1 : 4 for ground application [39]. Research on LU is very limited. Generally, it has been reported that where dry urea functions effectively the fluid urea should perform equally well or better due to having advantage of greater application uniformity over dry granular urea [40].

HNRGN has been marketed since the beginning of the 1990s; it is considered as one of the most efficient direct-applied N sources. HNRGN contains several forms of N and sulfur (S) as well as trace amounts of chlorophyll building elements such as iron (Fe), magnesium (Mg), manganese (Mn), and zinc (Zn). HNRGN also contains several proprietary enhancements. The product is very low in free ammonia and has been especially developed to minimize N loss and increased plant uptake. HNRGN has a reduced salt index and, therefore, is less corrosive compared to UAN [41].

The interest of crop growers in liquid N fertilizers is sustained by the pressing need to improve the efficiency of their farming operations and the successful marketing efforts by fertilizer industry and dealers. Many wheat growers in the Northern Great Plains, including the state of Montana, are already using fluid products or considering including them in their nutrient management program. These growers are in need of up-to-date and unbiased information about currently marketed liquid N fertilizers. Overall, opinion emphasized in most scientific reports could be summarized as follows: liquid N fertilizers could be successfully utilized; however, based on the products' labels, their application is limited due to potential leaf burn, where substantial N rates must be applied to satisfy crop needs.

2. Objectives

The objectives of this study were (i) to compare the efficacy of liquid N fertilizers (UAN, liquid urea, and HNRGN) applied to spring wheat and (ii) to determine the optimum N rate and dilution ratio of liquid fertilizers and the threshold at which spring wheat grain yield is reduced due to leaf burn.

3. Materials and Methods

This field study was conducted in 2012 and 2013 at three locations: two drylands, at Western Triangle Agricultural Research Center (WTARC, near Conrad, MT (48.309794,

−111.924684)) and in a cooperating producer's field (Jack Patton, Choteau County, MT (47.973032, −111.222696)), and one irrigated land, at Western Agricultural Research Center (WARC, near Corvallis, MT (46.328179, −114.089873)). Hard red spring wheat (cv. Choteau) was direct-seeded into plots measuring 1.5 by 7.6 m at the seeding rate of 1.8 million plants per hectare. Small plot drill with Conserva Pak™ openers manufactured by Swift Machining (Washougal, WA) was used to establish the research plots.

Appropriate weed and pest management control were employed when necessary. Treatment structure is reported in Table 1. At seeding, urea was applied in a band with the seed at $90 \, kg \, N \, ha^{-1}$ to all treatments except for the unfertilized check plot. At Feekes 5 growth stage (early tillering), $45 \, kg \, N \, ha^{-1}$ was applied utilizing an all-terrain vehicle- (ATV-) mounted stream-bar equipped sprayer. Three liquid N sources, UAN, LU, and HNRGN, and three dilution ratios of fertilizer%/water%, 100/0, 66/33, and 33/66, were evaluated. Because HNRGN contains Fe, Mg, Mn, and Zn, soil analysis was used to ensure that any of these nutrients were not deficient and can be corrected prior to top-dressing application. Similarly, because HNRGN contains S, plant samples were taken prior to top-dressing application to determine possible S deficiency and correct it as needed. At maturity, spring wheat was harvested with Hage 125 plot combine in 2012 and Wintersteiger Classic plot combine in 2013.

The field work activities are detailed in Tables 2 and 3. The harvested grain was dried in the drying room for 14 days at the temperature of 35°C. Then, the by-plot grain yield was determined utilizing scale. The subsamples (400 g) were analyzed by the Agvise Laboratories (Northwood, ND) for total N content utilizing near infrared reflectance (NIR) spectroscopy with a Perten DA 7250 NIR analyzer (Perten Instruments, Inc., Springfield, IL). The effects of N source and the dilution ratio (fertilizer/water) on N uptake (NUp), N use efficiency (NUE), spring wheat grain yield (GY), and grain protein (GP) content and protein yield (PY) were assessed. Grain N uptake was calculated by multiplying grain yield by total N concentration. N use efficiency was determined using the difference method [42] by deducting the total N uptake in wheat from the N-unfertilized treatment (check plot) from total N uptake in wheat from fertilized plots and then divided by the rate of N fertilizer applied. The analysis of variance was conducted using the PROC GLM procedure in SAS v9.3 (SAS Institute, Inc., Cary, NC). Mean separation was performed using the Orthogonal Contrasts method at a significance level of 0.05.

4. Results

4.1. Growing Season 2012

4.1.1. Grain Yield, Grain Protein Content, and Protein Yield. In 2012, GYs were higher at Conrad (5373 to $6456 \, kg \, N \, ha^{-1}$) and Corvallis (5406 to $6422 \, kg \, N \, ha^{-1}$) compared to Choteau (2092 to $3033 \, kg \, N \, ha^{-1}$). At all three locations, GYs were the highest with HNRGN. At dryland sites, the best GYs were

TABLE 1: Treatment structure, Choteau, Conrad, and Corvallis, 2012 and 2013.

Trt	Preplant N fertilizer (urea) rate, kg N ha^{-1}	Top-dressing N fertilizer source	Top-dressing N fertilizer rate, kg N ha^{-1}	Top-dressing N fertilizer/water ratio, %
1	0	—	—	—
2	90	UAN	45	100/0
3	90	UAN	45	66/33
4	90	UAN	45	33/66
5	90	LU	45	100/0
6	90	LU	45	66/33
7	90	LU	45	33/66
8	90	HNRGN	45	100/0
9	90	HNRGN	45	66/33
10	90	HNRGN	45	33/66

TABLE 2: Field activities and growing conditions, Conrad, Choteau, and Corvallis, 2012.

Field activity	Choteau	Conrad	Corvallis
Seeding date	April 24	April 18	April 15
Variety	Choteau	Choteau	Choteau
Seeding rate: seeds/ha	1.8 million	1.8 million	1.8 million
Herbicide	Bronate, Axial XL	Bronate, Axial XL	Bronate, Axial XL
Herbicide date	June 12	May 17	June 16
Sensing date	June 15	June 8	June 5
Top-dressing date	June 14	June 8	June 5
Harvest date	August 21	August 17	August 8
Average soil temperature °C	14.85	14.85	14.75
Average air temperature °C	14.25	14.25	15.75
Soil series	Scobey Clay Loam	Scobey Clay Loam	Burnt Fork Silt Loam
Soil N, kg ha^{-1}	31.7	39.5	31.7
Soil P, ppm	17	23	18
Soil K, ppm	287	423	345
Organic matter%	2.6	2.9	2.9
Soil pH	7.8	7.7	7.7

TABLE 3: Field activities and growing conditions, Conrad, Choteau, and Corvallis, 2013.

Field activity	Choteau	Conrad	Corvallis
Seeding date	May 1	April 26	April 20
Variety	Choteau	Choteau	Choteau
Seeding rate: seeds/ha	1.8 million	1.8 million	1.8 million
Herbicide	Supremacy, Axial XL	Supremacy, Axial XL	Supremacy, Axial XL
Herbicide date	June 7	May 29	May 20
Sensing date	June 25	June 24	June 22
Top-dressing date	June 25	June 24	June 22
Harvest date	August 23	August 19	August 12
Average soil temperature °C	15.10	15.10	15.25
Average air temperature °C	12.60	12.60	13.45
Soil series	Scobey Clay Loam	Scobey Clay Loam	Burnt Fork Silt Loam
Soil N, kg ha^{-1}	42.5	46	42.5
Soil P, ppm	21	25	21
Soil K, ppm	361	398	321
Organic matter%	2.7	2.2	2.8
Soil pH	7.8	7.7	7.8

TABLE 4: Treatment structure, Choteau, Conrad, and Corvallis, 2012 and 2013.

| Trt | Mean spring wheat grain yield, kg ha^{-1} | | | | | |
| | 2012 | | | 2013 | | |
	Choteau	Conrad	Corvallis	Choteau	Conrad	Corvallis
1	2529 (bcd)	5373 (c)	5629 (abc)	3564 (ab)	3698 (c)	1856 (b)
2	2118 (ed)	5979 (ab)	6012 (abc)	3477 (ab)	4001 (bc)	2125 (ab)
3	2233 (cde)	5810 (bc)	5710 (abc)	3403 (b)	4062 (bc)	2186 (ab)
4	2092 (e)	5837 (bc)	6348 (ab)	3490 (ab)	3954 (c)	2132 (ab)
5	2576 (bc)	6046 (ab)	5406 (c)	3537 (ab)	4593 (ab)	2361 (a)
6	2582 (bc)	6194 (ab)	5420 (bc)	3544 (ab)	4842 (a)	2139 (ab)
7	2690 (ab)	6207 (ab)	5548 (abc)	3880 (a)	4728 (a)	1957 (ab)
8	2811 (ab)	6382 (ab)	6422 (a)	3746 (ab)	4768 (a)	1957 (ab)
9	2616 (bc)	6369 (ab)	6288 (abc)	3356 (b)	5104 (a)	2246 (ab)
10	3033 (a)	6456 (a)	6147 (abc)	3867 (a)	5057 (a)	2334 (ab)

Means in the same column followed by the same letter are not significantly different at $p < 0.05$.

TABLE 5: Mean spring wheat grain protein content and protein yield, Choteau, Conrad, and Corvallis, 2012.

| Trt | Mean spring wheat grain protein content, % | | | Mean spring wheat protein yield, kg ha^{-1} | | |
	Choteau	Conrad	Corvallis	Choteau	Conrad	Corvallis
1	13.8 (c)	10.8 (c)	13.4 (f)	391 (d)	649 (b)	845 (b)
2	17.2 (a)	12.8 (b)	14.4 (bcde)	409 (d)	856 (a)	966 (ab)
3	16.8 (ab)	13.2 (ab)	13.9 (def)	422 (cd)	862 (a)	888 (ab)
4	17.0 (ab)	13.1 (ab)	14.2 (cde)	398 (d)	858 (a)	1010 (ab)
5	16.7 (ab)	13.2 (ab)	15.1 (a)	482 (bc)	897 (a)	916 (ab)
6	16.8 (ab)	13.7 (a)	15.0 (ab)	485 (bc)	947 (a)	907 (ab)
7	16.5 (b)	13.1 (ab)	14.9 (abc)	496 (b)	908 (a)	923 (ab)
8	16.9 (ab)	13.1 (ab)	13.8 (ef)	533 (ab)	934 (a)	989 (ab)
9	17.1 (a)	13.2 (ab)	14.6 (abcd)	501 (b)	943 (a)	1027 (a)
10	16.8 (ab)	12.9 (b)	14.0 (def)	572 (a)	929 (a)	963 (ab)

Means in the same column followed by the same letter are not significantly different at $p < 0.05$.

achieved with HNRGN at 33/66 dilution ratio (treatment 10), while at Corvallis (irrigated) the highest GY was obtained with undiluted HNRGN (treatment 8) (Table 4).

The GPs ranged from 13.8 to 17.2% at Choteau (highest among the three locations). The GP values ranged from 10.8 to 13.7% at Conrad and from 13.4 to 15.1% at Corvallis. At Choteau, application of HNRGN at 66/33 dilution ratio has produced the highest GPs (treatment 9), as well as application of undiluted UAN (treatment 2). Application of LU at 66/33 dilution ratio (treatment 6) has produced the best GP at Conrad. Similar results were noted for Corvallis, where comparable GPs were achieved with the application of undiluted LU (treatment 5) and LU at 66/33 dilution ratio (Tables 5 and 6).

In 2012, at Choteau, the best PY value of 572 kg N ha^{-1} was associated with HNRGN application at 33/66 ratio (treatment 10). The lowest PY values were obtained with UAN. At Conrad and Corvallis, the differences among the treatments were not as pronounced as at Choteau. Although at Conrad the differences were not statistically significant, the general trend was that HNRGN and LU resulted in higher PY values compared to UAN. At Corvallis, the highest PY value of

1027 kg ha^{-1} was noted for treatment 9 (HNRGN at 66/33 ratio), closely followed by treatment 4 (UAN at 33/66 ratio) (Tables 5 and 6).

4.1.2. N Uptake and Nitrogen Use Efficiency. In 2012, more pronounced differences between treatments in terms of NUp were observed at Choteau. The highest NUp values of 99 and 91 kg N ha^{-1} were observed for HNRGN applied at 33/66 ratio and undiluted, respectively. At Conrad, the differences between the treatments were not significant; treatments 6 (LU at 66/33 ratio) and 9 (HNRGN at 66/33 ratio) resulted in higher NUp values of 163 and 161 kg N ha^{-1}, respectively. Treatment 9 also produced the highest NUp at the irrigated site (Corvallis), followed by treatment 4 (UAN applied at 33/66 dilution ratio) (Tables 7 and 8).

At WTARC and Corvallis, no significant differences in NUEs associated with N source were observed in 2012. At Choteau, significantly greater NUE values were observed for HNRGN (treatments 10 and 8, followed by treatment 9), compared to LU and UAN. In general, similar trend was observed at Conrad, but the differences were not statistically

TABLE 6: Mean spring wheat grain protein content and protein yield, Choteau, Conrad, and Corvallis, 2013.

Trt	Mean spring wheat grain protein content, %			Mean spring wheat protein yield, kg ha^{-1}		
	Choteau	Conrad	Corvallis	Choteau	Conrad	Corvallis
1	12.5 (c)	10.6 (d)	16.0 (a)	446 (c)	394 (e)	296 (a)
2	14.9 (ab)	13.4 (ab)	15.2 (ab)	518 (b)	538 (d)	322 (a)
3	15.0 (ab)	13.5 (a)	14.5 (bc)	510 (b)	548 (cd)	317 (a)
4	15.3 (a)	13.3 (abc)	14.8 (bc)	535 (ab)	527 (d)	315 (a)
5	15.2 (ab)	13.1 (bc)	13.3 (c)	536 (ab)	603 (bcd)	314 (a)
6	14.7 (b)	13.3 (abc)	14.7 (bc)	521 (b)	641 (ab)	314 (a)
7	15.0 (ab)	13.1 (c)	14.4 (c)	584 (a)	617 (abc)	281 (a)
8	15.0 (ab)	13.4 (ab)	14.4 (bc)	564 (ab)	640 (ab)	281 (a)
9	15.4 (a)	13.4 (ab)	14.1 (bc)	517 (b)	684 (a)	316 (a)
10	15.2 (ab)	13.5 (ab)	14.8 (bc)	587 (a)	680 (a)	345 (a)

Means in the same column followed by the same letter are not significantly different at $p < 0.05$.

TABLE 7: Mean spring wheat N uptake and NUE, Choteau, Conrad, and Corvallis, 2012.

Trt	N uptake, kg N ha^{-1}			NUE, %		
	Choteau	Conrad	Corvallis	Choteau	Conrad	Corvallis
1	67 (d)	111 (b)	145 (b)	—	—	—
2	71 (d)	147 (a)	166 (ab)	1.8 (d)	23.5 (a)	13.8 (a)
3	73 (cd)	148 (a)	153 (ab)	3.0 (cd)	24.1 (a)	4.8 (a)
4	68 (d)	147 (a)	174 (ab)	0.5 (d)	23.8 (a)	18.7 (a)
5	83 (bc)	154 (a)	157 (ab)	10.0 (bc)	28.1 (a)	8.1 (a)
6	83 (bc)	163 (a)	156 (ab)	10.3 (bc)	33.8 (a)	7.1 (a)
7	85 (b)	156 (a)	158 (ab)	11.3 (b)	29.4 (a)	8.8 (a)
8	91 (ab)	160 (a)	169 (ab)	15.8 (ab)	32.2 (a)	16.3 (a)
9	86 (b)	161 (a)	176 (a)	12.0 (b)	33.3 (a)	20.6 (a)
10	99 (a)	159 (a)	165 (ab)	20.3 (a)	31.8 (a)	13.3 (a)

Means in the same column followed by the same letter are not significantly different at $p < 0.05$.

TABLE 8: Mean spring wheat N uptake and NUE, Choteau, Conrad, and Corvallis, 2013.

Trt	N uptake, kg N ha^{-1}			NUE, %		
	Choteau	Conrad	Corvallis	Choteau	Conrad	Corvallis
1	76 (c)	67 (e)	50 (a)	n/a	n/a	n/a
2	89 (b)	91 (d)	55 (a)	28.6 (b)	37.7 (bc)	15.9 (a)
3	87 (b)	94 (cd)	54 (a)	27.4 (b)	38.9 (bc)	15.0 (a)
4	92 (ab)	90 (d)	54 (a)	31.2 (ab)	36.0 (c)	14.3 (a)
5	92 (ab)	103 (bcd)	54 (a)	31.3 (ab)	47.7 (abc)	22.4 (a)
6	89 (b)	110 (ab)	54 (a)	28.6 (b)	53.6 (a)	14.5 (a)
7	100 (a)	105 (abc)	48 (a)	38.6 (a)	49.8 (ab)	12.0 (a)
8	97 (ab)	110 (ab)	48 (a)	35.8 (ab)	53.5 (a)	9.9 (a)
9	89 (b)	117 (a)	54 (a)	28.7 (b)	60.0 (a)	14.6 (a)
10	101 (a)	117 (a)	59 (a)	39.4 (a)	59.5 (a)	19.1 (a)

Means in the same column followed by the same letter are not significantly different at $p < 0.05$.

significant. At Corvallis, treatment 9 produced the best NUE of 20.6% (Tables 7 and 8).

4.2. Growing Season 2013

4.2.1. Grain Yield, Grain Protein Content, and Protein Yield. In general, in 2013, the GYs were lower at Conrad (3698 to 5104 kg N ha^{-1}) and Corvallis (1856 to 2361 kg N ha^{-1}) and higher at Choteau (3356 to 3867 kg N ha^{-1}), compared to 2012. Like in the first growing season, the highest GYs at dryland sites were observed for treatments that received HNRGN: at 33/66 dilution ratio at Choteau (treatment 10) and at 66/33 dilution ratio at Conrad (treatment 9). Treatment 10 at Conrad was the second best with the GY of 5057 kg N ha^{-1}. At the irrigated Corvallis location, application of undiluted LU resulted in the yielding the highest GY, closely followed by treatments 10 and 9 (Table 4).

Like in 2012, the highest GP values were observed again at Choteau (12.5–15.4%) in 2013. The highest GPs at Choteau were achieved with the application of HNRGN at 66/33 dilution ratio followed by treatment 4, UAN applied at 33/66 dilution ratio. At Conrad, application of HNRGN at 33/66 ratio (treatment 10) and UAN at 66/33 ratio (treatment 3) resulted in the highest GP values. At Corvallis, the highest GPs were noted for the unfertilized check plot (treatment 1) and with the application of undiluted UAN (treatment 2) (Tables 5 and 6).

In 2013, at all three locations, HNRGN has performed the best compared to other N sources. At Choteau, treatment 10 (HNRGN at 33/66 ratio) resulted in the highest PY value of 587 kg ha^{-1}. Application of HNRGN at 66/33 and 33/66 ratios (treatments 9 and 10) produced the highest PY values of 684 and 680 kg ha^{-1}, respectively. At the irrigated location (Corvallis), treatment 10 has also resulted in the highest PY, although the differences among the treatments were not significant (Tables 5 and 6).

4.2.2. N Uptake and Nitrogen Use Efficiency. In 2013, although no statistically significant differences in NUp values associated with N application source were observed at the irrigated Corvallis location, the highest NUp was noted for treatment 10 (HNRGN at 33/66 ratio). The same trend was observed at both dryland sites (Choteau and Conrad), where treatment

10 resulted in the highest NUp values of 101 and 117 kg N ha^{-1}, respectively (Tables 7 and 8).

In the second year of the study, the NUE values were higher compared to 2012 at all locations. Like in 2012, the differences between the treatments at Corvallis were not significant; higher NUEs were noted for treatments 5 (undiluted LU) and 10 (HNRGN at 33/66 ratio). Similarly, at the other two sites, the highest NUEs of 39.4 and 60% were observed for HNRGN treated plots at Choteau and Conrad, respectively (Tables 7 and 8).

4.2.3. Effect of Liquid N Fertilizer Dilution on GY, GP, PY, NUp, and NUE. The dilution ratios had no effect on GY, GP, PY, NUp, and NUE at any of the site-years in this study (data not shown).

5. Discussion

Although the effect of N source on GY was more pronounced in 2012, compared to 2013, in both growing seasons, HNRGN resulted in higher yields compared to UAN. At dryland locations, LU performed as well as HNRGN. At the irrigated location, there was little difference in yield associated with N product. In 2012, at Corvallis, lower GY but higher GP was observed with LU, compared to other N sources. The GP contents obtained in this study were excellent and ranged from 10.6% to 17.2%. Evaluation of product effect on PY and NUE allowed us to assess how efficiently N products were taken up, assimilated, and utilized to produce both yield and quality (protein). Protein yield is a valuable characteristic, especially for spring wheat in Montana. Protein yield was clearly higher with HNRGN at both dryland sites in 2012 and in 2013. Even where the differences were not statistically significant, over 35 kg ha^{-1} advantage in PY accumulation was observed with HNRGN compared to UAN. The effect of N source on NUE was very pronounced in favor of HNRGN at dryland locations in both growing seasons. The lowest NUE values were observed with UAN; LU produced intermediate results. The irrigated location had similar NUEs for all products, except for 2012, when LU resulted in lower (not statistically significant) NUE values.

Although various degrees of leaf burn were obvious during postapplication in the majority of the experimental plots, the wheat plants have recovered within next 2-3 weeks. The physical damage caused to the plants did not result in any significant yield or quality penalties. The dilution ratios had no effect on GY, GP, PY, NUp, and NUE at any of the site-years in this study (data not shown). Our results suggest that it is feasible to apply undiluted liquid N products to spring wheat when a stream bar sprayer is used without negatively impacting crop yield or quality. This statement is especially true for noncorrosive products like LU and HNRGN [43]. However, growers should be advised not to exceed the application rate of 45 kg N ha^{-1} (rate evaluated in this study) when applying undiluted liquid N fertilizers.

Over the 2008–2016 period, urea (and, thus, LU) and UAN averaged \$.24 and \$0.28 per kg of N [44]. On the other hand, HNRGN is typically about 20% more expensive, compared to UAN [45]. For this study, at the time of N fertilizer application, the costs were virtually the same for LU and UAN per unit of N, whereas HNRGN costed almost 25% more compared to both LU and UAN [46].

Many personal communications with Montana wheat growers have shown that they see LU as a very good N source choice. Popularity of LU is growing due to noncorrosive qualities. Several growers indicated that they produce their own LU on-site by dissolving dry granular urea in water. Results of our study suggested that choice of liquid N fertilizer might be more important in dryland cropping systems, compared to irrigated ones, with positive results obtained with LU at Choteau and Conrad experimental sites located in the heart of Golden Triangle: Montana's key dryland wheat producing region. In conclusion, taking into account agronomic and economic factors, LU can be recommended as the most suitable liquid N fertilizer source for spring wheat cropping systems of the Northern Great Plains.

Competing Interests

The authors declare that they have no competing interests.

References

[1] G. Vocke and O. Liefert, "USDA-ERS. Wheat: Background. Briefing Rooms. Economic Research Service," 2009, http://www.ers.usda.gov/briefing/wheat/background.htm.

[2] P. Thompson, *Oklahoma Ag in the Classroom. Wheat Facts*, Oklahoma State University Extension, 2015, http://oklahoma4h.okstate.edu/aitc/lessons/extras/facts/wheat.html.

[3] R. D. Taylor and W. W. Koo, Agribusiness & Applied Economics. Center for Agricultural Policy and Trade Studies, Department of Agribusiness and Applied Economics North Dakota State University, No. 738 Outlook of the U.S. and World Wheat Industries, 2015–2024, 2015, http://ageconsearch.umn.edu/bitstream/201310/2/AAE738.pdf.

[4] R. Engel, "Winter wheat response to available nitrogen and water. Montana State University Extension," Fertilizer Facts 4, 1993, http://landresources.montana.edu/fertilizerfacts/documents/FF4WWNwater.pdf.

[5] M. L. Gupta and R. Khosla, "Precision nitrogen management and global nitrogen use efficiency," in *Proceedings of the 11th International Conference on Precision Agriculture*, K. Harald and G. M. P. Butron, Eds., Indianapolis, Ind, USA, 2012.

[6] W. R. Raun and G. V. Johnson, "Improving nitrogen use efficiency for cereal production," *Agronomy Journal*, vol. 91, no. 3, pp. 357–363, 1999.

[7] S. H. Chien, L. I. Prochnow, and H. Cantarella, "Recent developments of fertilizer production and use to improve nutrient efficiency and minimize environmental impacts," *Advances in Agronomy*, vol. 102, pp. 267–322, 2009.

[8] V. Smil, "Global population and the nitrogen cycle," *Scientific American*, vol. 277, no. 1, pp. 76–81, 1997.

[9] O. Walsh, W. Raun, A. Klatt, and J. Solie, "Effect of delayed nitrogen fertilization on maize (*Zea mays* L.) grain yields and nitrogen use efficiency," *Journal of Plant Nutrition*, vol. 35, no. 4, pp. 538–555, 2012.

[10] F. J. Hay, *Energy-Efficient Use of Fertilizer and Other Nutrients in Agriculture*, 2012, http://www.extensionorg/pages/62014/energy-efficient-use-of-fertilizer-and-other-nutrients-in-agriculture#.U6sC70BinSR.

[11] W. M. Stewart, D. W. Dibb, A. E. Johnston, and T. J. Smyth, "The contribution of commercial fertilizer nutrients to food production," *Agronomy Journal*, vol. 97, no. 1, pp. 1–6, 2005.

[12] USDA, *Farm Production Expenditures 2014 Summary*, 2015, http://usda.mannlib.cornell.edu/usda/nass/FarmProdEx//2010s/2015/FarmProdEx-08-04-2015.pdf.

[13] J. W. Johnson, "Most Asked Agronomic Questions. The Ohio State University Extension Bulletin 760," 1999, http://agcrops.osu.edu/sites/agcrops/files/imce/fertility/Most%20Asked%20Agronomic%20Questions,%20Bulletin%20760,%20Chapter%2013_%20Miscellaneous.pdf.

[14] G. Silva, All Fertilizers are not Created Equal, Michigan State University Extension, 2016, http://msue.anr.msu.edu/news/all_fertilizers_are_not_created_equal.

[15] The Mosaic Company, "Fluid and dry fertilizers. Fluids and solids are equal agronomically," 2013, http://www.cropnutrition.com/efu-fluid-dry-fertilizers#overview.

[16] C. J. Watson, R. J. Stevens, R. J. Laughlin, and P. Poland, "Volatilization of ammonia from solid and liquid urea surface-applied to perennial ryegrass," *The Journal of Agricultural Science*, vol. 119, no. 2, pp. 223–226, 1992.

[17] E. Lombi, M. J. McLaughlin, C. Johnston, R. D. Armstrong, and R. E. Holloway, "Mobility and lability of phosphorus from granular and fluid monoammonium phosphate differs in a calcareous soil," *Soil Science Society of America Journal*, vol. 68, no. 2, pp. 682–689, 2004.

[18] R. E. Holloway, I. Bertrand, A. J. Frischke, D. M. Brace, M. J. Mclaughlin, and W. Shepperd, "Improving fertiliser efficiency on calcareous and alkaline soils with fluid sources of P, N and Zn," *Plant and Soil*, vol. 236, no. 2, pp. 209–219, 2001.

[19] M. J. McLaughlin, T. M. McBeath, R. Smernik, S. P. Stacey, B. Ajiboye, and C. Guppy, "The chemical nature of P accumulation in agricultural soils—implications for fertiliser management and design: an Australian perspective," *Plant and Soil*, vol. 349, no. 1-2, pp. 69–87, 2011.

[20] C. E. Boyer, W. B. Brorsen, J. B. Solie, D. B. Arnall, and W. R. Raun, "Economics of preplant, topdress, and variable rate nitrogen application in winter wheat," in *Proceedings of the Agricultural and Applied Economics Association*, Denver, Colo, USA, 2010.

[21] O. S. Walsh, R. J. Christiaens, and A. Pandey, "Foliar-applied nitrogen fertilizers in spring wheat production," Crops & Soils Magazine, 2013, https://dl.societies.org/publications/cns/articles/46/4/26/.

[22] O. S. Walsh, A. Pandey, and R. J. Christiaens, "Liquid N fertilizer evaluation in spring wheat," in *Proceedings of the Western Nutrient Management Conference*, Reno, Nev, USA, 2015.

[23] D. F. Leikam, Fluid Fertilizers: Properties and Characteristics, Fluid Fertilizer Foundation, 2012, http://www.fluidfertilizer.com/Forum%20Presentations/2010/2010%20Indianapolis%20Presentations/Dale%20Leikam%20-%20Fluid%20Fertilizer%20Basics.pdf.

[24] T. L. Jensen, How Do Lower-Rate Liquid Starter Fertilizers Compare to Traditional Seed-Row Fertilizer Blends in the Northern Great Plains? Plant Nutrition Today, 2011, http://www.ipni.net/ipniweb/pnt.nsf/5a4b8be72a35cd46852568d9001a18da/c9e7e746bd19dbc5852578fb004e5ce7/$FILE/No%202%20IPNI%20PNT%20FALL2011.pdf.

[25] F. G. Fernandez, Applying Nitrogen after Planting, 2010, http://web.extension.illinois.edu/state/newsdetail.cfm?NewsID=17983.

[26] Needham AG Technologies Llc, "Stream Bars for Uniform Liquid Fertilizer Application," 2016, http://www.needhamag.com/innovative_product_sales/stream_bars_for_uniform_liquid_fertilizer_application.php.

[27] J. Edwards, B. Arnall, and H. Zhang, Methods for Applying Topdress Nitrogen to Wheat, Oklahoma Cooperative Extension Service, 2004, http://pods.dasnr.okstate.edu/docushare/dsweb/Get/Document-6506/PSS-2261web.pdf.

[28] B. D. Arnall, J. Mullock, and B. Seabourn, "Can protein levels be economically increased?" *Fluid Journal*, vol. 77, no. 20, pp. 1–4, 2012.

[29] D. Beegle, Topdressing Wheat with Nitrogen, Pennsylvania State University Extension, 2014, http://extension.psu.edu/plants/crops/news/2014/04/topdressing-wheat-with-nitrogen.

[30] D. W. Franzen, Fertilizing Winter Wheat, North Dakota State University Extension, 2015, https://www.ag.ndsu.edu/pubs/plantsci/soilfert/sf1448.pdf.

[31] P. W. Stahlman, R. S. Currie, and M. A. El-Hamid, "Nitrogen carrier and surfactant increase foliar herbicide injury in winter wheat (*Triticum aestivum*)," *Weed Technology*, vol. 11, no. 1, pp. 7–12, 1997.

[32] R. Beck, *In Season N Applications to Wheat*, South Dakota State University Extension, 2014, http://igrow.org/agronomy/wheat/in-season-n-applications-to-wheat/.

[33] J. Ransom, "Nitrogen Application Studies. North Dakota State University Extension," 2013, https://www.ag.ndsu.edu/small-grains/presentations/2013-best-of-the-best-in-wheat-and-soy-bean/ransom-nitrogen.

[34] J. Wiersma and A. Sims, "Late Season Applications of Nitrogen in Spring Wheat, University of Minnesota Extension," 2014, http://blog-crop-news.extension.umn.edu/2014/07/late-season-applications-of-nitrogen-in.html.

[35] W. R. Raun and H. Zhang, *Nitrogen Fertilizer Sources, their Potential Losses and Management Tips*, Oklahoma State University Extension, 2006, http://www.nue.okstate.edu/Anhydrous_Ammonia/PT2006-5%20FertilizerSource2.doc.

[36] T. L. Wesley, R. E. Lamond, V. L. Martin, and S. R. Duncan, "Effects of late-season nitrogen fertilizer on irrigated soybean yield and composition," *Journal of Production Agriculture*, vol. 11, no. 3, pp. 331–336, 1998.

[37] simplot.com, Urea Ammonium Nitrate, 2012, http://techsheets.simplot.com/Plant_Nutrients/Urea_Ammon_Nitrate_Solution_32_0_0.pdf.

[38] fetizona.com, "Liquid Urea," 2015, http://www.fertizona.com/pop-up-liquidurea.html.

[39] B. Brown and L. Long, "Response of 'Ute' to Rate and Source of Foliar N," in *Proceedings of the 39th Annual Far West Regional Conference*, pp. 111–116, Bozeman, MT, USA, 1988.

[40] C. A. Jones, R. T. Koenig, J. W. Ellsworth, B. D. Brown, and G. D. Jackson, *Management of Urea Fertilizer to Minimize Volatilization*, Montana State University Extension, Bozeman, Mont, USA, 2007, http://cru.cahe.wsu.edu/CEPublications/eb173/eb173.pdf.

[41] tirms.net, Power-Line. 21% Urea Solution, 2007, http://pdf.tirmsdev.com/Web/538/39212/538_39212_LABEL_English_.pdf?download=true.

[42] C. J. Overdahl, G. W. Rehm, and H. L. Meredith, Fertilizer Urea, Fluid Urea, University of Minnesota Extension, 1991, http://www.extension.umn.edu/distribution/cropsystems/dc0636.html.

[43] agroliquid.com, "High NRG-N: A Different Kind of Nitrogen Product," 2013, http://1amtlr2w7hkt2eg60up2ibg4.wpengine.netdna-cdn.com/wp-content/uploads/2013/05/Product-Catalog-High-NRG-N-California.pdf.

[44] G. E. Varvel and T. A. Peterson, "Nitrogen fertilizer recovery by corn in monoculture and rotation systems," *Agronomy Journal*, vol. 82, no. 5, pp. 935–938, 1990.

[45] G. Schnitkey, Averages and Seasonality of Prices for Nitrogen Fertilizers, University of Illinois Extension, 2016, http://farmdoc-daily.illinois.edu/2016/04/averages-and-seasonality-of-prices-nitrogen.html.

[46] B. Battel, P. Kaatz, M. Nagelkirk, and J. Vincent, *Thumb Ag Research & Education 2013 Field Trials*, Michigan State University Extension, East Lansing, Mich, USA, 2013, http://msue.anr.msu.edu/uploads/236/43436/2013_Tare_Book.pdf.

10

Response of Boron and Light on Morph-Physiology and Pod Yield of Two Peanut Varieties

Md. Quamruzzaman,[1] **Md. Jafar Ullah,**[1] **Md. Fazlul Karim,**[1] **Nazrul Islam,**[2] **Md. Jahedur Rahman,**[2] **and Md. Dulal Sarkar**[2]

[1]*Department of Agronomy, Sher-e-Bangla Agriculture University, Dhaka 1207, Bangladesh*
[2]*Department of Horticulture, Sher-e-Bangla Agriculture University, Dhaka 1207, Bangladesh*

Correspondence should be addressed to Md. Quamruzzaman; bdquamu@gmail.com

Academic Editor: Sudhakar Srivastava

Boron is an important micronutrient that enhances vegetative growth and yield of crops, like peanut. Light also plays an important role in pegging of peanut. There has been little information regarding the application of boron and light in peanut in Bangladesh. Therefore, a field experiment was conducted to study the response of boron and light on morph-physiology and pod yield of two peanut varieties. Treatments considered two peanut varieties, *namely*, Dhaka-1 and BARI Chinabadam-8, three levels of boron (B), *namely*, 0-kg B ha^{-1} (B_0), 1-kg B ha^{-1} (B_1), and 2-kg B ha^{-1} (B_2), and two levels of light, *namely*, normal day light ($\approx$12 h light) and normal day light + 6 h extended red light at night ($\approx$18 h light). Result revealed that days to first-last emergence and days to first-50% flowering took shorter times and vegetative growth, pods dry weight plant^{-1}, pod yield, and germination were markedly increased with the application of boron. Vegetative growth and germinations were significantly increased in light, but the lowest leaf area, pods dry weight plant^{-1}, and pod yield were found in light. Without germination, the highest vegetative growth, reproductive unit, and pod yield were observed from BARI Chinabadam-8. Days to first-last emergence, days to first-50% flowering, and number of branches plant^{-1} were found linearly related to pod yield.

1. Introduction

Peanut (*Arachis hypogaea* L.) is one of the most important oil seed crops throughout the world [1]. Boron (B) is a micronutrients required by plants in a very small quantity [2] which are rapidly becoming deficient in soils [3]. Boron is an essential element needed for normal growth and development of peanut plant [4–7]. Boron makes the stigma receptive and sticky, makes pollen grain fertile, and enhances the pollination [8]. It regulates carbohydrate metabolism and plays role in seed formation [9]. Application of boron in soil significantly increases the growth and yield of groundnut [10, 11]. But boron deficiency problems for crop production have been identified [12] because application of boron in crops is limited at farmer's field [13]. To overcome this problem and to specify the optimum doses of boron in peanuts a little bit of research has been found. So, more research is needed regarding on application of boron at farmer's fields in Bangladesh.

Therefore, it is important to study the effect of boron on morph-physiology and pod yield of peanut.

Light plays an important role in the vegetative and reproductive growth in peanut. The quantity, quality, and direction of light are perceived by several different photosensory systems that together regulate nearly all stages of plant development, presumably in order to maintain photosynthetic efficiency [14]. However, the number of flowers markedly reduces if less light is received by the peanut plants [15]. Total numbers of pegs and pods and therefore yield are lower in long day photoperiods, but vegetative production is higher in long day photoperiod [16, 17]. In peanut, for light supplementation, peg to pod conversation rate and yield are lower [18], but light stress can lead to ROS (Reactive Oxygen Species) accumulation and antioxidant enzymes activation in plant [19]. Little or no research studies were conducted in Bangladesh to find out the impact of light on peanut. Therefore, the present studies were conducted to find out the effect of boron

TABLE 1: (a) Soil test results of the experimental filed (mean of two years). (b) Monthly record of air temperature, relative humidity, and rainfall of the experimental site during the period of March to July 2014 and 2015 (mean of two years).

(a)

Element	*Levels in the soil plot
pH	5.9
Total nitrogen	0.071%
Exchangeable K	0.31 meq/100 g soil
Exchangeable Ca	6.36 meq/100 g soil
Exchangeable P	14.04 μg/g soil
Exchangeable S	15.16 μg/g soil
B	0.30 μg/g soil
Sand	27%
Silt	43%
Clay	30%
Organic matter	0.78%

*Soil was tested at Soil Resource Development Institute (SRDI) Laboratory, Farmgate, Dhaka, Bangladesh.

(b)

Month	Air temperature (°C)		Relative humidity (%)		Rainfall (mm) (total)
	Maximum	Minimum	Maximum	Minimum	
March	37.4	20.2	80.2	32.4	3.80
April	39.4	19.4	80.2	39.2	65.60
May	38.2	19.3	89.2	40	202
June	37.2	17.4	88.4	46.3	282.7
July	35.6	18.2	88.2	55.4	107.8

Source. Sher-e-Bangla Agricultural University mini weather station, Dhaka 1207, Bangladesh.

and duration of light on morph-physiology and pod yield of peanut.

2. Materials and Methods

2.1. Experimental Site. The experiment was conducted at the Central Experimental Farm, Sher-e-Bangla Agricultural University, Dhaka 1207, Bangladesh, during March to July 2014, and the same experiment was conducted during Mach to July 2015 in the same plot. Soil of the experimental field was analyzed before the studies were conducted and means of two years were recorded (Table 1(a)). The experimental filed was located at 23°41′N latitude and 90°22′E longitude at a height of 8.6 m above the sea level belonging to the agroecological zone "AEZ-28" of Madhupur Tract [20]. The environmental factor, that is, mean air temperature, relative humidity, and rainfall in 2014 and 2015 of the experimental site, was also recorded (Table 1(b)).

2.2. Methods of Soil Nutrient Elements and
Particle Size Analysis

2.2.1. pH. pH was determined by Jenway 3570 pH meter using soil and water ratio 1 : 2.5.

2.2.2. Total Nitrogen. Micro Kjeldahl method was used for determining total nitrogen.

2.2.3. Exchangeable Potassium and Calcium. For these two elements soil extraction was made by using 1 M ammonium acetate solution and K; Ca was measured directly from the soil extract in the flame photometer.

2.2.4. Exchangeable Phosphorous. Phosphorous was extracted with 0.3 M NH_4F according to Bray and Kurtz method.

2.2.5. Exchangeable Sulphur. Sulphur was determined turbid metrically using acid seed solution and turbid metric reagent with soil filtrate. Reading was taken on Perkin Elmer Lambda 11 (2.2) UV/VIS Spectrometer at 535 nm.

2.2.6. Boron. Extraction of boron was made by using 0.01 M $CaCl_2$. The extract was then processed with buffer solution and azomethine-H reagent. The concentration of boron was measured in spectrophotometer.

2.2.7. Sand, Silt, and Clay. Hydrometer method was used to analyze the percentage of sand, silt, and clay.

2.2.8. Organic Matter. Total organic carbon was determined with LECO-C-200 carbon analyzer. Organic matter content of individual soil sample was determined by multiplying the presence of carbon by the factor 1.724.

TABLE 2: Effect of boron and light on days to emergence and days to flowering in two peanut varieties (mean of two trials).

Treatment	Days to 1st emergence	Days to last emergence	Days to 1st flowering	Days to 50% flowering
Boron (B)				
B_0	7.83^{a^z}	17.83^a	28.91^a	34.83^a
B_1	7.42^b	17.42^a	28.75^a	33.83^b
B_2	6.75^c	16.50^b	27.75^c	32.58^c
Light (L)				
L	—	—	—	—
L_0	—	—	—	—
Variety (V)				
V_1	7.72^a	16.89^b	28.83	34.11
V_2	6.94^b	17.61^a	28.11	33.39
Significance (P)				
B	<0.001	<0.001	<0.001	<0.001
L	—	—	—	—
V	<0.001	<0.001	<0.001	<0.001

B = boron; L = light; V = variety; ^{z}means, column having the same letter(s) are insignificant and different letter(s) statistically significant, P = probability. Means were separated by Tukey's test at $P \leq 0.05$, $B_0 = 0$ kg B ha^{-1} (control), $B_1 = 1$ kg B ha^{-1}, $B_2 = 2$ kg B ha^{-1}, L = normal day light + 6 h extended red light at night (≈18 h light), L_0 = normal day light (≈12 h light), V_1 = Dhaka-1, and V_2 = BARI Chinabadam-8.

2.3. Plant Material and Treatments. Two peanut varieties were used in this experiment, *namely*, Dhaka-1 (Maizchar Badam) and BARI Chinabadam-8. The seeds of the groundnuts were collected from Bangladesh Agricultural Research Institute (BARI), Gazipur, Bangladesh. The experiment was laid out in a $2 \times 3 \times 2$ factorial design with three replications. The experimental unit was 4 m^2 (2 m × 2 m) plot. The first factor was the two peanut varieties, *namely*, Dhaka-1 (V_1) and BARI Chinabadam-8 (V_2); second factor was the three levels of boron, *namely*, 0 kg B ha^{-1} (B_0), 1 kg B ha^{-1} (B_1), and 2 kg B ha^{-1} (B_2) and third factor was duration of light, that is, normal day light (≈12 h light) (L_0) and normal day light + 6 h extended red light at night (≈18 h light) (L). In both years, to extend the photoperiod, one month after seed sowing (after seedling emergence), artificial lightening was used by florescence bulb from 1800 h to 2400 h at 30–50,000 lux, measured by lux meter.

2.4. Field Preparation and Data Recorded. The recommended doses of organic manure and inorganic fertilizer were also used for the present experiment. Cow dung, urea, triple superphosphate, muriate of potash, gypsum, and zinc sulphate were applied at 10 t ha^{-1}, 25 kg ha^{-1}, 160 kg ha^{-1}, 75 kg ha^{-1}, 170 kg ha^{-1}, and 4 kg ha^{-1}, respectively. The crop was harvested at maturity stage (114 days after planting (DAP) for 1st EXPT and 120 days after planting for 2nd EXPT); in the meantime randomly three plants of each plots were uprooted and different reproductive data were recorded at 60 DAP and 90 DAP and at harvest.

2.5. Data Analysis. Data recorded in 2014 and 2015 cropping seasons were mean together on account of nonsignificant interaction between year and treatment. Mean data of two trials, days to 1st and last emergence, days to 1st and 50% flowering, plant height, number of branches plant^{-1}, shoot dry weight, leaf area, pods dry weight plant^{-1}, and pod yield were analyzed using SPSS (version 20.0) and the means were separated using Tukey's test at $P \leq 0.05$. Pearson correlation was also analyzed using statistical computer software SPSS (version 20.0).

3. Results and Discussion

3.1. Days to Emergence. Boron had a significant impact on days to groundnut seed emergence. From the three levels of boron, when B was applied at 2 kg ha^{-1}, seed took shorter times for days to 1st and last emergence in both the varieties compared to that of control (Table 2). BARI Chinabadam-8 took shorter time to first emergence, but in case of last emergence Dhaka-1 took shorter times. This might be due to the application of boron because boron is the important micronutrient that helps to facilate early germination and faster growth of the hypocotyl [21]. Rerkasem [22] reported that low boron is responsible for poor seed germination and/or seedling establishment in peanut.

In both studies, the extended light was used after 30 days of seed sowing and the effect of light on days to seed emergence could not be observed.

3.2. Days to Flowering. The application of boron at 2 kg ha^{-1} facilitated 2 days early of 1st flowering and 3 days early of 50% flowering (Table 2). Result revealed that the application of B at 2 kg ha^{-1}, 1 kg ha^{-1} and control had a significant variation of days to flowering because application of B had pronounced influence on flowering [23]. Singh et al. [23] reported that B application caused 2-3 days of early flowering. There was an evidence that application of boron reduced days to 50% flowering by 4-5 days over control in peanut [24].

Though light treatment showed 1 day early of flowering (50% flowering), in spite of light treatment, we could not

TABLE 3: Effect of boron and light on plant height and number of branches plant^{-1} in two peanut varieties (mean of two trials).

Treatment	Plant height at 30 DAP	Plant height at 60 DAP	Plant height at 90 DAP	Plant height at harvest	Number of branches plant^{-1} at 30 DAP	Number of branches plant^{-1} at 60 DAP	Number of branches plant^{-1} at 90 DAP	Number of branches plant^{-1} at harvest
Boron (B)								
B_0	14.10^{a^z}	41.48^c	81.22^c	95.17^c	1.57^b	8.36^c	7.56^b	7.42^b
B_1	13.71^{ab}	42.64^b	84.30^b	99.87^b	1.79^b	9.06^b	7.86^b	7.95^{ab}
B_2	13.23^b	43.78^a	88.02^a	103.49^a	2.08^a	10.39^a	8.75^a	8.61^a
Light (L)								
L	—	46.63	90.79	109.08	—	9.91	8.39	8.54
L_0	—	38.64	78.24	89.84	—	8.63	7.72	7.44
Variety (V)								
V_1	14.48	44.74	86.86	102.90	1.40	8.78	7.54	7.06
V_2	12.88	40.53	82.17	96.12	2.22	9.76	8.57	8.93
Significance (P)								
B	0.004	<0.001	<0.001	<0.001	<0.001	<0.001	0.001	0.002
L	—	<0.001	<0.001	<0.001	—	<0.001	0.010	<0.001
V	<0.001	<0.001	<0.001	<0.001	<0.001	<0.001	<0.001	<0.001

DAP = day after planting, B = boron; L = light; V = variety; ^{z}means, columns having the same letter(s) are insignificant and different letter(s) statistically significant, P = probability. Means were separated by Tukey's test at $P \leq 0.05$, B_0 = 0 kg B ha^{-1} (control), B_1 = 1 kg B ha^{-1}, B_2 = 2 kg B ha^{-1}, L = normal day light + 6 h extended red light at night (≈18 h light), L_0 = normal day light (≈12 h light), V_1 = Dhaka-1, and V_2 = BARI Chinabadam-8.

report any findings because light treatment was imposed after 30 days of seed sowing.

3.3. Plant Height.
Plant height increased significantly with the application of boron at all the sampling dates except 30 DAP. At 30 DAP control produced highest plant height; this might be due to slow release of available form of boron from boric acid and probably environmental factor was involved to get highest plant height from control treatment. In the rest of sampling dates boron at 2 kg ha^{-1} showed the best result for both of the varieties compared to control treatment, but Dhaka-1 showed the best result over BARI Chinabadam-8 (Table 2). This might be due to the fact that boron helped in cell elongation and meristematic tissue development in plant [25]. It was reported that plant height increased with the application of boron in peanut [10].

The significant increasing trend of plant height was also obtained from light treatment for both of the varieties. Artificial light showed the highest plant height than control (Table 2). Light had a positive effect on cell development and plant growth rate significantly influenced by light in peanut [26]. Wynne and Emery [27] stated that long day photoperiod produced taller plant than the short day photoperiod.

3.4. Number of Branches Plant^{-1}.
The effect of boron on the number of branches plant^{-1} was significantly higher for both varieties. The highest number of branches plant^{-1} was obtained from 2 kg B ha^{-1} compared to 1 kg B ha^{-1} and control (Table 3). The increasing trend of number of branches plant^{-1} is due to the fact that B helped in side branching and

it also promoted the vegetative growth of peanut [11]. The similar result was reported that number of branches plant^{-1} increased with application of boron in peanut [10, 28].

Light had a positive effect on number of branches plant^{-1}. Additional light helped to increase the number of branches plant^{-1} in peanut (Table 3). This might be due to the fact that light helped in cell elongation and cell development and plays crucial role in increasing the vegetative growth in peanut [18]. Wynne and Emery [27] have also reported that vegetative growth increased in long day treatment.

3.5. Shoot Dry Weight Plant^{-1}.
With the increase of dose of boron, a significant increment in shoot dry weight plant^{-1} was observed from the present study. Maximum shoot dry weight was recorded from B at 2 kg ha^{-1} and BARI Chinabadam-8 showed the best result over the Dhaka-1 variety (Table 4). Boron had a positive effect on vegetative growth as like plant height and number of branches. And as a result shoot dry weight might be increased due to the application of boron in peanut plant [10]. Harris and Brolman [5] also reported that shoot dry weight increased with the application of boron in peanut.

With the supplementation of artificial light, shoot dry weight markedly increased and the highest shoot dry weight was recorded from light treatment (Table 4). The reason behind the result might be due to the fact that extended photoperiod helped in increasing the vegetative growth [29]. Since vegetative growth was higher in light treatment, shoot dry weight might also increase for supplementation of light. The present finding is consistent with the finding of Nigam et al. [18].

TABLE 4: Effect of boron and light on shoot dry weight and leaf area in two peanut varieties (mean of two trials).

Treatment	Shoot dry weight at 30 DAP	Shoot dry weight at 60 DAP	Shoot dry weight at 90 DAP	Shoot dry weight at harvest	Leaf area at 30 DAP	Leaf area at 60 DAP	Leaf area at 90 DAP	Leaf area at harvest
Boron (B)								
B_0	1.57^{b^z}	27.73^c	86.92^c	110.33^c	24.41^c	125.69^c	127.16^c	98.94^c
B_1	1.79^b	33.31^b	113.42^b	135.42^b	33.84^a	144.02^a	141.87^a	106.32^a
B_2	2.08^a	38.84^a	130.42^a	148.08^a	30.47^b	129.60^b	134.59^b	103.02^b
Light (L)								
L	—	36.68	118.83	136.61	—	129.16	131.10	101.69
L_0	—	29.91	101.67	125.94	—	137.04	137.98	103.83
Variety (V)								
V_1	1.40	28.26	100.61	98.83	28.53	118.44	124.50	99.00
V_2	2.22	38.33	119.89	163.72	30.62	147.77	144.59	106.52
Significance (P)								
B	<0.001	<0.001	<0.001	<0.001	<0.001	<0.001	<0.001	<0.001
L	—	<0.001	<0.001	<0.001	—	<0.001	<0.001	<0.001
V	<0.001	<0.001	<0.001	<0.001	<0.001	<0.001	<0.001	<0.001

DAP = day after planting, B = boron; L = light; V = variety; ^{z}means, columns having the same letter(s) are insignificant and different letter(s) are statistically significant, P = probability. Means were separated by Tukey's test at $P \leq 0.05$, B_0 = 0 kg B ha^{-1} (control), B_1 = 1 kg B ha^{-1}, B_2 = 2 kg B ha^{-1}, L = normal day light + 6 h extended red light at night (≈18 h light), L_0 = normal day light (≈12 h light), V_1 = Dhaka-1, and V_2 = BARI Chinabadam-8.

3.6. *Leaf Area Plant*$^{-1}$. Leaf area increased with the application of boron in case of both varieties. Leaf area significantly increased for B at 1 kg ha^{-1} and the lowest was found from control (Table 4); probably in boron deficiency soil, supplementation of boron helped in the leaf expansion of peanut [30]. Kabir et al. [10] also reported that the leaf area increased with the application of boron.

Leaf area of both varieties increased gradually with the advancement of growth stage that was up to 90 DAP and then decreased at harvest. This might be due to the fact that vegetative growth was highest up to 90 DAP and then photosynthates diverted to pod development because we found highest pod dry weight at harvest. Data (Table 4) showed that leaf area was significantly lower in light treatment compared to control. The lowest leaf area was observed from light treatment and this result is not supported by the report of Nigam et al. [18]. Imposition of light did not increase the leaf area; probably photosynthates increased leaf thickness instead of leaf area. In this study leaf thickness was not monitored.

3.7. *Pod Dry Weight Plant*$^{-1}$. Pod dry weight plant^{-1} was adversely affected in control and with the application of varying levels of boron in peanut a significant increment in pod dry weight plant^{-1} was found. Result showed that 2 kg B ha^{-1} produced the highest value of pod dry weight plant^{-1} compared to that of control and BARI Chinabadam-8 produced maximum pod dry weight (Table 5) because B helped in flowering, pod retaining and increased pod weight [31]. The present finding agreed with the findings of Quamruzzaman et al. [32] and Chitdeshwari and Poongothai [33].

Lowest pod dry weight was recorded in the light treatment for both varieties (Table 5). Extended photoperiod had limited impact on reproductive growth as per Bagnall and King [16]. Quamruzzaman et al. [32] stated similar findings.

3.8. *Pod Yield*. Significant pod yield variations were observed from the varying boron levels and maximum yield was recorded from B at 2 kg ha^{-1} whereas BARI Chinabadam-8 gave best result as compared with Dhaka-1 (Table 5). Boron had a positive effect on the reproductive development of peanut and significantly increased the pod yield [34]. Naiknaware et al. [35] reported that application of boron increased the number of pegs and pods and finally it helped to get the maximum pod yield of peanut.

Light plays a vital role in pod yield of peanut. In case of imposition of artificial light, pod yield was decreased (Table 5). This might be due to the fact that extended photoperiod limits the reproductive development of groundnut [29]. The present finding is consistent with the findings of Ansari et al. [31] and Wynne and Emery [27].

3.9. *Germination Percentage*. After harvesting seeds were stored in normal store condition and after 3 months the germination percentage was checked for both of studies.

Germination percentage showed significant variation due to different levels of boron application (Figure 1). Data revealed that 2 kg B ha^{-1} showed the highest germination percentage (90.67%) and control showed the lowest germination percentage (82.00%) for both varieties. Boron is responsible for vigorous seedling [22]. The present finding is consistent with the findings of Gupta and Solanki [36].

TABLE 5: Effect of boron and light on pod dry weight plant^{-1} and pod yield in two peanut varieties (mean of two trials).

Treatment	Pod dry weight plant^{-1} at (g) 60 DAP	Pod dry weight plant^{-1} at (g) 90 DAP	Pod dry weight plant^{-1} at (g) Harvest	Pod yield (t/ha)
Boron (B)				
B_0	1.97$^{c^z}$	19.17^c	35.00^c	1.63^c
B_1	2.36^b	27.42^b	49.58^b	1.83^b
B_2	3.58^a	33.42^a	54.09^a	2.16^a
Light (L)				
L	3.18	28.44	48.05	1.74
L_0	2.10	24.89	44.39	2.02
Variety (V)				
V_1	0.81	20.39	43.06	1.65
V_2	4.47	32.94	49.39	2.10
Significance (P)				
B	<0.001	<0.001	<0.001	<0.001
L	<0.001	<0.001	<0.001	<0.001
V	<0.001	<0.001	<0.001	<0.001

DAP = day after planting, B = boron; L = light; V = variety; ^{z}means, columns having the same letter(s) are insignificant and different letter(s) statistically significant, P = probability. Means were separated by Tukey's test at $P \leq 0.05$, B_0 = 0 kg B ha^{-1} (control), B_1 = 1 kg B ha^{-1}, B_2 = 2 kg B ha^{-1}, L = normal day light + 6 h extended red light at night ($\approx$18 h light), L_0 = normal day light ($\approx$12 h light), V_1 = Dhaka-1, and V_2 = BARI Chinabadam-8.

TABLE 6: Pearson correlation coefficient (r) among the days to emergence, days to flowering, and pod yield of peanut (mean of two trials).

		Days to emergence 1st	Last	Days to flowering 1st	50%	Number of branches at 30 DAP	60 DAP	90 DAP	HV	Pod yield (t/ha)
Days to emergence	1st	1								
	Last	0.037	1							
Days to flowering	1st	0.224	−0.127	1						
	50%	0.120	−0.242	0.644**	1					
Number of branches at	30 DAP	−0.273	0.298	−0.684**	−0.600**	1				
	60 DAP	−0.380*	−0.001	−0.564**	−0.667**	0.603**	1			
	90 DAP	−0.289	0.075	−0.671**	−0.635**	0.662**	0.652**	1		
	HV	−0.190	0.226	−0.609**	−0.627**	0.834**	0.666**	0.609**	1	
Yield (t/ha)		−0.317	0.601**	−0.570**	−0.520**	0.670**	0.280	0.423*	0.505**	1

Notes. DAP = days after planting; HV = harvesting.
$^*P < 0.05$ and $^{**}P < 0.01$ (means were separated by Tukey's test).

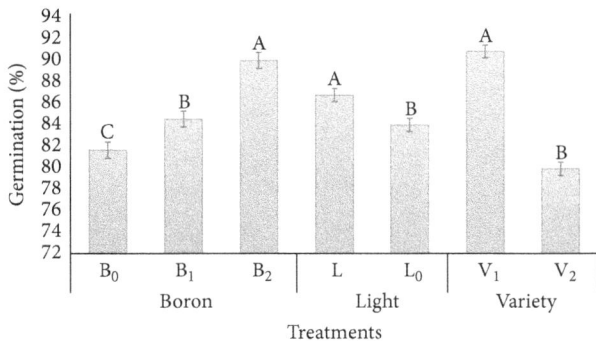

FIGURE 1: Effect of boron and light on germination percentage of two peanut varieties (mean of two trials). B = boron; L = light; V = variety; B_0 = 0 kg B ha^{-1}, B_1 = 1 kg B ha^{-1}, B_2 = 2 kg B ha^{-1}, L = normal day light + 6 h extended red light at night ($\approx$18 h light), L_0 = normal day light ($\approx$12 h light), V_1 = Dhaka-1, and V_2 = BARI Chinabadam-8. Means were separated by Tukey's test at $P \leq 0.05$. Vertical bars represent the standard error of the means. A, B, and C are statistically significant among the treatment means.

Germination percentage of peanut showed statistically significant variations with the imposition of light for both varieties. It was observed that light treatment showed highest germination percentage (87.33%) compared to control treatment (84.44%) (Figure 1). Little or no information is available regarding this finding. This might be due to the fact that crop cultivated under artificial light helped to get viable seed as well as vigorous seedling.

3.10. Coefficient of Determination. Some significant correlation among days to 1st emergence, days to 1st and 50% flowering, number of branches plant^{-1} at 30 DAP, 60 DAP, and 90 DAP and at harvest was found out. Correlation of coefficient (Table 6) and coefficient of determination showed that with decrease of days to 1st emergence, 1st flowering, and 50% flowering, the pod yield of peanut was increased. On the contrary, with the increase in number of branches plant^{-1} at all the sampling dates, the pod yield of peanut was increased (Figures 2–5).

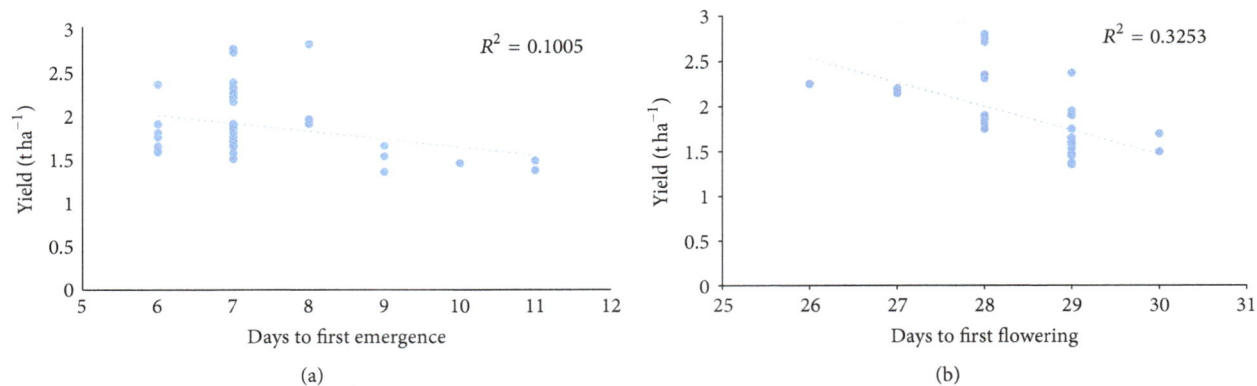

FIGURE 2: Relationship between days to first emergence and first flowering on yield of peanut (mean of two trials).

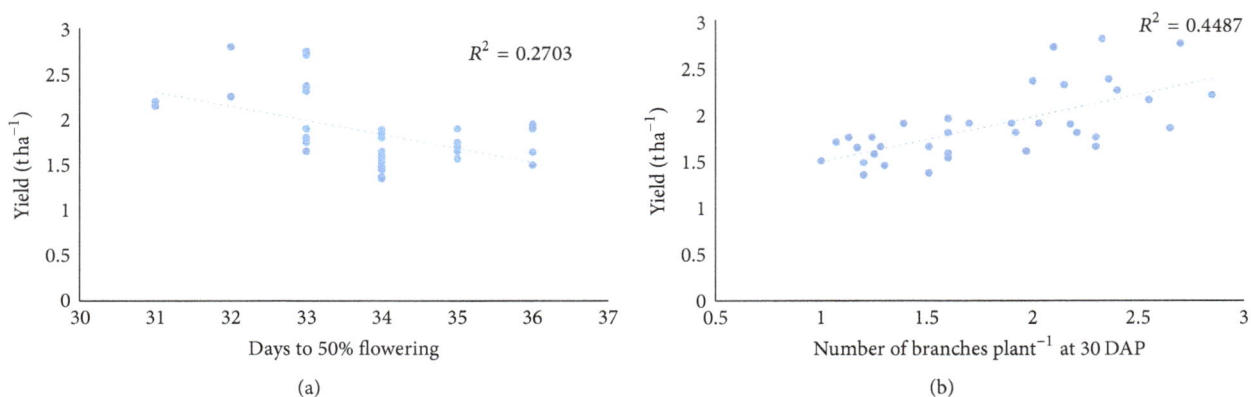

FIGURE 3: Relationship between days to 50% flowering and number of branches/plant at 30 DAP with yield of peanut (mean of two trials).

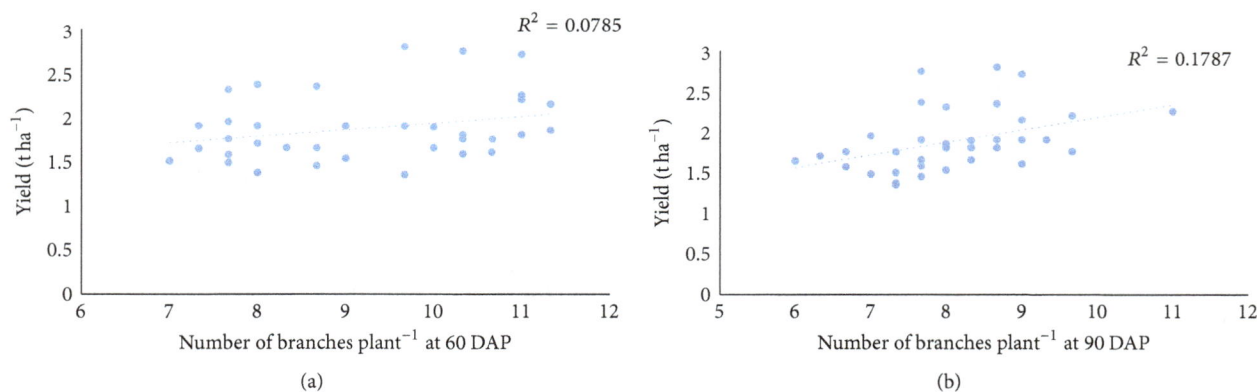

FIGURE 4: Relationship between number of branches/plant at 60 DAP and number of branches/plant at 90 DAP with yield of peanut (mean of two trials).

3.11. Correlation (r). A significant correlation was found out among the days to first-last emergence, days to first-50% flowering, and number of branches plant^{-1}. Correlation of coefficient showed that boron had a positive effect on growth and reproductive unit whereas growth and reproductive unit are positively correlated with yield of peanut (Table 6).

4. Conclusion

The present investigation indicated that the application of boron on soil has a positive effect on vegetative growth, yield performance, and germination percentage of peanut. Light treatment showed the best result for plant height, number

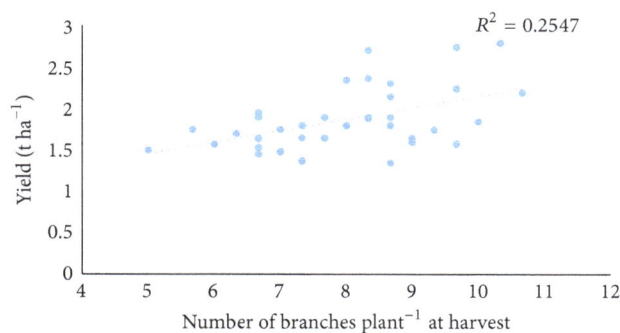

FIGURE 5: Relationship between number of branches/plant during harvest and yield of peanut (mean of two trials).

of branches plant, shoot dry weight plant, and germination percentage, but light had a negative effect on leaf area, pods dry weight plant, and pod yield. It was also observed that BARI Chinabadam-8 produced the best result for all studied parameters except plant height. Therefore, it can be concluded that the application of boron and supplementation of artificial light helped to increase the vegetative growth, yield, and germination behavior of peanut.

Competing Interests

The authors declare that there is no conflict of interests regarding the publication of this paper.

Acknowledgments

The authors are thankful to Shabiha Sultana, Deputy Director (Budget), Sher-e-Bangla Agricultural University, Dhaka, Bangladesh, for helping to arrange initial fund and support for the present experiment. The authors also thank the Ministry of Science and Technology, Government of Bangladesh, for providing the National Science and Technology (NST) fellowship for this experiment.

References

[1] F. Onemli, "Impact of climate change on oil fatty acid composition of peanut (*Arachis hypogaea* L.) in three market classes," *Chilean Journal of Agricultural Research*, vol. 72, no. 4, pp. 483–488, 2012.

[2] A. M. Wahab and A. Mohamed, "Effect of some trace elements on growth, yield and chemical constituents of *Trachyspermum ammi* L. (AJOWAN) plants under Sinai conditions," *Research Journal of Agricultural and Biology Sciences*, vol. 4, no. 6, pp. 717–724, 2008.

[3] M. Tahir, A. Tanveer, T. H. Shah, N. Fiaz, and A. Wasaya, "Yield response of wheat (Triticum aestivum L.) to boron application at different growth stages," *Pakistan Journal of Life and Social Sciences*, vol. 7, no. 1, pp. 39–42, 2009.

[4] G. J. Gascho and J. G. Davis, "Soil fertility and plant nutrition," in *Advances in Peanut Science. Stillwater*, H. T. Pattee and H. T. Stalker, Eds., pp. 383–418, American Peanut Research and Education Society, 1995.

[5] H. C. Harris and J. B. Brolman, "Comparison of calcium and boron deficiencies of peanut I. Physiological and yield differences," *Agronomy Journal*, vol. 58, no. 6, pp. 575–578, 1966.

[6] H. C. Harris and J. B. Brolman, "Comparison of calcium and boron deficiencies of the peanut II. Seed quality in relation to histology and viability," *Agronomy Journal*, vol. 58, no. 6, pp. 578–582, 1966.

[7] H. C. Harris and J. B. Brolman, "Effect of imbalance of boron nutrition on the peanut," *Agronomy Journal*, vol. 58, no. 6, pp. 97–99, 1966.

[8] M. S. Kaisher, M. A. Rahman, M. H. A. Amin, A. S. M. Amanullah, and A. S. M. Ahsanullah, "Effects of sulphur and boron on the seed yield and protein content of mungbean," *Bangladesh Research Publications Journal*, vol. 3, no. 4, pp. 1181–1186, 2010.

[9] BARC (Bangladesh Agricultural Research Council), *Fertilizer Recommendation Guide—2005*, BARC, Dhaka, Bangladesh, 2005.

[10] R. Kabir, S. Yeasmin, A. K. M. Mominul Islam, and M. A. Rahman Sarkar, "Effect of phosphorus, calcium and boron on the growth and yield of groundnut (*Arachis hypogea* L.)," *International Journal of Bio-Science and Bio-Technology*, vol. 5, no. 3, pp. 51–60, 2013.

[11] R. Singaravel, V. Parasath, and D. Elayaraja, "Effect of organics and micronutrients on the growth, yield of groundnut in coastal soil," *International Journal of Agriculture Sciences*, vol. 2, no. 2, pp. 401–402, 2006.

[12] S. Ahmed and M. B. Hossain, "The problem of boron deficiency in crop production in Bangladesh," in *Boron in Soils and Plants: Proceedings of the International Symposium on Boron in Soils and Plants held at Chiang Mai, Thailand, 7–11 September, 1997*, vol. 76 of *Developments in Plant and Soil Sciences*, pp. 1–5, Springer, Amsterdam, The Netherlands, 1997.

[13] S. Nasreen, M. Siddiky, R. Ahmed, and R. Rannu, "Yield response of summer country bean to boron and molybdenum fertilizer," *Bangladesh Journal of Agricultural Research*, vol. 40, no. 1, pp. 71–76, 2015.

[14] R. P. Hangarter, "Gravity, light and plant form," *Plant, Cell and Environment*, vol. 20, no. 6, pp. 796–800, 1997.

[15] F. R. Cox, "Effect of quantity of light on the early growth and development of the peanut," *Peanut Science*, vol. 5, no. 1, pp. 27–30, 1978.

[16] D. J. Bagnall and R. W. King, "Response of peanut (*Arachis hypogea*) to temperature, photoperiod and irradiance 2. Effect on peg and pod development," *Field Crops Research*, vol. 26, no. 3-4, pp. 279–293, 1991.

[17] H. T. Stalker and J. C. Wynne, "Photoperiodic response of peanut species," *Peanut Science*, vol. 10, no. 2, pp. 59–62, 1983.

[18] S. N. Nigam, R. C. Nageswara Rao, and J. C. Wynne, "Effects of temperature and photoperiod on vegetative and reproductive growth of groundnut (*Arachis hypogaea* L.)," *Journal of Agronomy and Crop Science*, vol. 181, no. 2, pp. 117–124, 1998.

[19] R. Mittler, "Oxidative stress, antioxidants and stress tolerance," *Trends in Plant Science*, vol. 7, no. 9, pp. 405–410, 2002.

[20] BBS (Bangladesh Bureau of Statistics), *Annual Survey Report*, Bangladesh Bureau of Statistics. Statistics Division, Government of the Peoples Republic of Bangladesh, Dhaka, Bangladesh, 2015.

[21] B. Rerkasem, R. W. Bell, and J. F. Loneragan, "Effects of seed and soil boron on early seedling growth of black and green gram (*Vigna mungo* and *V. radiata*)," in *Plant Nutrition—Physiology*

and Applications, M. L. van Beusichem, Ed., vol. 41 of *Developments in Plant and Soil Sciences*, pp. 281–285, Springer, Amsterdam, The Netherlands, 1990.

[22] B. Rerkasem, "Boron deficiency in food legumes in Northern Thailand," *Thai Journal of Soils and Fertilizers*, vol. 16, pp. 130–152, 1989.

[23] A. L. Singh, R. S. Jat, and J. B. Misra, "Boron fertilization is a must to enhance peanut production in India," in *Proceedings of the International Plant Nutrition Colloquium XVI*, pp. 1–5, University of California, Sacramento, Calif, USA, August 2009.

[24] M. A. Ansari, N. Prakash, I. M. Singh, and P. K. Sharma, "Efficacy of boron sources on groundnut production under North East Hill Regions," *Indian Research Journal of Extension Education*, vol. 14, no. 2, pp. 123–126, 2016.

[25] K. Mengel and E. A. Kirkby, *Principles of Plant Nutrition*, International Potash Institute, Bern, Switzerland, 3rd edition, 1982.

[26] S. N. Nigam, R. C. Rao, J. C. Wynne, J. H. Williams, M. Fitzner, and G. V. Nagabhushanam, "Effect and interaction of temperature and photoperiod on growth and partitioning in three groundnut (*Arachis hypogaea* L.) genotypes," *Annals of Applied Biology*, vol. 125, no. 3, pp. 541–552, 1994.

[27] J. C. Wynne and D. A. Emery, "Response of intersubspecific peanut hybrids to photoperiod," *Crop Science*, vol. 14, no. 6, pp. 878–880, 1974.

[28] X. Y. Luo, Y. C. Peng, and B. Y. Wang, "Effect of boron fertilizer on yield and quality of groundnut," *Journal of Zhejiang Agricultural Sciences*, vol. 1, pp. 30–32, 1990.

[29] D. L. Ketring, "Light effects on development of an indeterminate plant," *Plant Physiology*, vol. 64, no. 4, pp. 665–667, 1979.

[30] B. Dell and L. Huang, "Physiological response of plants to low boron," *Plant and Soil*, vol. 193, no. 1-2, pp. 103–120, 1997.

[31] M. A. Ansari, N. Prakash, I. M. Singh, P. K. Sharma, and P. Punitha, "Efficacy of boron sources on productivity, profitability and energy use efficiency of groundnut (*Arachis hypogaea*) under North East Hill Regions," *Indian Journal of Agricultural Sciences*, vol. 83, no. 9, pp. 959–963, 2013.

[32] M. Quamruzzaman, M. J. Ullah, M. J. Rahman, R. Chakraborty, M. M. Rahman, and M. G. Rasul, "Organoleptic assessment of groundnut (*Arachis hypogaea* L.) as influenced by boron and artificial lightening at night," *World Journal of Agricultural Science*, vol. 12, no. 1, pp. 1–6, 2016.

[33] T. Chitdeshwari and S. Poongothai, "Yield of groundnut and its nutrient uptake as influenced by Zinc, Boron and Sulphur," *Agricultural Science Digest*, vol. 23, no. 4, pp. 263–266, 2003.

[34] D. Jena, S. C. Nayak, B. Mohanty, B. Jena, and S. K. Mukhi, "Effect of boron and boron enriched organic manure on yield and quality of groundnut in boron deficient Alfisol," *Environment and Ecology*, vol. 27, no. 2, pp. 685–688, 2009.

[35] M. D. Naiknaware, G. R. Pawar, and S. B. Murumkar, "Effect of varying levels of boron and sulphur on growth, yield and quality of summer groundnut (*Arachis hypogea* L.)," *International Journal of Tropical Agriculture*, vol. 33, no. 2, part 1, pp. 471–474, 2015.

[36] U. Gupta and H. Solanki, *Boron: Impact on Seed Germination and Growth of Solanum melongena L.: Plant Nutrient Relation*, LAP LAMBERT Academic Publishing, 2012.

11

Commodity Systems Assessment Methodology of Postharvest Losses in Vegetable Amaranths: The Case of Tamale, Ghana

Mildred Osei-Kwarteng,[1] Joseph Patrick Gweyi-Onyango,[2] and Gustav Komla Mahunu[1]

[1]*Department of Horticulture, Faculty of Agriculture, University for Development Studies, P.O. Box TL 1882, Tamale, Ghana*
[2]*Department of Agricultural Science and Technology, Kenyatta University, P.O. Box 4384400100, Nairobi, Kenya*

Correspondence should be addressed to Mildred Osei-Kwarteng; misokwart@yahoo.com

Academic Editor: Allen Barker

A semistructured questionnaire based on the commodity system assessment methodology (CSAM) was used to determine postharvest losses in vegetable amaranths (VA). Fifty producers and retailers were randomly selected from five and four major VA producing areas and markets, respectively, and interviewed. Data obtained were subjected to descriptive statistical analyses. The survey revealed that absence of laws, regulation, incentives, and inadequate technical information affected the production of VA. The utmost preproduction challenge was poor quality seeds with poor seed yield (35%), low viability (19%), and nontrueness (46%). It was noted that some cultural practices including planting pattern and density, irrigation, and fertiliser use had effects on postharvest losses. Some postharvest practices used were cleaning with water, trimming, sorting, and grading. Usually the produce was transported to marketing centers by cars and motor cycle trailers. Generally poor temperature management after harvest was a big challenge for the postharvest handling of VA. The potential of vegetable amaranths as a commodity in the study area can be enhanced by providing the necessary institutional support, incentives, and use of good management practices along the value chain. An interdisciplinary approach and quantification of losses along the chain are recommended for any future study.

1. Introduction

In Ghana, it is estimated that about 20–50% of vegetables produced are lost due to preharvest and postharvest factors, including cultural practices (e.g., fertilisation, water supply, and harvesting method) and poor postharvest handling [1, 2]. In developing countries (e.g., Benin), postharvest losses in vegetable amaranths can be as a high as 79–89.5% due to damage and decay [3, 4] resulting from poor handling and storage facilities [5].

Commodity systems assessment methodology (CSAM) is a postharvest loss assessment procedure that was originally developed and modified over years [6, 7]. This method evolved as a result of the perceived need for a systematic approach to identify, prioritize, and resolve postharvest problems from planning to product distribution to ensure that all factors affecting a given commodity are considered along the value chain. The CSAM covers two major aspects of the value chain, preharvest (preproduction and production) and postharvest (harvesting and marketing) [6, 7].

An important feature of this methodology is that, with 26 components (Figure 1), it permits an analysis of a whole commodity system, thus facilitating the identification and prioritization of problems throughout the system. The CSAM was employed to assess postharvest losses of about 16 crops including leafy vegetables in South Asia (India) and sub-Saharan Africa (Benin, Ghana, and Rwanda) and for other crops worldwide as a first step towards the reduction of food losses and improve the efficiency and productivity of the value chain of these commodities [7, 8].

Amaranths (*Amaranthus* spp. L.) are widely grown in the tropics as one of the most important leafy vegetables in the lowlands of Africa and Asia but have a short postharvest shelf life of 1-2 days [9]. Grubben and Denton [10] noted the economic value of amaranths as the main African leafy vegetable ranking high in terms of quantity and area under cultivation. They are recognized as an easy-to-grow, extremely productive, and nutritious vegetable and probably the highest yielding leafy vegetable in the tropics. The protein, vitamin A, vitamin C, calcium, iron, and zinc content in the leaves

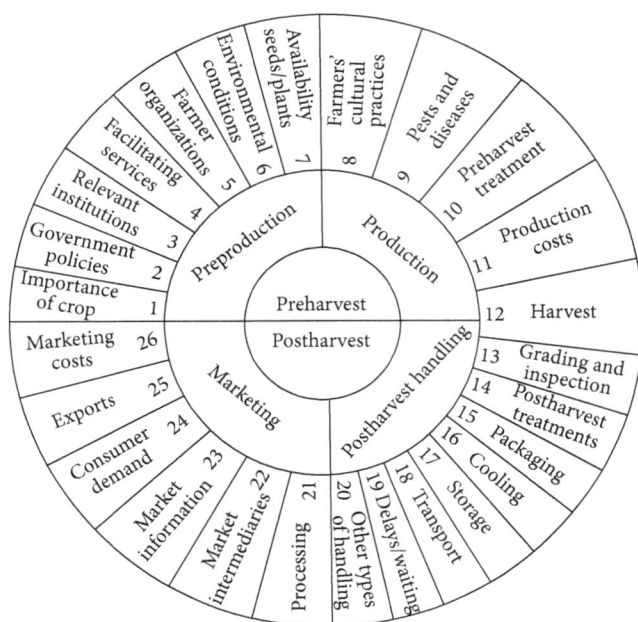

FIGURE 1: Components of the CSAM [6, 7].

per 100 g edible portion are 3.2–4.6 g, 1.7–5.7 mg, 36–64 mg, 270–582 mg, 2.4–8.9 mg, and 0.7–1.5 mg, respectively [11]. Their excellent nutritional values make them important for human nutrition, both in rural areas for home consumption and as cheap green vegetable in cities [10]. It was also noted by Osei-Kwarteng et al. [12] as the indigenous leafy vegetable (ILV) with the highest market share in Tamale metropolitan area. The above qualities enhance their potential for nutrition sensitive agriculture and therefore the need for systematic analysis of the postharvest losses of vegetable amaranths. Moreover, ILVs that are adapted to the agroecological zones of Northern Ghana are to be promoted towards food security, eradication of malnutrition, and poverty reduction. A simplified survey was conducted to assess the contribution of the preproduction, production, postharvest, and marketing activities on the postharvest losses of vegetable amaranths. The output from the CSAM will synthesize information which can be used to improve the value chain of vegetable amaranths and food security in the study area.

2. Material and Methods

The survey was conducted in Tamale (Figure 2), the northern regional capital of Ghana, which lies between latitude 9°15′ and 9°05′N and longitude 0°45′ and 1°0′W and at an altitude of 185 m above sea level and its suburban areas. The investigation was conducted between May and June 2012. Two separate semistructured questionnaires based on the components of the commodity systems assessment methodology [CSAM, [6]] were used in interviewing 50 producers and 50 retailers of vegetable amaranths. Simple random sampling technique [13] was used in selecting the respondents for the survey. This study concentrated on preharvest (preproduction and production) and postharvest (postharvest handling and

marketing) components of the CSAM for the assessment of losses in vegetable amaranth production. The selected respondents were from five major vegetable producing areas (Datoyili, Bulpiela, Gumbihini, Manguli, and Sangani) and four major markets (Tamale Central, Lamashegu, Kukuo, and Aboabo markets; Figure 2).

Some of the issues considered in the survey instrument were (i) preproduction: relative importance of the crop, governmental policies, relevant institutions, facilitating services, producer/shipper organizations, environmental conditions, and availability of planting materials; (ii) production: farmers' general cultural practices, pest and diseases, preharvest treatments, and production cost; and (iii) postharvest: harvest, grading, sorting and inspection, postharvest treatments, packaging, cooling, storage, transport, delay or waiting, agroprocessing, market intermediaries, market information, consumer demand, and exports. However the study did not quantify the losses within the product value chain.

Data obtained from the survey were subjected to descriptive statistical analyses using the Statistical Package for Social Sciences (version 17.0) software.

3. Results and Discussion

3.1. Preproduction. Vegetable amaranth is the most cultivated indigenous vegetable in Tamale metropolitan area as reported by Osei-Kwarteng et al. [12]. The reasons why producers cultivated vegetable amaranths included high demand (33%), high profit (32%), low initial capital (25%), and fast growth (20%). The importance of the crop is further deepened by its relative contribution to the local diet and livelihoods of the people. These are corroborated by the report of Ngugi et al. [14] which stated that, in urban centers in Kenya, there is high demand for African indigenous vegetables (AIVs) due to their nutritional and medicinal values.

The survey noted the nonexistence of laws, regulation, or incentives supporting the production of vegetable amaranths due to the fact that (1) they are relatively less important in the crop production system in Ghana and consequently ignored by development and government agencies [15], (2) there is less research focus [15], and (3) information on the quantities cultivated and their urban demand structure is inadequate [16, 17] since they are minor crops that contribute little to the livelihoods and nutrition of people. Shackleton et al. [18] reported that policies on ALVs are absent or weak in sub-Saharan Africa (SSA) countries because they are a subset of agricultural crops and are considered under general agricultural policies. They can only have specific policy considerations when their advantages are evident by numerous research [18]. Hence, it is not surprising to find an absence of policy on vegetable amaranths in Ghana in spite of their high market share among the indigenous leafy vegetables in the Tamale metropolitan area [12]. Moreover, relevant government and nongovernment agencies work on major mandated crops or principal fruits and vegetables such as tomatoes, pepper, onions, okra, and eggplant. A review on the food and agriculture policies in Ghana a decade ago confirmed lack of comprehensive national policy document on underutilized plant species including amaranths [19].

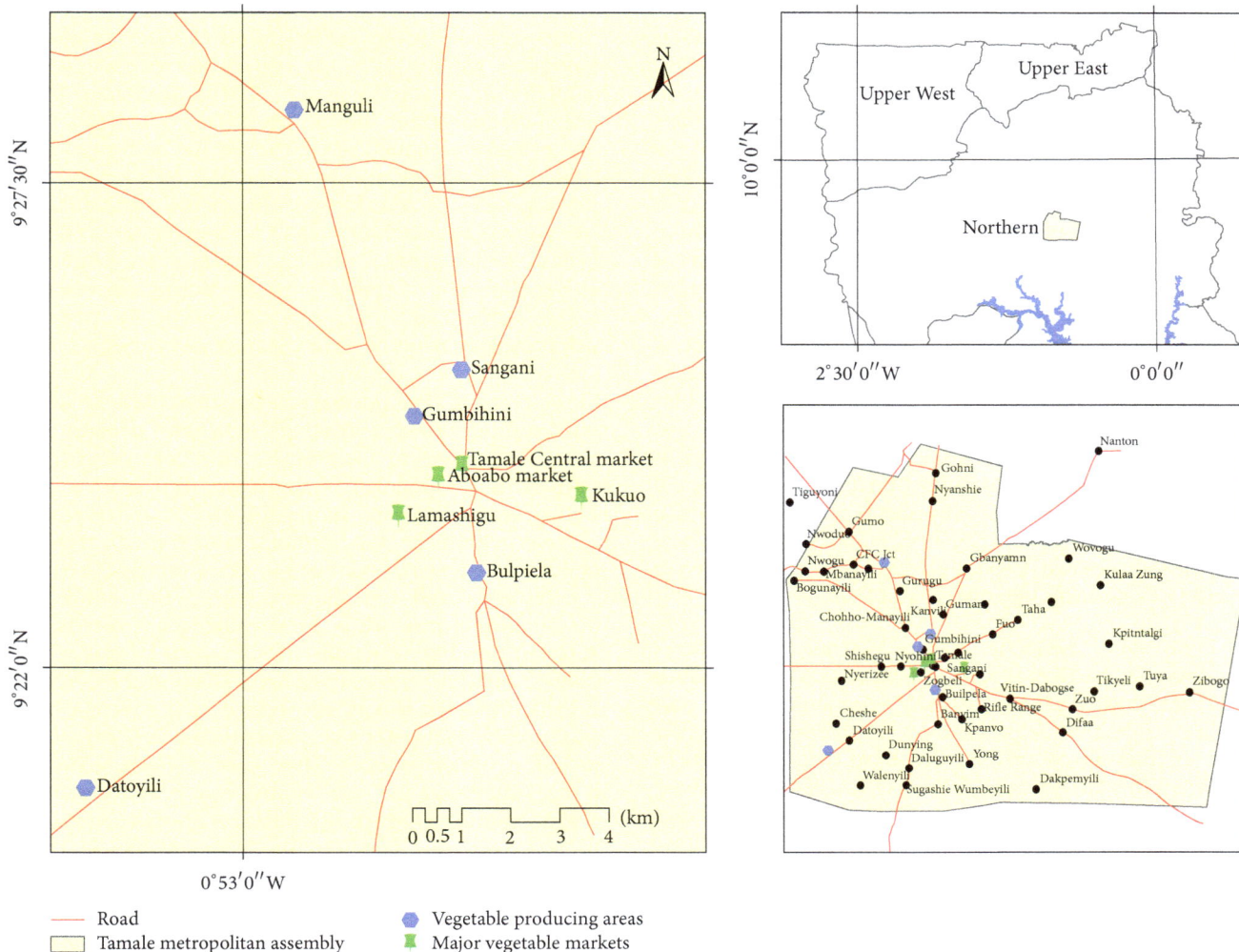

FIGURE 2: Map of Tamale metropolitan area showing major vegetable producing areas and markets.

However, the study showed that the support from nongovernmental organizations (NGOs) is noteworthy as producers (24%) cited the German Technical Service (GIZ) for technical support whiles less of that came from the Ministry of Food and Agriculture (4%). In the Upper East Region (Bongo district), the World Vision International provided extension support for indigenous leafy vegetables (ILVs) production while Adventist Development and Relief Agency (ADRA), Catholic Diocese and Action Aid Ghana, supported infrastructure development (dams) for the production of ILVs including vegetable amaranths [15]. TRAX Ghana also trained ILVs farmers in the Bongo district on soil and water conservation technologies [15]. The support given by the NGOs is based on the importance (commercial potential) of ILVs including amaranths to the livelihoods of the people in the communities.

Results from the investigation also indicated that producers obtained seeds from either the local market (35%) or their own farms (65%). However, it was noted that the quality of the seeds was not reliable due to poor seed yield (35%), low viability (19%), and being not true to type (46%).

Similarly, in Upper East Region 10–80% of the farmers obtained seeds for ILVs including amaranths from the market [15] and our findings on amaranth seeds corroborate their observation. Hence farmers suggested that the availability of quality seeds will moderate this challenge. The source of seeds for amaranth production affirms the existence of two parallel seed systems in Ghana (traditional/informal and formal) [20]. Absence of improved and standard seeds for ILVs including amaranths was also observed in the South African Development Community region (Tanzania, Zambia, and Botswana) [17] and Kenya [21]. Generally, the nonexistence of improved seeds for the cultivation of African ILVs is a major challenge to the broad production of ILVs in Africa including Ghana, as farmers use their own seed from unselected planting materials mostly without any quality and standard criteria and technical supervision [20, 21]. Additionally, farmers own and purchased seeds which may be poorly handled at storage and consequently affect the vigour and germination percentage [20]. These seed constraints indicate the need for research and improvement of seed sources or delivery system [19] for vegetable amaranths.

The environmental and other constraints that were recognized by the producers as affecting the quality of their produce included insect damage (48%), water shortage (6%), low soil fertility (18%), and low yields (28%). Producers could not identify the type of insect pest that damaged their produce but reported holes in the leaves which reduced the quality of leaves. ILVs are reported to be tolerant to local pest and disease [22] and in Southeast Asia [23] amaranths are also noted to have no serious insect pest, disease, and nematodes [24] attack. However, there is an expected pest build-up in hot and humid climates when an intensive monoculture is regularly practiced [18]. Moreover, in some West African countries insect pests and disease have been found on amaranths [23]. Hence in regular production, producers must practice safe and effective integrated pest management methods [22] before or after production to reduce postharvest loses. An avoidable loss of 20% in vegetable amaranths was estimated in Kenya as caused by insect pests [22]. Inadequate water supply during production induces flowering and reduces the leave yield in vegetable amaranth [25]. Vegetable amaranths do best on fertile well-drained deep soils although they are adapted to a variety of soils including marginal soils [25].

3.2. Production. The study revealed that all the producers were conversant with production practices that affect produce quality. They listed planting density and pattern, weed, pest and disease control, irrigation, and fertiliser use as familiar practices. They also observed insect pests that made holes in the leaves. This knowledge is based on the experience gained over the years. Conversely, they could not identify viral and fungal infections. Previous reports on the postharvest losses of 16 commodities including leafy vegetables showed that farmers used some production practices that increased the challenges of postharvest quality and losses of these crops [26]. Vegetable amaranths are usually grown commercially as a sole crop and intercropped with food crops in home gardens [27]. Most farmers plant at a density of 180 plants/m^2 for single harvest by uprooting the whole plant but a plant density of 100 plts/m^2 is strategic for good quality produce. If the desire is several harvests during the season then vegetable amaranths should be planted at 20 plts/m^2 [27]. Generally good cultural practices specific for vegetable amaranths are recommended for good quality produce at harvest for immediate consumption or further handling.

Ninety-two percent (92%) of the producers harvested themselves in the morning by cutting with sharp knives or by uprooting. Harvesting with sharp knives avoids severe abrasions or bruises that occur with blunt knives which opens up entry points for microorganisms and hastens the rate of produce decay [7] while physical damage causes 3-4 times more moisture loss than undamaged produce [8]. Other types of equipment used during harvesting include baskets, basins, sacks, and fiber and fertiliser bags. Majority of farmers (54%) used the number of nodes and branches as an indicator of maturity. Additional determinants of maturity were plant height (27%) and canopy spread (19%). Vegetable amaranths can be harvested between 30 and 55 days after sowing at a height of 60 cm when leaves are young and slender [25]. Harvesting at the right physiological age prevents the rapid loss of quality of the produce, since at the appropriate harvest maturity, leaves are less prone to moisture loss and subsequent wilting [8]. Leafy vegetables harvested at early hours (0400 hrs) and late hours (2000 hrs) contained higher water potential which reduces the rate of water loss. Moreover, produce harvested late in the day has high sugar levels due to the photosynthesis in the day [8]. At late harvesting, leafy vegetables become fibrous and lose their tenderness [7]. Poor packaging materials such as baskets, basins, and sacks easily make the produce prone to damage from bruises and further allow entry of microorganism during transport and storage. Additionally, leafy vegetables are easily stressed in such packaging material during storage and transport [7].

3.3. Postharvest. The postharvest practices used by farmers were cleaning (66%), trimming (34%), and cooling either under shady trees in the ambient air (25%) or with cool water (75%) at ambient temperature. The problem with cooling under ambient air conditions is that temperature increases within the piled produce with time. In this case, the respiration rate of the piled produce increases resulting in water and weight loss and thus reduces the market value [8]. Cooling of produce is done immediately after harvest on farms. Two basins of cool water are used for washing the produce after harvesting; roots are washed in the first basin and the harvested shoots are washed in the second basin. Subsequently cool water is sprinkled on the day's harvest and stacked in containers for transportation. Washing all produce in the same basin may contaminate the produce with soil or plant pathogens which will facilitate postharvest diseases [8]; hence for good sanitation practices, produce should be washed in clean water. At the market, produce is also cooled again by sprinkling water on display. Unsold produce is also sprinkled with water to maintain the freshness for the next day. In spite of the above cooling practices, stacking or piling of produce in containers (woven baskets and pans) during transport or storage renders the whole practice ineffective. Temperature management after harvesting is to avoid a high ambient temperature which reduces the quality of produce and the market value after harvest [8]. The high mean day time temperatures (28–43°C [28]) of the Tamale metropolitan area demand a good cooling system for the harvested produce.

Retailers stored produce in their stalls, while producers stored produce either at shady places (46%) or in their kitchen or hut (54%) at ambient temperature. In these storage situations the produce experiences fluctuating temperature which affects the stability of the quality and acceptability of the produce by the consumers. Freshly harvested vegetable amaranths can be maintained at a temperature of 0–2°C and relative humidity of 95–100% for 14 days [29]. Vegetable amaranths packaged under modified atmospheric conditions (active bags) and stored at 5°C at 75% relative humidity could be stored for 23 days with about 55% retention of ascorbic acid compared to the control sample stored at room temperature (25°C) [5].

Middlemen transports produce by using cars (57%) and motorcycle trailers (motor king in Ghana; 43%). Producers transports produce by either head porting (34%) or motorcycle trailers (65%). During transport fresh produce is stressed by packing them tightly so as to have more room for more produce. Nevertheless the atmospheric conditions (e.g., high temperature and low relative humidity) during transport also increases the respiration rates which result in weight loss of the produce as tissues lose turgor and succulence [30].

Delay in transporting produce to marketing centers can be at least 2 hours (43%) and a maximum of 6 hours (36%). Fifty-eight percent (58%) of middlemen registered delay in transportation. Delays were attributed to vehicular breakdown (30%), heavy rains (16%), and sheer delay by drivers, retailers, and producers (54%). Because harvested produce are transported under high ambient temperatures and also piled, such delays are to be avoided.

Retailers were very particular about grading and sorting in the market for better pricing. This indicates that the consumers are not only particular about price but also some standard for quality produce [7]. Thirty-two percent of retailers stated the inadequacy of labour for grading and sorting when they have large volumes of produce, while 68% did not see that as a problem because they are sorted and graded by themselves. Few (6%) of handlers (6%) are well trained in sorting and grading but the retailers first demonstrate how their hired labour should do the activity.

There was no processing method for harvested produce other than the drying and grinding of dried leaves in the study area. Drying is not mostly done because consumers have preference for fresh produce for their soups and sauces. All actors emphasized the urgent need for simple innovative agroprocessing methods during the glut period. Traditional drying in the sun can be improved by using affordable solar drying methods which avoid possible incomplete or uneven drying that allows microbial growth and prevents loss of nutritional quality (vitamins and minerals) [30, 31]. Processing freshly harvested vegetable amaranths can reduce postharvest losses of about 20–50% within the value chain [31]. Processing methods such as solar drying and freeze drying are recommended for the upgrading of vegetable amaranths into supermarkets in Tamale.

Twenty percent of retailers indicated that they are able to handle produce for 2 days while the majority (80%) can do so in a day because of the perishable nature of the amaranth leaves. Vegetable amaranths displayed in the market have a short postharvest shelf life due to the hot environmental conditions and no cooling facilities.

Packaging materials for transport and storage of produce are basins (30%), baskets (28%), polyethylene bags (26%), and sacks (16%) which are not highly conducive for the purpose. Therefore the need for an appropriate packaging material will protect the produce from damage, preserve the quality, and enable good ventilation to avoid the accumulation of heat and/or unwanted gases in the produce [29].

3.4. Marketing. Seventy-six percent of retailers indicated that producers set prices at the beginning of the production period to ensure that price per unit weight is similar throughout the production area. This implies that producers are well organized and take decisions that make them benefit from their produce. Producers do not have any formal means of getting market information other than visiting the markets themselves. However, they are mostly informed on trends and consumer demand in the market by the retailers and middlemen who mostly come to their farms to harvest purchased plots or harvestable produce. Retailers also decide on market prices from the prevailing farm gate price.

Retailers (64%) indicated that consumers preferred the produce bundled while 36% desired loosely packed portions. Out of the retailers who indicated consumer preference for bundled produce, 52% observed that consumers normally like smaller bundle sizes compared to 48% who like bigger bundles. Fresh dark green leaves were much preferred. The smaller bundle sized produce is mostly for home or household use while the bigger bundle sized produce is for commercial cooking purposes. Bigger bundle sized produce is more prone to postharvest losses as there is less air circulation within the produce because it is mostly piled in basins or baskets during sales.

Almost all retailers (98%) expressed unmet demand and oversupply of produce in some occasions. Unmet demands occurred during the dry season when produce is scarce due to water shortage (33%) and pest and diseases (35%). Thirty-two percent of retailers had excess supply of the produce in the wet season resulting in cheaper farm gate prices but maintained their prices and rather increased the size of the bundles for the same price. This suggests the need to process into other forms to avoid losses in the wet season. Consumers rather looked out for high quality produce (undamaged and dark green) during the wet seasonal glut. Producers confirmed that amaranth leaves are not exported; as such there is no existing regulation on its trade. Producers, marketers, and consumers of vegetable amaranths wished that there could be simple postharvest treatments. Nevertheless, they preferred nonchemical and inexpensive treatments.

4. Conclusion

The study assessed the contribution of current practices on postharvest losses of vegetables amaranths using a commodity system assessment methodology in the Tamale metropolis, Ghana. Our investigation revealed absence of laws, regulation, and incentives for vegetable amaranth production. There is also deficiency in quality seed for production since seeds are mostly obtained from the producer's field and open markets and not subjected to any standard quality criteria. Producers had a fair understanding of the influence of some cultural practices (e.g., planting pattern and density, weed, pest and disease control, irrigation, and fertiliser use) on postharvest losses. Further, produce is transported at ambient temperature to market centers by cars or motorcycle trailers. Retailers are particular in grading and sorting of the produce to enhance pricing.

There are observed challenges at all stages of the product chain and the potential of vegetable amaranths as a commodity in the study area can be enhanced by providing the necessary institutional support and incentives. This work provides first-hand information for development planners and government and nongovernment institutions that are seeking to improve a popular indigenous leafy vegetable in urban communities such as the Tamale metropolitan area. Finally, the quantification of losses at each level of the chain is recommended.

Conflicts of Interest

The authors declare no conflicts of interest for this publication and that data used are from their own investigation.

Acknowledgments

The authors are grateful to Dr. Lisa Kitinoja and her Postharvest Education Foundation team for the revision of the first draft of the manuscript.

References

[1] Y. S. Gonzalez, Y. Dijkxhoorn, I. Koomen et al., "Vegetable business opportunities in Ghana," The GhanaVeg Sector Reports, 2016, http://ghanaveg.org/.

[2] I. K. Arah, H. Amaglo, E. K. Kumah, and H. Ofori, "Preharvest and postharvest factors affecting the quality and shelf life of harvested tomatoes: a mini review," International Journal of Agronomy, vol. 2015, Article ID 478041, 6 pages, 2015.

[3] H. Affognon, C. Mutungi, P. Sanginga, and C. Borgemeister, "Unpacking postharvest losses in sub-Saharan Africa: a meta-analysis," World Development, vol. 66, pp. 49–68, 2015.

[4] L. Kitinoja, "Identification of appropriate postharvest technologies for improving market access and incomes for small horticultural farmers in sub-Saharan Africa and south Asia," WFLO Grant Final Report, 2010.

[5] J. A. Nyaura, D. N. Sila, and W. O. Owino, "Postharvest stability of vegetable amaranths (Amaranthus dubius) combined low temperature and modified atmospheric packaging," Food Science and Quality Management, vol. 30, pp. 66–72, 2014.

[6] J. La Gra, A Commodity Systems Assessment Methodology for Problem and Project Identification, Postharvest Institute for Perishables, University of Idaho, IICA, AFHB, Moscow, Idaho, USA, 1990.

[7] J. La Gra, L. Kitinoja, and K. Alpizar, Commodity Systems Assessment Methodology for Value Chain Problem and Project Identification: A First Step in Food Loss Reduction, Inter-American Institute for Cooperation on Agriculture, San Isidro, Costa Rica, 2016.

[8] A. L. Acedo Jr., Postharvest Technology for Vegetables, AVDRC—ADB Postharvest Projects RETA 6208/6376, AVDRC Publication no. 10-733, AVDRC—The World Vegetable Center, Tainan, Taiwan, 2010.

[9] J. Norman, Tropical Vegetable Crops, Arthur H. Stockwell Ltd, Devon, UK, 1992.

[10] G. J. H. Grubben and O. A. Denton, Eds., Plant Resources of Tropical Africa 2: Vegetables, PROTA Foundation, Wageningen, The Netherlands; Backhuys Publishers, Leiden, The Netherlands; CTA, Wageningen, The Netherlands, 2004.

[11] E. G. Achigan-Dako, O. E. D. Sogbohossou, and P. Maundu, "Current knowledge on Amaranthus spp.: research avenues for improved nutritional value and yield in leafy amaranths in sub-Saharan Africa," Euphytica, vol. 197, no. 3, pp. 303–317, 2014.

[12] M. Osei-Kwarteng, G. K. Mahunu, and E. Yeboah, "Baseline survey on the seasonal market share of exotic and indigenous

[13] J. W. Creswell, Educational Research: Planning, Conducting, and Evaluating Quantitative and Qualitative Research (4th edn.), Pearson Education, Inc., Boston, Mass, USA, 2012.

[14] I. K. Ngugi, R. Gitau, and J. Nyoro, Access to High Value Markets by Smallholder Farmers of African Indigenous Vegetables in Kenya, Regoverning Markets Innovative Practice Series, IIED, London, UK, 2007.

[15] ICRA/CBUD, Indigenous Leafy Vegetables in the Upper East Region of Ghana: Opportunities and Constraints for Conservation and Commercialisation, Working Document Series 102, ICRA/CBUD, Kumasi, Ghana, 2002.

[16] F. I. Smith and P. Eyzaguirre, "African leafy vegetables: their role in the World Health Organization's global fruit and vegetables initiative," African Journal of Food, Agriculture, Nutrition and Development, vol. 7, no. 3, pp. 1–17, 2007.

[17] E. Lyatuu and L. Lebotse, Eds., Marketing of Indigenous Leafy Vegetables and How Small-Scale Farmers Can Improve Their Incomes, Agricultural Research Council, Dar es Salaam, Tanzania, 2010.

[18] C. M. Shackleton, M. W. Pasquini, and A. W. Drescher, "African indigenous vegetables in urban agriculture: recurring themes and policy," in African Indigenous Vegetables in Urban Agriculture, C. M. Shackleton, M. W. Pasquini, and A. W. Drescher, Eds., pp. 271–283, Earthscan, Routledge, UK, 2009.

[19] Council for Scientific and Industrial Research Ghana/Global Facilitation Unit for Underutilized Species, "Underutilized plant species strategies and policies. Analysis of existing national policies and legislation that enable or inhibit the wider use of underutilized plant species for food and agriculture in Ghana," Report 2007, http://www.underutilizedspecies.org/record_details_id_836.html.

[20] P. M. Etwire, I. D. K. Atokple, S. S. J. Buah, A. L. Abdulai, A. S. Karikari, and P. Asungre, "Analysis of the seed system in Ghana," International Journal of Advance Agricultural Research, vol. 1, pp. 17–13, 2013.

[21] M. O. Abukutsa-Onyango, "Seed production and support systems for African leafy vegetables in three communities in western Kenya," African Journal of Food Agriculture Nutrition and Development, vol. 7, no. 3, pp. 1–16, 2007.

[22] M. O. Oluoch, G. N. Pichop, D. Silué, M. O. Abukutsa-Onyango, M. Diouf, and C. M. Shackleton, "Production and harvesting systems for African indigenous leafy vegetables," in African Indigenous Vegetables in Urban Agriculture, C. M. Shackleton, M. W. Pasquini, and A. W. Drescher, Eds., pp. 145–170, Earthscan, London, UK, 2009.

[23] G. J. H. Grubben, The Cultivation of Amaranths as a Tropical Leaf Vegetable with Special Reference to South-Dahomey, Communication 67 of the Department of Agricultural Research Koninklijk Instituut voor de Tropen, Amsterdam, Netherlands, Royal Tropical Instistute, 1976.

[24] J. E. Knott and J. Deanon, Vegetable production in South-East Asia, University of the Philippines Press, Metro Manila, Philippines, 1967.

[25] Production Guidelines, "Amaranthus," Department of Agriculture, Forestry and Fisheries. 2010, http://www.nda.agric.za/docs/Brochures/Amaranthus.pdf.

[26] WFLO, "Identification of appropriate postharvest technologies for improving market access and incomes for small horticultural farmers in Sub-Saharan Africa and South Asia,"

World Food Logistics Organisation (WFLO), Appropriate Postharvest Technology Planning Project. 2010, http://www.ucce.ucdavis.edu/files/datastore/234-1848.pdf.

[27] G. J. H. Grubben, "Amaranthus cruentus L.," in *Record from PROTA4U. PROTA (Plant Resources of Tropical Africa)*, G. J. H. Grubben and O. A. Denton, Eds., PROTA, Wageningen, The Netherlands, 2004, http://www.prota4u.org/search.asp.

[28] Klimatafel von Tamales/Ghana, "Baseline climate means (1961–1990) from stations all over the world (in German)," Deutscher Wetterdienst.

[29] L. Kitinoja and A. A. Kader, *Small-Scale Postharvest Handling Practices: A Manual for Horticultural Crops*, Postharvest Horticulture Series No. 8E, University of California Davis, Postharvest Technology and Information Center, 4th edition, 2003.

[30] K. M. Maquire, H. T. Sabarez, and D. J. Tanner, "Postharvest preservation and storage," in *Handbook of Vegetable Preservation and Processing*, Y. H. Hui, S. Ghazala, D. M. Graham, K. D. Murrell, and W.-K. Nip, Eds., Marcel Dekker Inc, New York, NY, USA, 2004.

[31] S. Gupta, B. S. Gowri, A. J. Lakshmi, and J. Prakash, "Retention of nutrients in green leafy vegetables on dehydration," *Journal of Food Science and Technology*, vol. 50, no. 5, pp. 918–925, 2013.

12

Growth Performance and Nutrient Uptake of Oil Palm Seedling in Prenursery Stage as Influenced by Oil Palm Waste Compost in Growing Media

A. B. Rosenani, R. Rovica, P. M. Cheah, and C. T. Lim

Department of Land Management, Faculty of Agriculture, Universiti Putra Malaysia (UPM), 43400 Serdang, Selangor, Malaysia

Correspondence should be addressed to A. B. Rosenani; rosenani.abubakar@gmail.com

Academic Editor: Iskender Tiryaki

The use of composted oil palm wastes in the oil palm nursery as an organic component of growing medium for oil palm seedlings seems promising in sustainable oil palm seedling production. This study was conducted to investigate the effects of six oil palm waste compost rates (0, 20, 40, 60, 80, and 100%) on the growth performance of oil palm seedling and nutrient uptake in the prenursery stage (0–3 months). The addition of oil palm compost reduced the soil bulk density (1.32 to 0.53 g cm^{-3}) and increased soil pH (4.7 to 5.1) of growth media. Oil palm waste compost treatment produced positive growth performance up to 70%. A regression analysis indicated in 72% of compost and topsoil mixture as a polybag growth medium was optimum in producing best growth performance of oil palm seedling in the prenursery stage. Foliar analysis implied highest nutrients uptake (N, P, K, Mg, Ca, Fe, Zn, and Cu) for seedlings grown in 60 to 100% compost media.

1. Introduction

Oil palm (*Elaeis guineensis*) was first introduced to Malaysia in 1870 as an ornamental plant. However, the cultivation of oil palm dates back to 1917 and only in the last 50 years was the rubber industry replaced by oil palm cultivation as the main commercial crop in the agriculture sector [1]. The total planted area had then rapidly increased from 1.5 million hectares in 1985 to 5.39 million hectares in 2014 [2]. Currently, Malaysia is the second largest palm oil producer (39%) and exporter (44%) in the world [2]. This industry is a significant contributor toward economic growth but it also contributes toward environmental pollution due to the large amount of by-product produced during the oil extraction process. It is estimated that 53 million tonnes of oil palm waste residue was generated every year with a 5% increment annually [3]. These oil palm wastes are rich in nutrients and present potential agronomic values. For instance, oil palm waste is suitable as raw material for composting to reduce the volume and recycle the nutrients. Currently, the oil palm empty fruit bunch (EFB) is applied in raw or unprocessed form to newly transplanted palms as mulch and nutrient source.

Topsoil has been conventionally used as the growing medium for oil palm seedling during nursery stage. However, it is not practical for long term due to the increasing nutrient demand of oil palm seedling and depletion of fertile soils. The production of high quality seedlings is dependent on good growing media. The physicochemical and biological properties of a growing medium will affect plant growth and directly influence roots growth. Furthermore, the planting medium must be porous and well drained to permit free roots penetration, secure anchorage, and have sufficient nutrients to support crop growth [4].

Amendment materials such as dry effluent and coir dust [5], organic-based substrate [6], and oil palm waste compost [7, 8] in soil growing media for oil palm seedling have been reported prior to this study. The positive effects of compost in raising seedlings include changes in the soil physical properties, exchange and buffering capacities, and being a direct source of nutrients for plants. Study by Aisueni and Omoti [9] in Nigeria found that application of composted EFB with POME at rate of 150 g per polybag increased seedling dry matter weight up to 71% at the main nursery stage. Previous study by Siregar et al. [10] conducted in Riau,

TABLE 1: Selected physical and chemical characteristics of growing media.

Treatments	Bulk density (g cm^{-3})	pH (H$_2$O)	C:N Ratio	C	N	P (%)	K	Ca	Mg	Mn	Zn (mg kg^{-1})	Cu
C0	1.32^a	4.68^e	23.89^d	2.75^f	0.12^f	0.24^e	0.56^a	0.11^d	0.09^d	18.25^c	22.25^d	11.75^{cb}
C20	1.24^b	4.85^d	26.05^{cd}	5.72^e	0.22^e	0.31^d	0.56^a	0.15^{cd}	0.13^d	22.00^{ab}	29.25^{cb}	13.00^b
C40	1.11^b	5.08^c	28.23^{bc}	10.65^d	0.38^d	0.35^d	0.57^a	0.18^c	0.15^d	24.25^a	25.50^{cd}	12.00^{cb}
C60	0.95^{bc}	5.13^a	31.33^b	18.47^c	0.59^c	0.49^c	0.53^a	0.22^b	0.22^c	19.50^{bc}	23.75^d	9.75^c
C80	0.75^{cd}	5.15^a	29.78^b	28.32^b	0.95^b	0.83^b	0.56^a	0.26^a	0.28^b	21.75^{ab}	32.25^b	13.75^b
C100	0.56^d	5.10^b	61.51^a	73.19^a	1.19^a	1.24^a	0.55^a	0.26^a	0.35^a	21.50^{ab}	46.25^a	23.00^a

Note: means with the same letter within row are not significantly different at $p < 0.05$ according to LSD test. C0, C20, C40, C60, C80, and C100 represent 0, 20, 40, 60, 80, and 100% of compost, respectively.

Indonesia, reported that mixing 7.5 kg EFB compost together with topsoil could replace the standard mineral fertilization in main nursery stage as the 12-month-old seedling grows as good as seedlings under normal fertilization practices. Ovie et al. [11] reported that, under water stress condition, the use of 300 g of EFB and poultry manure compost (4 : 1 ratio) in oil palm seedlings planting media resulted in significant improvement of plant vegetative growth and soil chemical properties when compared to seedling practices under conventional main nursery establishment.

Earlier reports of oil palm waste application on oil palm seedling were mainly on raw EFB during the main nursery stage. In this study, a different type of oil palm waste, the oil palm mesocarp, was utilized and focused on the prenursery stage (rather than the main nursery stage) to produce high quality seedlings. Thus, the aims of this study were to investigate the effects of oil palm waste compost which consisted of pressed oil palm fruit mesocarp and palm oil mill effluent (POME) as a component of polybag medium on growth performance of oil palm seedlings and nutrients uptake during the prenursery stage (0–3 months) in the double stage nursery system.

2. Materials and Methods

This study was conducted under shelter in Universiti Putra Malaysia (2°59′59N, 101°42′25E) with air temperature of 24–33°C. Oil palm waste of pressed oil palm fruit mesocarp and POME were obtained from Golden Hope Plantation Berhad and composted in Universiti Putra Malaysia. Oil palm seeds, GH500, a cross-breed product between Elite Deli Dura and second generation of Pesifera, BM119, were used in this study. Polybag media were prepared by mixing compost with the Serdang series topsoil (Ultisol, Tipik Lutualemkuts) at the ratio of 0 (C0), 20 (C20), 40 (C40), 60 (C60), 80 (C80), and 100% (C100). The characteristics of the topsoil and oil palm waste compost were similar to C0 (topsoil only without compost) and C100 (solely compost), respectively. The experiment was carried out in a randomized complete block design (RCBD) with six replications.

Planting distance between each of the poly bags in the block was 30 cm and was 50 cm between each of the blocks. One germinated oil palm seed (GH500) was sown in each polybag of 15 cm × 23 cm with equal volume of media.

The seedlings were watered twice a day and weeding was done manually. A compound fertilizer, N-P-K-Mg (14 : 7 : 9 : 2.5), was applied weekly (1.0 g per polybag) from week 7 to week 12 after sowing.

Plant height was recorded during the 12 weeks of prenursery period at weeks 1, 2, 4, 6, 8, 10, and 12. All seedlings were left to grow until the 12th week (Figure 4). At the 12th week, chlorophyll content was recorded using a chlorophyll meter (SPAD 502 Plus Chlorophyll Meters). The seedlings were harvested 2 cm above ground for the shoot weight, while the roots were carefully removed from the polybag to record root weight. Fresh weight was recorded and the plant samples were oven-dried at 55°C until constant weight was achieved. Then, seedling (excluded root) was ground using a tissue grinder (IKA Labortechnik, MF10 Basic) for macro- and micronutrients analysis. Total N and selected nutrients were determined using Kjeldahl method [12] and dry ashing method [13], respectively. Concentrations of N, P, and K were measured by Auto-Analyzer (LACHAT Instrument, QuikChem FIA+ 8000 series), while Ca, Mg, Fe, Cu, Zn, Mn, and Fe were measured via Atomic Adsorption Spectrometer (AAS; PerkinElmer Analyst 400), respectively.

Each medium treatment was analyzed for bulk density, pH (1 : 4, soil : water), and macro- and micronutrient contents as shown in Table 1. Bulk density was measured according to the core method [14]. Total N was determined using Kjeldahl method [12], total organic carbon (TOC) was determined with Walkey and Black [15], available P was determined according to Bray and Kurtz [16], and selected nutrients (K, Mg, Ca, Fe, Cu, Zn, and Mn) were determined with dry ashing method [13]. N and P concentrations were measured by Auto-Analyzer (LACHAT Instrument, QuikChem FIA+ 8000 series). Other elements (K, Mg, Ca, Fe, Cu, Zn, and Mn) were determined by Atomic Adsorption Spectrometer (AAS; PerkinElmer Analyst 400).

Analysis of variance (ANOVA) was performed for all parameters with the statistical analysis system, SAS version 9.2 (SAS Institute, Inc., Cary, NC, USA). Differences between treatments were analysed using LSD at $p < 0.05$. Regression analyses (with ANOVA) were performed with Sigma Plot version 11.0 (Systat Software, San Jose, CA) to determine the relationships between oil palm waste compost and the chemical characteristics of the media and the growth parameters and nutrient contents of oil palm seedling. Initial characteristics of the treated growing media characteristic were analysed

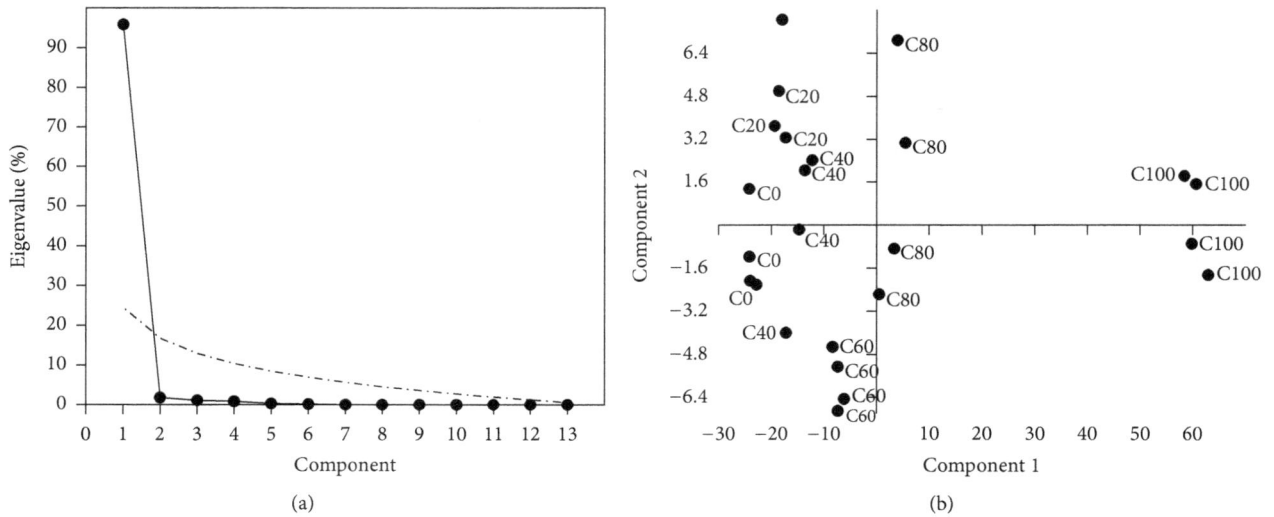

FIGURE 1: (a) Scree plot of eigenvalue. (b) Principal components analysis (PCA) of initial physicochemical properties of growing media (–·– broken stick. C0, C20, C40, C60, C60, C80, and C100 represent 0, 20, 40, 60, 80, and 100% of compost, resp.).

with PCA (PAST version 3.07, PAST: Paleontological Statistics Software, Paleontological Association) to determine the most prominent factor induced by compost addition on the overall alteration of the growing media.

3. Results

3.1. Physicochemical Characteristics of Growing Media. The pH value of growing media significantly increased from 4.68 in C0 and peaked at 5.15 in C80. Thus, the pH value for mixture of compost and topsoil of C60 and C80 was significantly higher than pure compost (C100). Mixing the compost with topsoil reduced the bulk density by up to 43% from the original topsoil (C0) and there was no significant difference between C80 and C100 (Table 1). In general, the optimum bulk density of potting growing media is about 0.40 g cm^{-3} [17]. The C content of growing media significantly increased with 100% compost (73.19%) as compared to control treatment, C0, with only 2.75% C (Table 1). Furthermore, N content of the treatments increased significantly with increasing amount of compost incorporated. The C/N ratio of C100 (61.51) was significantly higher than the other treatments (23–31) due to higher presence of C associated with increasing amount of compost. However, C100 was not suitable as potting media due to the high C/N ratio as the ideal range was recommended to be between 20 and 40 [18]. There was no significant difference between C40 and C80 (Table 1). Correspondingly, P, Ca, and Mg contents of soil-compost mixtures increased significantly when compared to C0 (Table 1). However, K content was not affected by the addition of compost and this could be attributed to leaching, commonly reported for K loss [19, 20]. In general, micronutrients (Mn, Zn, and Cu) concentrations were increased with the addition of compost to the soil (Table 1), which might be contributed by the compost itself (C100).

TABLE 2: Factor pattern, eigenvalues, and percentages of the variance explained by each factor, obtained from the first PCA performed on the physicochemical properties of growing media.

	PC 1	PC 2	PC 3
Soil pH	0.003	−0.010	−0.028
Bulk density (g cm^{-3})	−0.008	0.008	0.033
C : N ratio	0.444	−0.213	0.829
C%	0.840	−0.230	−0.474
N%	0.012	−0.006	−0.055
P%	0.012	0.001	−0.031
K%	0.000	0.002	−0.002
Ca%	0.001	−0.002	−0.010
Mg%	0.003	−0.001	−0.011
Cu (mg/kg)	0.140	0.376	0.171
Zn (mg/kg)	0.270	0.834	−0.025
Mn (mg/kg)	0.010	0.221	−0.026
Na (mg/kg)	0.061	0.127	0.228
Eigenvalue	841.176	15.539	9.631
% variance	95.845	1.7705	1.0974

PCA of the chemical properties of growth media was generated (Table 2). The physicochemical properties of growing media (soil pH, bulk density, and total C, N, P, K, Ca, Mg Zn, Cu, Mn, and Na) were regarded as independent factor and PCA was conducted using variance-covariance matrix. The eigenvalue scree plot (Figure 1(a)) showed one factor analysis. Besides, PC 1 explained 95.85% of the total variance and was primary associated with C content (Table 2). The PCA score plot revealed that the growing media properties were categorized according to the volume of compost content as C100 was clearly separated from other treatments (Figure 1(b)). This was attributed to the high C content in C100 (73.17%).

TABLE 3: Effect of compost addition in growing media on nutrients uptake by oil palm seedlings.

Treatments	N	P	K	Ca	Mg	Cu	Zn	Mn	Fe
	mg plant^{-1}						μg plant^{-1}		
C0	12.60[b]	8.39[c]	11.43[d]	3.00[b]	2.83[b]	25.86[b]	53.35[b]	110.12[a]	509.79[b]
C20	16.80[b]	11.78[bc]	13.52[cd]	3.09[b]	4.45[b]	28.42[b]	49.67[b]	110.63[a]	373.19[c]
C40	19.30[b]	14.53[b]	15.66[c]	3.42[b]	4.28[b]	28.10[b]	57.63[b]	121.83[a]	436.94[bc]
C60	29.66[a]	25.72[a]	19.70[b]	5.12[a]	6.44[a]	44.48[a]	86.73[a]	142.56[a]	739.40[a]
C80	33.68[a]	24.21[a]	24.01[a]	5.71[a]	6.53[a]	52.50[a]	92.00[a]	135.35[a]	675.38[a]
C100	32.48[a]	27.76[a]	21.38[ab]	4.89[a]	6.44[a]	43.91[a]	85.15[a]	105.74[a]	752.78[a]

Note: means with the same letter within row are not significantly different at $p < 0.05$ according to LSD test. C0, C20, C40, C60, C80, and C100 represent 0, 20, 40, 60, 80, and 100% of compost, respectively.

Therefore, C content largely affected the physicochemical properties in growth media.

3.2. Macro- and Micronutrients Uptake by Oil Palm Seedlings. The nutrients uptake of oil palm seedlings is shown in Table 3. In general, there was no significant difference between topsoil (C0) and topsoil-compost mixtures up to 40% (C20–40). However, the macronutrients uptake increased significantly at C60 and there was no further significant rise subsequently. Likewise, Cu, Zn, and Fe uptakes imitated the pattern of macronutrients uptake; no significant difference was detected at low compost rate (C0–40) but it increased significantly at C60 and remained stagnant successively (C60–100).

3.3. Oil Palm Seedlings Growth. Based on the regression studies, the shoot dry weight, root dry weight, plant height, and SPAD reading of oil palm seedlings were related to the amount of compost added (Figure 2). Hence, the shoot dry weight was best fitted into a positive quadratic relationship ($p \leq 0.001$) with compost ($R^2 = 0.725$). It peaked at C80 and declined afterwards. Meanwhile, the root dry weight was also best fitted into a positive quadratic relationship ($p \leq 0.01$) with compost ($R^2 = 0.496$) and the highest root dry weight was detected at C100. The plant height was directly proportional or positive linearly related ($p \leq 0.001$) with compost ($R^2 = 0.534$). Meanwhile, the SPAD was best fitted into a positive quadratic relationship ($p \leq 0.001$) with compost ($R^2 = 0.717$) and peaked at C80. Addition of compost up to 72% in growing media resulted in progressive seedlings growth, but shoot dry weight started to decrease when more than 72% of compost was added (Figure 2(a)). SPAD value also showed a strong correlation ($p = 0.01$) to the seedlings nitrogen uptake (Figure 3).

4. Discussion

4.1. Effects of Oil Palm Waste Compost on the Characteristics of Growth Media. Addition of oil palm waste compost decreased the bulk density of topsoil-compost growing medium. This could be attributed to the dilution effect [21] as bulk density of compost was significantly lower than topsoil and higher amount of compost. Thus, compost addition improved soil aeration and pore space. Bronick and Lal [22] had shown that organic amendments such as compost

have a dilution effect in lowering bulk density. Organic applications (e.g., compost) increased soil C content and also improved soil aggregation and macroporosity and lowered the penetration resistance, contributing to better crop growth [23, 24]. This further highlighted the importance of compost in growing medium to alleviate soil C content and improved soil physical properties.

Adding compost also improved the chemical properties of the growing medium as shown by various studies, which reported the ability of compost to reduce soil acidity [25–27]. This could be attributed to the presence of complex humic acid in compost [28] which provided binding sites for exchangeable bases and decreased the availability of polycations like Fe^{2+} and Al^{3+} [29]. Thus, addition of compost-C rich in negative functionalities could affect the pH value of the growth media. Besides, the high content of alkaline elements (excess cations), Ca and Mg, was concomitant with increasing compost which could react with the free H^+. This explained the liming effect of compost and decarboxylation of organic anions during decomposition, improving the pH of growing media [30].

Furthermore, compost could also supply the seedlings with additional N, P, Ca, Mg, and Zn, functioning as a source of nutrients beside applied fertilizers. Another plausible mechanism was compost retaining the nutrients in the growing media by surface adsorption and increased nutrient recovery [31]. Vegetal compost was shown to be effective in removing Zn via sorption process [32]. Potassium was mostly leached away as it existed in ionic form (K^+) and can be easily dissolved.

4.2. Effects of Oil Palm Waste Compost on Seedling Growth and Nutrients Uptake. The oil palm seedling growth and physiology improved with the application of compost. This was majorly attributed to the alteration of physicochemical properties of the growing medium. Multiple regression study showed that root and shoot growth and macronutrients uptake were highly significantly related to the pH of the growing media (Table 4). Previous studies also reported plants, specifically lettuce and ornamental plant, showing similar response when the growing medium improved from acidic to slightly acidic [33, 34]. This highlighted the effect of pH on the nutrient availability as macronutrients like P were heavily affected by the pH value and subsequently

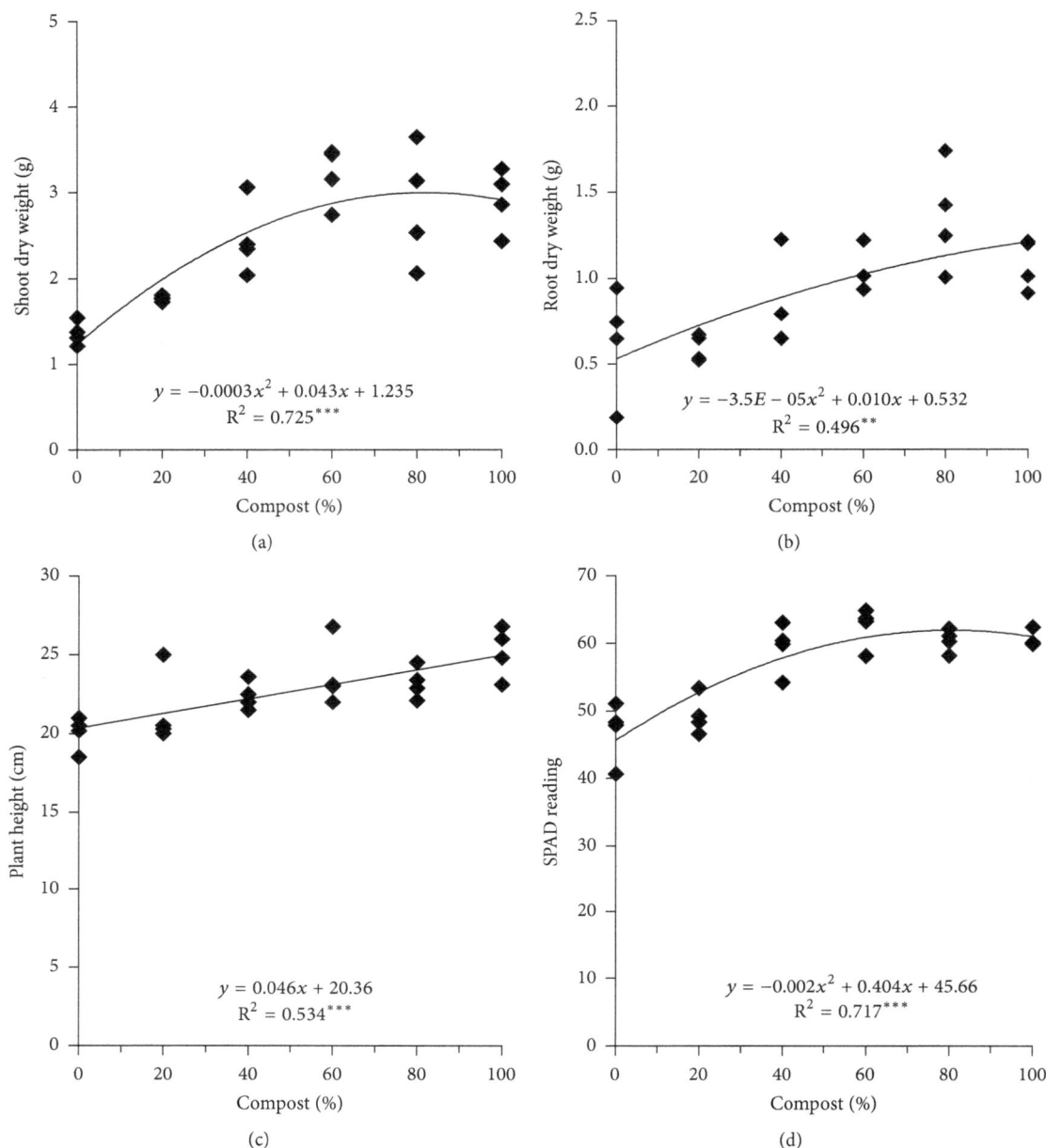

FIGURE 2: Effect of compost application rates on (a) shoot dry weight, (b) root dry weight, (c) plant height, and (d) chlorophyll meter readings (SPAD value) of oil palm seedlings. Note: **,***significant at the level of 0.01 and 0.0001 probability level, respectively.

the nutrients uptake of oil palm seedlings. At low pH, soil P reacted with Al and Fe oxides forming insoluble Al or Fe phosphate through ligand bridging and anionic repulsion rendering P unavailable [35]. According to Budianta et al. [36], available P concentrations increased to 73.82% when oil palm waste derived compost was applied to soil, while the soil exchangeable Al decreased. This could be due to the presence of the negative surface functional groups (phenolic, carboxylic, carbonyl, and alcohol) on compost immobilizing Al or Fe and releasing P back to the soil.

Furthermore, the seedling root growth was clearly associated with the bulk density, reflected by the increasing pore

space and soil aeration. Addition of composted agricultural waste had been reported to significantly improve root growth [37, 38]. The nutrients uptake increased with root growth as shown by the correlation matrix (Table 5). Thus, lower bulk density and alleviated pH due to compost addition were ideal to promote root growth and increased nutrients uptake and biomass.

Apart from promoting nutrient uptake, compost addition also influenced nutrient availability and retention in the growing medium. The surface of organic matter like compost is rich in negative functionalities such as phenolic, carboxylic, carbonyl, and alcohol which serve as exchange

TABLE 4: Relationship between shoot and root dry weight, nitrogen (N), phosphorous (P), potassium (K), calcium (Ca), and magnesium (Mg) uptake and soil pH as affected by compost addition in growing media.

Interaction between soil pH and	Equation	R^2
Shoot dry weight, g^{-1}	$y = 4.233x^2 - 37.99x + 86.42$	0.86^{***}
Root dry weight, g^{-1}	$y = 6.163x^2 - 59.37x + 143.5$	0.78^{***}
N uptake, $mg\,plant^{-1}$	$y = 109.8x^2 - 1039x + 2469.86$	0.84^{***}
P uptake, $mg\,plant^{-1}$	$y = 94.63x^2 - 894.6x + 2123.6$	0.84^{***}
K uptake, $mg\,plant^{-1}$	$y = 71.15x^2 - 677.4x + 1623.6$	0.85^{***}
Ca uptake, $mg\,plant^{-1}$	$y = 24.58x^2 - 236.6x + 572.3$	0.85^{***}
Mg uptake, $mg\,plant^{-1}$	$y = 20.68x^2 - 195.2x + 463.6$	0.88^{***}

Note: *** significant at the level of 0.0001 probability level.

TABLE 5: Correlation matrix between dry shoot and root, nitrogen (N), phosphorous (P), potassium (K), calcium (Ca), and magnesium (Mg) in soil and plant nutrient uptake as affected by compost addition in growing media.

	Parameter	Root	Shoot	N	P	K	Ca	Mg	N	P	K	Ca	Mg
		Dry weight (g)		Uptake ($mg\,plant^{-1}$)					Soil before treatment (%)				
Root	Dry weight (g)	1.00	0.57^{**}	0.81^{***}	0.68^{**}	0.84^{***}	0.78^{***}	0.40^{*}	0.68^{**}	0.58^{**}	-0.22^{ns}	0.79^{***}	0.63^{***}
Shoot				0.80^{***}	0.84^{***}	0.74^{***}	0.86^{***}	0.67^{**}	0.70^{***}	0.61^{**}	-0.27^{ns}	0.73^{***}	0.67^{**}
N					0.87^{***}	0.92^{***}	0.88^{***}	0.55^{**}	0.83^{***}	0.75^{***}	-0.31^{ns}	0.86^{***}	0.84^{***}
P						0.82^{***}	0.82^{***}	0.57^{**}	0.84^{***}	0.77^{***}	-0.38^{ns}	0.83^{***}	0.77^{***}
K	Uptake ($mg\,plant^{-1}$)						0.88^{***}	0.55^{**}	0.82^{***}	0.72^{***}	-0.17^{ns}	0.86^{***}	0.80^{***}
Ca								0.61^{**}	0.74^{***}	0.64^{**}	-0.27^{ns}	0.77^{***}	0.71^{***}
Mg									0.50^{*}	0.44^{*}	-0.31^{ns}	0.54^{**}	0.45^{*}

Notes: *,**,*** significant at the level of 0.05, 0.01, and 0.0001 probability level, respectively; ns: not significant at $p < 0.05$.

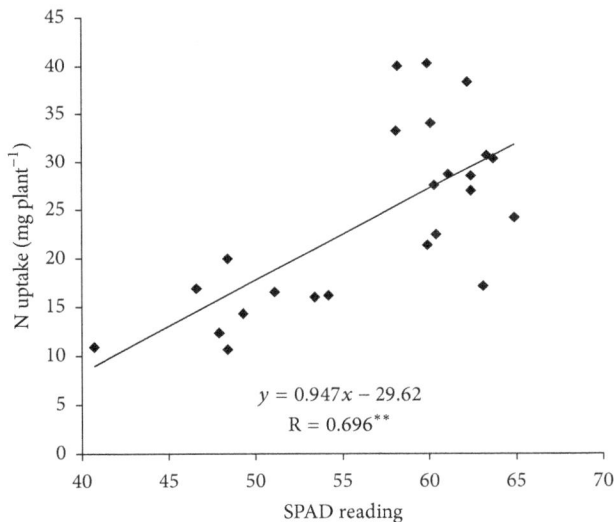

FIGURE 3: Relationship between SPAD value and plant nitrogen (N) uptake. Note: ** significant at the level of 0.01 probability level.

$$y = 0.947x - 29.62$$
$$R = 0.696^{**}$$

FIGURE 4: Growth of seedlings at the 12th week in prenursery stage.

sites that ultimately increased the CEC of growing media [39]. Subsequently, it also increased dry matter production and enabled efficient nutrients usage [39–43]. This finding was consistent with Palanivell et al. [39], which reported that the application of saw dust based compost as growing media increased N, P, K, Ca, Mg, and Na uptake by maize when compared to conventional fertilizer.

However, growth of oil palm seedling is negatively affected by compost application, solely due to the high C : N ratio, causing N immobilization. However, C : N value would decrease gradually when microorganisms start breaking down the compost and slowly release plant available nutrients to the soil throughout the seedlings growth period. Similar results have been reported by [33] whereby addition of composted green waste of more than 70% caused reduction in plant growth and root morphology, probably due to the presence of phytotoxic substance and negative changes in chemical and physical properties of growth media.

5. Conclusions

Prenursery polybag medium amended with oil palm waste compost up to 70% increased oil palm seedling growth.

Meanwhile, 72% of the compost mixed with topsoil could produce the best planting material with respect to the high DMW production, oil palm seedlings growth, and development as well as greater nutrient uptake. Further study is required to investigate whether the amount of chemical fertilizer employed during nursery stage could be reduced, thus cutting operational cost by making use of oil palm waste compost as polybag growth medium.

Conflict of Interests

The authors declare that there is no conflict of interests regarding the publication of this paper.

Acknowledgments

The authors would like to thank Universiti Putra Malaysia for providing the facilities and Golden Hope Plantation Bhd., Banting, for the oil palm waste that was used to produce the compost.

References

[1] Y. Basiron, "Palm oil production through sustainable plantations," *European Journal of Lipid Science and Technology*, vol. 109, no. 4, pp. 289–295, 2007.

[2] MPOB, "Oil palm planted area by state as at December 2014," 2014, http://bepi.mpob.gov.my/images/area/2014/Area_summary.pdf.

[3] M. A. A. Mohammed, A. Salmiaton, W. A. K. G. Wan Azlina, M. S. Mohammad Amran, A. Fakhru'L-Razi, and Y. H. Taufiq-Yap, "Hydrogen rich gas from oil palm biomass as a potential source of renewable energy in Malaysia," *Renewable and Sustainable Energy Reviews*, vol. 15, no. 2, pp. 1258–1270, 2011.

[4] M. M. Khan, M. A. Khan, M. Abbas, M. J. Jaskani, M. A. Ali, and H. Abbas, "Evaluation of potting media for the production of rough lemon nursery stock," *Pakistan Journal of Botany*, vol. 38, no. 3, pp. 623–629, 2006.

[5] M. T. Hashim, K. H. Yeow, and Y. C. Poon, "Recent development in nursery practice—potting media," in *Proceedings of the International Oil Palm/Palm Oil Conferences: Progress & Prospects. Conference 1, Agriculture*, A. H. B. Hassan, Ed., pp. 369–371, Palm Oil Research Institute of Malaysia and Incorporated Society of Planters, Kuala Lumpur, Malaysia, June 1987.

[6] M. Khairil, A. A. Kamarozaman, I. Arifin, A. Ramadzan, and N. Nasir, "Evaluation of several planting media for oil palm (ElaiesGuineensis) seedlings in main nursery," in *Proceedings of the Soil Science Conference of Malaysia*, Kota Bharu, Malaysia, April 2012.

[7] O. O. Adeoluwa and G. O. Adeoye, "Potential of oil palm empty fruit bunch (EFB) as fertilizer in oil palm (*Elaeis guineensis* L. Jacq.) nurseries," in *Proceedings of the 16th IFOAM Organic World Congress*, Modena, Italy, June 2008, http://orgprints.org/view/projects/conference.html.

[8] E. G. Uwumarongie-Ilori, B. B. Sulaiman-Ilobu, O. Ederion et al., "Vegetative growth performance of oil palm (*Elaeis guineensis*) seedlings in response to inorganic and organic fertilizers," *Greener Journal of Agricultural Science*, vol. 2, pp. 26–30, 2012.

[9] N. O. Aisueni and U. Omoti, "The role of compost in sustainable oil palm production," in *Cutting-Edge Technologies for Sustained Competitiveness: Proceedings of the 2001 PIPOC International Palm Oil Congress, Agriculture Conference, Kuala Lumpur, Malaysia, 20–22 August 2001*, Nigerian Institute for Oil Palm, Ed., pp. 536–541, Malaysian Palm Oil Board (MPOB), 2001.

[10] F. A. Siregar, S. Salates, J. P. Caliman, and Z. Liwang, "Enhancing oil palm industry development through environmentally friendly technology," in *Proceedings of the Chemistry & Technology Conference*, Indonesian Oil Palm Research Institute (IOPRI), Ed., Empty Fruit Bunch Compost: Processing and Utilities, pp. 225–234, Nusa Dua, Indonesia, July 2002.

[11] S. Ovie, M. O. Ekabafe, A. Nkechika, and O. N. Udegbunam, "Influence of composted oil palm bunch waste on soil pH, nitrogen, organic matter status and growth of oil palm seedlings under water stress condition," *Continental Journal of Agronomy*, vol. 1, pp. 1–15, 2014.

[12] J. M. Bremner and C. S. Mulvaney, "Nitrogen-total," in *Methods of Soil Analyses, Part 2. Chemical and Mineralogical Properties*, A. L. Page, R. H. Miller, and D. R. Keeney Madison, Eds., pp. 610–615, American Society of Agronomy and Soil Science Society of America, 1982.

[13] D. E. Baker, G. W. Gorslinc, C. G. Smith, W. I. Thomas, W. E. Grube, and J. L. Ragland, "Techniques for rapid analysis of corn leaves for eleven elements," *Agronomy Journal*, vol. 56, pp. 133–136, 1964.

[14] M. De Boodt and O. Verdonck, "The physical properties of the substrates in horticulture," *Acta Horticulturae*, vol. 26, no. 5, pp. 37–44, 1972.

[15] D. W. Nelson and L. E. Sommers, "Methods of soil analysis. Part 2. Chemical and microbiological properties," in *Total Carbon, Organic Carbon, and Organic Matter*, A. L. Page, R. M. Miller, and D. R. Keeney, Eds., vol. 9 of *Agronomy Monograph*, pp. 542–560, ASA and SSSA, Madison, Wis, USA, 2nd edition, 1982.

[16] R. H. Bray and L. T. Kurtz, "Determination of total, organic, and available forms of phosphorus in soils," *Soil Science*, vol. 59, no. 1, pp. 39–46, 1945.

[17] M. Abad, P. Noguera, and S. Burés, "National inventory of organic wastes for use as growing media for ornamental potted plant production: case study in Spain," *Bioresource Technology*, vol. 77, no. 2, pp. 197–200, 2001.

[18] M. Abad, P. F. Martínez, M. D. Martínez, and J. Martínez, "Agronomic evaluation of growing media," *Actas de Horticultura*, vol. 11, pp. 141–154, 1993.

[19] B. Ulén, "Leaching of plant nutrients and heavy metals during the composting of household wastes and chemical characterization of the final product," *Acta Agriculturae Scandinavica B: Soil & Plant Science*, vol. 47, no. 3, pp. 142–148, 1997.

[20] S. G. Sommer, "Effect of composting on nutrient loss and nitrogen availability of cattle deep litter," *European Journal of Agronomy*, vol. 14, no. 2, pp. 123–133, 2001.

[21] S. C. Gupta, R. H. Dowdy, and W. E. Larson, "Hydraulic and thermal properties of a sandy soil as influenced by incorporation of sewage sludge," *Soil Science Society of America Journal*, vol. 41, no. 3, pp. 601–605, 1977.

[22] C. J. Bronick and R. Lal, "Soil structure and management: a review," *Geoderma*, vol. 124, no. 1-2, pp. 3–22, 2005.

[23] A. Rivenshield and N. L. Bassuk, "Using organic amendments to decrease bulk density and increase macroporosity in compacted soils," *Arboriculture & Urban Forestry*, vol. 33, no. 2, pp. 140–146, 2007.

[24] I. Celik, H. Gunal, M. Budak, and C. Akpinar, "Effects of long-term organic and mineral fertilizers on bulk density

and penetration resistance in semi-arid Mediterranean soil conditions," *Geoderma*, vol. 160, no. 2, pp. 236–243, 2010.

[25] G. V. Giannakis, N. N. Kourgialas, N. V. Paranychianakis, N. P. Nikolaidis, and N. Kalogerakis, "Effects of municipal solid waste compost on soil properties and vegetables growth," *Compost Science & Utilization*, vol. 22, no. 3, pp. 116–131, 2014.

[26] P. J. Valarini, G. Curaqueo, A. Seguel et al., "Effect of compost application on some properties of a volcanic soil from central South Chile," *Chilean Journal of Agricultural Research*, vol. 69, no. 3, pp. 416–425, 2009.

[27] T. J. Butler and J. P. Muir, "Dairy manure compost improves soil and increases tall wheatgrass yield," *Agronomy Journal*, vol. 98, no. 4, pp. 1090–1096, 2006.

[28] L. R. Bulluck III, M. Brosius, G. K. Evanylo, and J. B. Ristaino, "Organic and synthetic fertility amendments influence soil microbial, physical and chemical properties on organic and conventional farms," *Applied Soil Ecology*, vol. 19, no. 2, pp. 147–160, 2002.

[29] Y. Chen, P. Gat, F. H. Frimmel, and G. Abbt-Braun, "Metal binding by humic substances and dissolved organic matter derived from compost," in *Soil and Water Pollution Monitoring, Protection and Remediation*, I. Twardowska, H. E. Allen, M. M. Häggblom, and S. Stefaniak, Eds., vol. 69 of *NATO Science Series*, pp. 275–297, Springer, Dordrecht, The Netherlands, 2006.

[30] J. M. Xu, C. Tang, and Z. L. Chen, "The role of plant residues in pH change of acid soils differing in initial pH," *Soil Biology and Biochemistry*, vol. 38, no. 4, pp. 709–719, 2006.

[31] E. P. Jouquet, E. Bloquel, T. T. Doan et al., "Do compost and vermicompost improve macronutrient retention and plant growth in degraded tropical soils?" *Compost Science and Utilization*, vol. 19, no. 1, pp. 15–24, 2011.

[32] O. Gibert, D. J. Pablo, J. L. Cortina, and C. Ayora, "Sorption studies of Zn (II) and Cu (II) onto vegetal compost used on reactive mixtures for in situ treatment of acid mine drainage," *Water Resources*, vol. 39, pp. 2827–2838, 2005.

[33] L. Zhang, X. Sun, Y. Tian, and X. Gong, "Composted green waste as a substitute for peat in growth media: effects on growth and nutrition of *Calatheainsignis*," *PLoS ONE*, vol. 8, no. 10, Article ID e78121, 2013.

[34] G. Y. Jayasinghe, "Sugarcane bagasses sewage sludge compost as a plant growth substrate and an option for waste management," *Clean Technologies and Environmental Policy*, vol. 14, no. 4, pp. 625–632, 2012.

[35] W. L. Lindsay, *Chemical Equilibiria in Soils*, John Wiley & Sons, New York, NY, USA, 1979.

[36] D. Budianta, P. K. S. A. Halim, N. S. Bolan, and R. J. Gilkes, "Palm oil compost reduces aluminum toxicity thereby increases phosphate fertilizer use efficiency in ultisols," in *Proceedings of the 19th World Congress of Soil Science*, pp. 221–223, Brisbane, Australia, August 2010.

[37] M. M. Khan, M. A. Khan, A. Mazhar, J. Muhammad, J. M. A. Ali, and H. Abbas, "Evaluation of potting media for the production of rough lemon nursery stock," *Pakistan Journal of Botany*, vol. 38, no. 3, pp. 623–629, 2006.

[38] S. B. Wilson, P. J. Stoffella, and D. A. Graetz, "Compost-amended media for growth and development of Mexican heather," *Compost Science & Utilization*, vol. 9, no. 1, pp. 60–64, 2001.

[39] P. Palanivell, K. Susilawati, O. H. Ahmed, and A. M. N. Muhamad, "Effects of crude humin and compost produced from selected waste on *Zea mays* growth, nutrient uptake and nutrient use efficiency," *African Journal of Biotechnology*, vol. 12, no. 13, pp. 1500–1507, 2013.

[40] C. Lazcano and J. Domínguez, "Soil nutrients," in *The Use of Vermicompost in Sustainable Agriculture: Impact on Plant Growth and Soil Fertility*, M. Miransari, Ed., p. 336, Nova Science, 2011.

[41] A. C. Petrus, O. H. Ahmed, A. M. N. Muhamad, H. M. Nasir, and M. Jiwan, "Effect of K-N-humates on dry matter production and nutrient use efficiency of maize in Sarawak, Malaysia," *TheScientificWorldJOURNAL*, vol. 10, pp. 1282–1292, 2010.

[42] M. D. Perez-Murcia, R. Moral, J. Moreno-Caselles, A. Perez-Espinosa, and C. Paredes, "Use of composted sewage sludge in growth media for broccoli," *Bioresource Technology*, vol. 97, no. 1, pp. 123–130, 2006.

[43] R. M. Atiyeh, S. Subler, C. A. Edwards, G. Bachman, J. D. Metzger, and W. Shuster, "Effects of vermicomposts and composts on plant growth in horticultural container media and soil," *Pedobiologia*, vol. 44, no. 5, pp. 579–590, 2000.

13

Coastal Mudflat Saline Soil Amendment by Dairy Manure and Green Manuring

Yanchao Bai,[1,2,3] **Yiyun Yan,**[1] **Wengang Zuo,**[1] **Chuanhui Gu,**[4] **Weijie Xue,**[1] **Lijuan Mei,**[1] **Yuhua Shan,**[1,5] **and Ke Feng**[1,5]

[1]*College of Environmental Science and Engineering, Yangzhou University, Yangzhou 225009, China*
[2]*State Key Laboratory of Soil and Sustainable Agriculture, Institute of Soil Science, Chinese Academy of Sciences, Nanjing 210008, China*
[3]*Institute of Biotechnology, Jiangsu Academy of Agricultural Sciences, Nanjing 210014, China*
[4]*Department of Geology, Appalachian State University, Boone, NC 28608, USA*
[5]*Jiangsu Collaborative Innovation Center for Solid Organic Waste Resource Utilization, Nanjing 210095, China*

Correspondence should be addressed to Yanchao Bai; byc529@gmail.com and Yuhua Shan; shanyuhua@outlook.com

Academic Editor: Mathias N. Andersen

Dairy manure or green manuring has been considered as popular organic amendment to cropland in many countries. However, whether dairy manure combined with green manuring can effectively amend mudflat saline soil remains unclear. This paper was one of first studies to fill this knowledge gap by investigating impact of dairy manure combined with green manuring on soil chemical properties of mudflat saline soil. Dairy manure was used by one-time input, with the rates of 0, 30, 75, 150, and 300 t ha^{-1}, to amend mudflat saline soil. Ryegrass, *Sesbania*, and ryegrass were chosen as green manures for three consecutive seasons, successively planted, and tilled, and maize was chosen as a test crop. The results indicated that one-time application of dairy manure enhanced fertility of mudflat saline soil and supported growth of ryegrass as the first season green manure. By the cycles of the green manuring, it rapidly improved the chemical properties of mudflat saline soil by decreasing soil salinity and pH and increasing soil organic carbon and available N and P, which promoted growth of maize. Dairy manure combined with green manuring can be applied for mudflat saline soil amendment, which provides an innovative solution for mudflat saline soil reclamation, dairy manure disposal, and resource recycling.

1. Introduction

Coastal mudflats located in the interaction zone between land and sea are found in many parts of the world [1]. According to statistics, there are 10000 km^2 coastal mudflat along the east coast of China. More than 6500 km^2 and 3500 km^2 mudflat mainly located in the Yangtze River Estuary of western Yellow Sea and the Yellow River Delta of western Bohai Sea, respectively. The mudflats can be important alternative sources for arable lands after being amended by large amount of organic fertilizers. About 1.1~1.2 million ha mudflats have been reclaimed to croplands in the past 50 years [2]. It is estimated that additional 2000 km^2 mudflats will be reclaimed to cropland in 2020s, according

to the policy of coastal development planning in Jiangsu province in China, and another 4400 km^2 in the Yellow River Delta are under discussion [3]. The newly reclaimed mudflats as typical saline soils are not suitable for cultivation due to high salinity and macronutrient deficiencies, which hinder the germination and growth of plants [4]. The key to improve the newly reclaimed mudflats is to increase the soil organic matter (OM) content for fertility enhancement and salinity reduction [5–7], which is usually achieved through instantaneous application of a great amount of organic matter because soil natural organic matter formation is extremely slow.

Dairy manure generated in dairy industry is rich in OM and inorganic nutrients and has caused serious environmental pollution and ecological safety concern. Therefore, mudflat saline soil that needs organic amendment may become a potential land source for safe disposal of massive dairy manure, which also has a great incentive in view of mudflat saline soil amendment, and nutrient recycling and reuse including OM, nitrogen (N), phosphorus (P), and other plant nutrients. Moreover, it is well known that cultivation of green manures plays an important role in soil quality and sustainability of agricultural systems and has been used to increase the soil fertility, as it adds organic carbon and nutrients to the soil [8]. Green manures were tilled into soil to increase soil OM content, which improved soil bulk density, soil porosity, soil structure, and water holding capacity [9, 10]. Input of biomass of green manures can facilitate mineralization, which enhanced soil fertility [11]. To amend saline agricultural soils, different green manure, such as straw, cotton, and maize, have been applied to increase soil OM [12, 13].

Past research on the application of dairy manure for soil amendment has mainly focused on farmlands [14–16], which showed that land application of dairy manure increased soil OM [17], yield of plants [18, 19]. However, the mudflat saline soil amended by dairy manure and green manure has received no attention. The effect and mechanism of dairy manure amendment on mudflat saline soil can be quite different because farmlands are different from mudflats in soil structure, nutrient, microbial flora, and so forth [1, 20].

In this study, we aimed to assess the method for mudflat saline soil amendment by dairy manure and green manuring. Dairy manure was used by one-time application to amend infertile mudflat saline soil and to support growth of green manure at the first season. Then, green manure tilled into mudflat saline soil can be decomposed and converted into soil OM. By the cycles of planting and tilling green manures, OM in mudflat saline soil would rapidly increase through build-up added organic matter. However, whether dairy manure and green manuring can effectively amend mudflat saline soil remains unclear. This paper was one of first studies to fill this knowledge gap by assessing the effects of dairy manure combined with green manuring on the change of organic carbon, salinity, pH, and N and P of mudflat saline soil, as well as the yield and metal uptake of maize (*Zea mays* L.) as a test crop. Our goals are to assess the effects of dairy manure combined with green manuring on mudflat saline soil chemical properties and crop yield, and to assure environmental safety for heavy metals in crops grown in mudflats.

2. Materials and Methods

2.1. Study Area and Experimental Materials. The experiment was conducted at the farm of Senmao Company Ltd. located in Rudong county, Jiangsu Province, China (E 121°24′04″, N 32°20′00″). The land, as reclaimed from coastal mudflats by building an artificial seawall in 2007 to prevent seawater intrusion, is under amendment since 2011. The distance from the site to the new Yellow Sea coastline (the artificial seawall) is approximately 1.2 km. The experimental mudflat soil was

TABLE 1: Basic chemical properties of the mudflat saline soil at the start of the experiment and dairy manure used in this study.

Parameters	Mudflat saline soil	Dairy manure
pH	8.92	7.74
Salinity (‰)	8.82	11.61
Organic matter ($g\,kg^{-1}$)	3.16	415.7
Total N ($N\,g\,kg^{-1}$)	0.209	32.3
Total P ($P\,g\,kg^{-1}$)	0.565	5.31
Alkaline N ($N\,mg\,kg^{-1}$)	17.48	358.1
Available P ($P\,mg\,kg^{-1}$)	17.86	111.9
Total Cd ($mg\,kg^{-1}$)	0.451	2.15
Total Cr ($mg\,kg^{-1}$)	13.08	41.7
Total Cu ($mg\,kg^{-1}$)	15.89	769.4
Total Mn ($mg\,kg^{-1}$)	152.8	133.9
Total Ni ($mg\,kg^{-1}$)	30.4	18.0
Total Zn ($mg\,kg^{-1}$)	49.9	146.7

typic saline soil, which belonged to the halaquepts group of aquepts in inceptisols based on US soil taxonomy. The experimental dairy manure was collected from the Dairy Farm of Rudong County in September 2011. The chemical properties of mudflat soil and dairy manure were shown in Table 1.

2.2. Experimental Design. The experiment was carried out in randomized complete block design with each plot of 4 m length and 4 m width. The average OM content in arable lands soil was about 0.5%–3.0% of total soil weight. Therefore, 30, 75, 150, and 300 t ha^{-1} (0.5%, 1.0%, 1.5%, and 3.0% organic matter of total soil weight, resp.) dairy manure amendment (DMA) rates on a dried weight basis by one-time application were used. The unamended soil was the control soil. Each of the five treatments had three replicates. The dairy manure was mixed uniformly with soil down to the depth of 20 cm by a rototiller on 20th October 2011. Ryegrass (*Lolium perenne* L.) as a popular high-quality and salt-tolerant green manure was chosen for the first season green manure, sowed 35 g for each plot in 25th October 2011, and tilled in 30th May 2012. *Sesbania (Sesbania cannabina)* as the second season green manure was sowed 120 g for each plot in 12th June 2012 and tilled in 20th September 2012. Ryegrass as the third season green manure was sowed in 30th October 2012 and tilled in 25th May 2013. Maize (*Zea mays* L.) as one of the most important grain crop in China was chosen as a test crop, sowed two seeds per hole with average row spacing of 0.50 m and plant spacing of 0.25 m in 10th July 2013, and harvested in 22th September 2013. Thinning was carried out 20 days after sowing (DAS), leaving one plant per hole. Weeding was carried out two times during the whole experiment, 20 and 48 DAS, through hand-hoeing. Plants were rain-fed, and no extra artificial irrigation was carried out. Soil and plant samples were collected for analysis in 30th May and 20th September 2012 and 25th May and 22th September 2013.

2.3. Soil Analysis. Soil samples for 0–20 cm depth were collected in quadruplicate from control, 30, 75, 150, and 300 t ha^{-1} DMA rates. For analysis of organic carbon (OC), 0.3 g air-dried sample through 0.150 mm mesh size was measured by the $K_2Cr_2O_7$ method. Soil salinity was measured by the gravimetric method. The pH of soil was measured in suspension of 1 : 5 (weight/volume) by pH meter (Model IQ150, Spectrum, USA). Alkaline N was determined by alkaline hydrolysis diffusion method. Available P was analyzed by sodium bicarbonate ($NaHCO_3$) extraction and subsequent colorimetric analysis. The analytical methods for the above soil chemical properties are described in detail by the Soil and Agro-Chemistry Analysis [21].

2.4. Plant Analysis. The biomasses of green manures for three consecutive seasons and maize were determined by weighting all biomass from each plot after harvesting. For estimation of aerial part growth, 10 plants of maize were sampled randomly from each plot and then washed with deionized water to remove soil particles adhering on them, separated into stem, leaves, husk, cob, and grain parts, deactivated at 105°C for 15 minutes, and oven-dried at 80°C until constant weight was achieved. The plant parts were then weighed separately and biomass accumulation was expressed as grams per plant. For extraction of metals (Cd, Cr, Cu, Mn, Ni, and Zn) in maize grain, 0.5 g oven-dried sample was digested in 10 mL tri acid mixture ($HNO_3/H_2SO_4/HClO_4$, 5 : 1 : 1) till transparent color appeared. Metal concentrations were determined after filtering the digested samples by using Atomic Absorption Spectrometer (Model SOLAAR M6, Thermo Elemental, Thermo Fisher Scientific Inc., USA).

2.5. Statistical Analysis. After tests of normality of data distribution and homogeneity of variance among treatments, analysis of variance (ANOVA) of monofactorial (DMA) for randomized complete block design was conducted using SPSS version 13 software. Then, the least significant difference (LSD) method at the 0.05 level of significance was performed to test the significant difference between the treatments.

3. Results

3.1. Biomass of Green Manures and Maize. The DMA significantly increased biomass of ryegrass, *Sesbania*, ryegrass, and maize grown in mudflat saline soil, and the biomass of all plants increased with increasing DMA rates (Figure 1). The biomass of three green manures increased by 35.9%–87.3%, 56.2%–119.8%, 77.0%–169.7%, and 141.3%–223.0% at 30, 75, 150, and 300 t ha^{-1} DMA rates, respectively, compared to the unamended soil. The dairy manure combined with green manuring significantly increased biomass of maize at all DMA rates ($p < 0.05$), and maize biomass in the mudflat saline soil increased with increasing DMA rate. There were increments in maize biomass by 44.4%, 117.1%, 142.2%, and 214.9% at 30, 75, 150, and 300 t ha^{-1} DMA rates, respectively, compared to 29.35 kg plot^{-1} in the unamended soil.

3.2. Soil Chemical Properties. The DMA significantly increased organic carbon concentration in mudflat saline

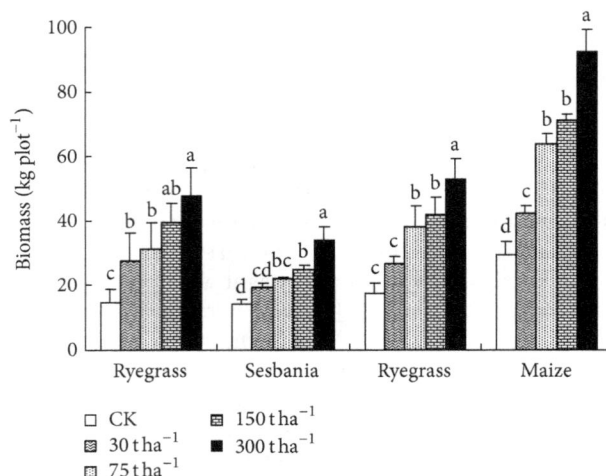

FIGURE 1: Effects of dairy manure amendment on biomass of green manures and maize. Vertical bars indicate standard deviations of the means. Columns with different letters show significant difference between dairy manure amendment rates at $p < 0.05$ by LSD's multiple range test.

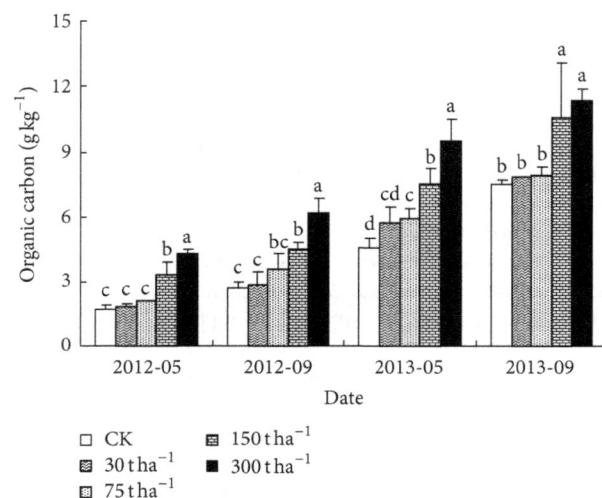

FIGURE 2: Effects of dairy manure amendment on organic carbon in mudflat saline soil. Vertical bars indicate standard deviations of the means. Columns with different letters show significant difference between dairy manure amendment rates at $p < 0.05$ by LSD's multiple range test.

soil, and the OC concentration in the topsoil at all DMA rates increased over time (Figure 2). The dairy manure combined with green manuring significantly increased OC concentration by 4.1%–26.7%, 5.4%–32.2%, 40.6%–92.2%, and 51.2%–151.1% at 30, 75, 150, and 300 t ha^{-1} DMA rates in the four experimental dates, respectively, compared to the unamended soil. The increments of OC concentration were 36.1%, 33.0%, 39.8%, and 19.8% at 30, 75, 150, and 300 t ha^{-1} DMA rates in September 2013, respectively, compared to May 2013.

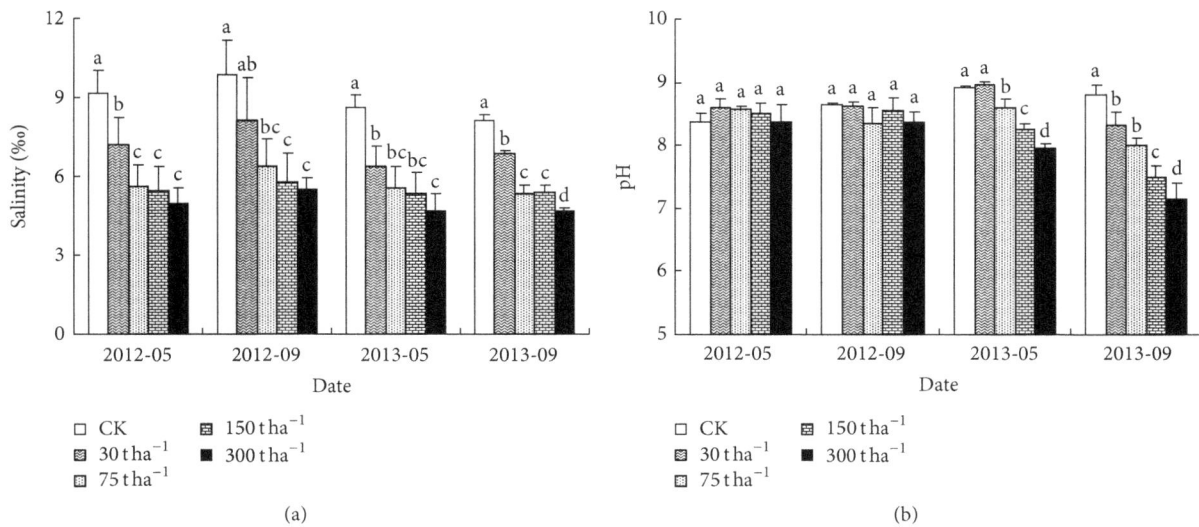

FIGURE 3: Effects of dairy manure amendment on salinity (a) and pH (b) in mudflat saline soil. Vertical bars indicate standard deviations of the means. Columns with different letters show significant difference between dairy manure amendment rates at $p < 0.05$ by LSD's multiple range test.

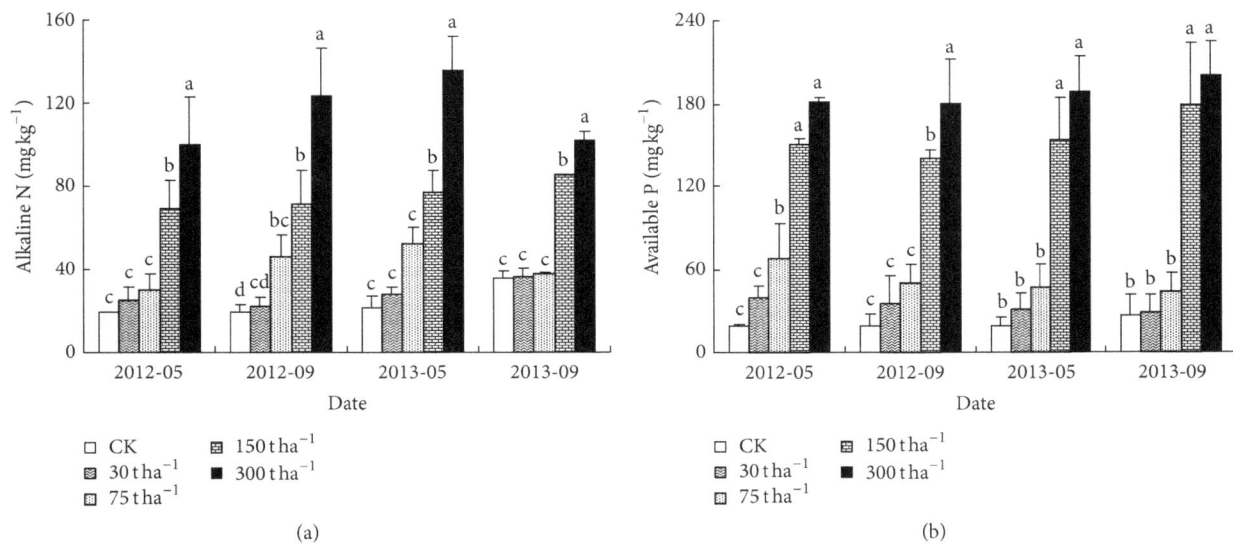

FIGURE 4: Effects of dairy manure amendment on concentrations of alkaline nitrogen (a) and available phosphorus (b) in mudflat saline soil. Vertical bars indicate standard deviations of the means. Columns with different letters show significant difference between dairy manure amendment rates at $p < 0.05$ by LSD's multiple range test. N, nitrogen; P, phosphorus.

The DMA significantly decreased salinity in mudflat saline soil, and salinity decreased with increasing DMA rates in the four experimental dates (Figure 3(a)). However, the green manuring did not significantly change soil salinity. The salinity decreased by 15.0%–26.1%, 34.0%–38.4%, 33.7%–41.3%, and 42.1%–45.9% at 30, 75, 150, and 300 t ha^{-1} DMA rates in the four experimental dates, respectively, compared to the unamended soil.

The application of dairy manure did not significantly affect pH in mudflat saline soil, and the pH decreased with increasing biomass of tilled green manures in May and September 2013 (Figure 3(b)). The pH decreased by 3.1%–5.3%, 7.1%–9.0%, 10.7%–14.7%, and 14.1%–18.8%

at 30, 75, 150, and 300 t ha^{-1} DMA rates in May and September 2013, respectively, compared to the unamended soil.

The dairy manure combined with green manuring significantly increased alkaline N and available P in mudflat saline soil, and the concentrations of alkaline N and available P in the soil increased with increasing DMA rates at the four experimental dates (Figure 4). The alkaline N and available P increased by 2.8%–38.7%, 5.9%–144.0%, 142.2%–280.1%, and 190.7%–544.2% and 6.3%–112.3%, 61.7%–264.7%, 568.8%–707.0%, and 648.0%–872.7% at 30, 75, 150, and 300 t ha^{-1} DMA rates in the four experimental dates, respectively, compared to the unamended soil.

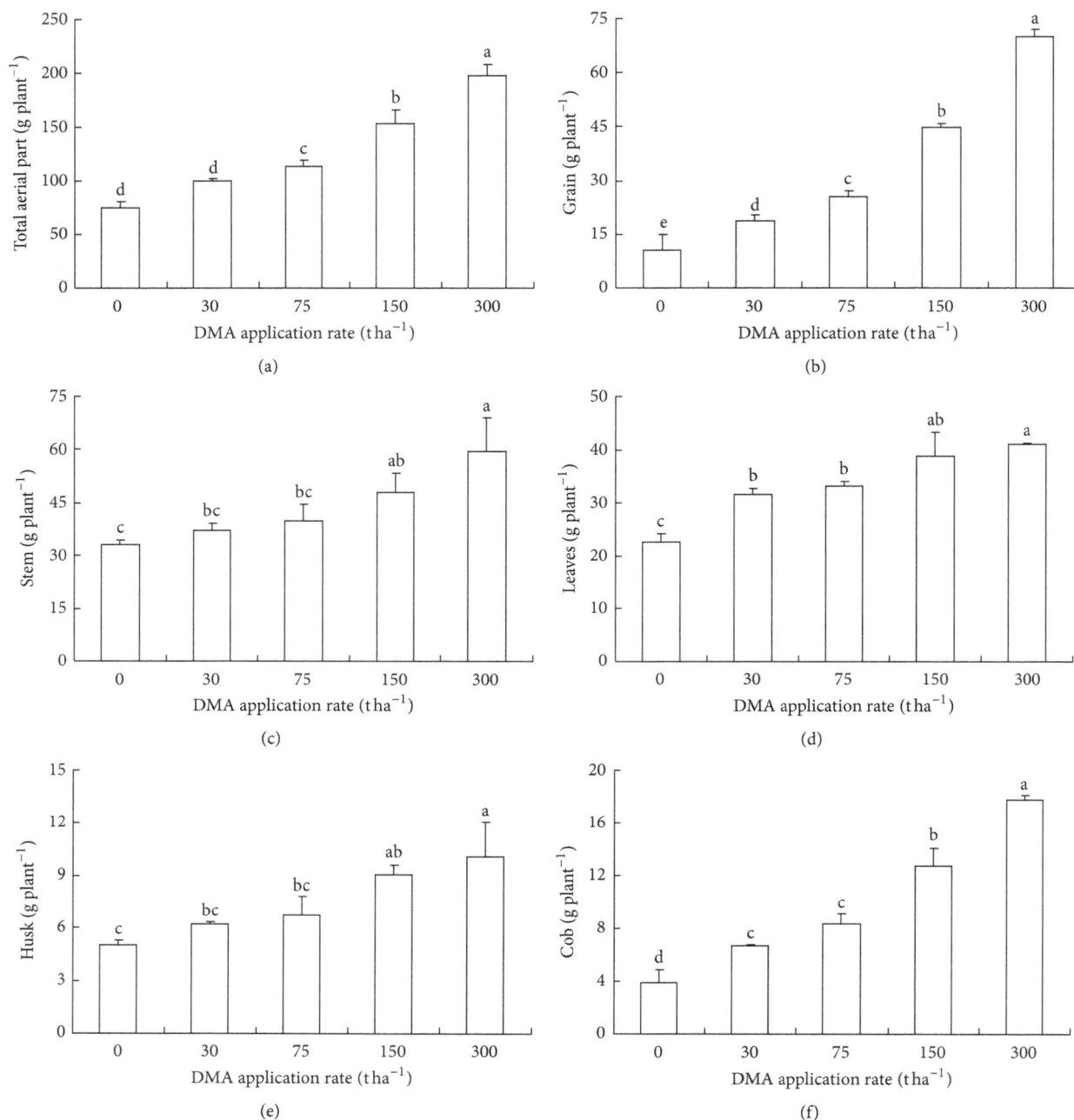

FIGURE 5: Effects of dairy manure amendment (DMA) on dry matter for total aerial part (a), grain (b), stem (c), leaves (d), husk (e), and cob (f) of maize grown in mudflat saline soil. Vertical bars indicate standard deviations of the means. Columns with different letters show significant difference between dairy manure amendment rates at $p < 0.05$ by LSD's multiple range test.

3.3. *Dry Matter of Maize.* Dry matter of total aerial part increased with increasing DMA rates and was significantly higher at all DMA soils than the unamended soil (Figure 5(a)). Grain yield of maize at 30, 75, 150, and 300 t ha^{-1} DMA rates increased by 78.6%, 141.6%, 323.5%, and 562.2%, respectively, as compared to 10.54 g plant^{-1} in the unamended soil (Figure 5(b)). Dry matter for stem, leaves, husk, and cob of maize in mudflat saline soil significantly increased with increasing DMA rates (Figures 5(c)–5(f)). Stem and husk dry matter at 150 and 300 t ha^{-1} DMA rates were 47.86,

59.58 g plant^{-1} and 9.01, 10.07 g plant^{-1}, respectively, which corresponded to significant increases of 45.1%, 80.6% and 78.8%, 99.7%, compared to 32.99 and 5.04 g plant^{-1} in the unamended soil. The increments in leaves and cob dry matter were 39.9%, 47.1%, 71.8%, and 81.7% and 70.0%, 114.6%, 225.4%, and 354.5% at 30, 75, 150, and 300 t ha^{-1} DMA rates, respectively, as compared to the unamended soil.

3.4. *Metals Uptake of Grain in Maize.* The effect of dairy manure combined with green manuring on metal

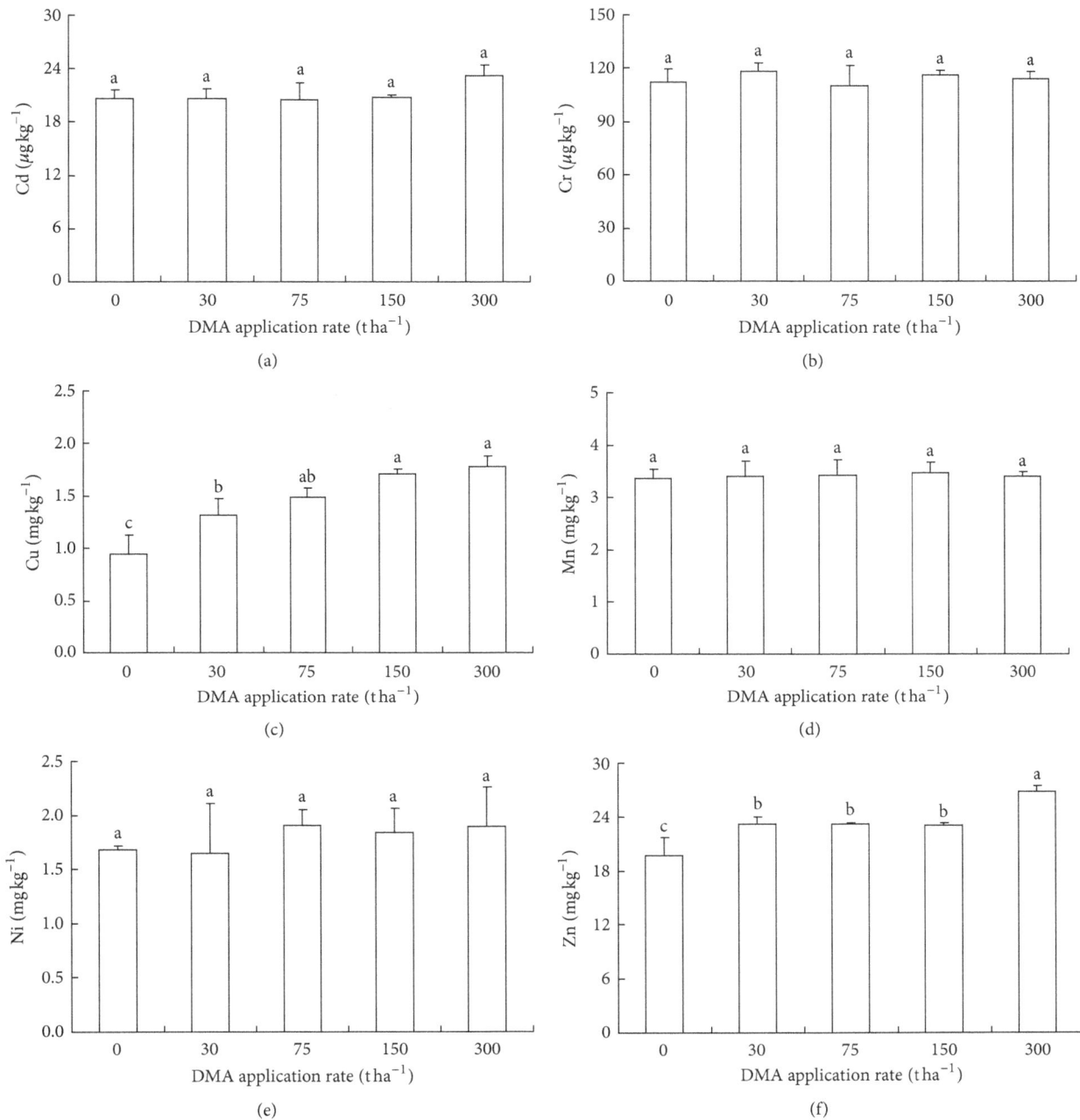

FIGURE 6: Effects of dairy manure amendment (DMA) on heavy metals concentration for Cd (a), Cr (b), Cu (c), Mn (d), Ni (e), and Zn (f) of maize grain. Vertical bars indicate standard deviations of the means. Columns with different letters show significant difference between dairy manure amendment rates at $p < 0.05$ by LSD's multiple range test. Cd, cadmium; Cr, chromium; Cu, copper; Mn, manganese; Ni, nickel; Zn, zinc.

concentrations in grain of maize was shown in Figure 6. Cu and Zn concentrations in grain of maize significantly increased with increasing DMA rates. The increments in Cu and Zn concentrations in grain of maize were 39.4%, 57.3%, 81.8%, and 88.3% and 17.3%, 17.3%, 16.8%, and 35.7% at 30, 75, 150, and 300 t ha^{-1} DMA rates, respectively, as compared to the unamended soil. Cd, Cr, Mn, and Ni concentrations in grain of maize showed no significant changes ($p > 0.05$) between all the treatments.

4. Discussion

The one-time application of dairy manure improved initial fertility formation for infertile mudflat saline soil and promoted growth of ryegrass as the first season green manure. The unamended mudflat saline soil is not suitable for ryegrass growth due to its high salinity and low organic matter content [13, 22]. The present study showed that the one-time application of dairy manure rapidly decreased soil salinity and increased soil organic matter in mudflat saline soil, which provided good conditions for growth of ryegrass as first green manure. In addition, mudflat saline soil was enriched with alkaline N, available P, and other nutrients from DMA, which provided sufficient nutrients for ryegrass growth as the first green manure. Previous studies also found that DMA increased biomass of sorghum-sudangrass [23], pasture grass, and ryegrass [24] grown in croplands.

Biomass of ryegrass, *Sesbania*, and ryegrass as green manures for three consecutive seasons, tilled into the mudflat saline soil, can be decomposed and converted into soil organic carbon, which subsequently improved the chemical properties of mudflat saline soil. In the present study, OC concentration in mudflat saline soil during the 2-year experimental period increased over time and also increased with increasing DMA rates, which were 3.6, 3.7, 5.2, and 5.7-fold increase at 30, 75, 150, and 300 t ha^{-1} DMA rates, respectively, compared to 1.705 g kg^{-1} in the unamended soil at the beginning of this study. In addition, OC concentration in the control soil also increased over time, which is due to added organic matter from green manures, but to a lesser extent during the 2-year experimental period by 3.4-fold increase compared to the control at the beginning of this study. The OC concentrations in mudflat saline soil at all DMA rates were following the increasing trend of tilled total biomass of green manures. Previous studies in cropland showed that tilled green manure increased soil OC concentration [25] and then improved soil aggregation status and decreased the bulk density [9], increased porosity [26], and saturated hydraulic conductivity [27]. Therefore, salinity reduction of mudflat saline soil might be attributed to the fact that OM enrichment by dairy manure combined with green manuring reduced soil bulk density, broke soil capillary, and suppressed salt solution rise through soil capillary [28, 29].

The pH of mudflat saline soil did not show significant variation by one-time application of dairy manure. Previous studies also found that soil pH did not change by dairy manure application in Calcareous soil [30] and Plano soil [31]. Other studies showed that dairy manure application increased pH value of alum shale soil [32] and Rosholt soil [31]. The various effects of DMA on soil pH might be due to original soil pH, OM content, and soil buffering capacity. In this study, the pH of mudflat saline soil significantly decreased after green manuring with *Sesbania*, and the decrement increased with tilling total biomass of *Sesbania*. It might be due to proton release from roots during cultivation of legumes and decomposition of organic N after legumes being tilled [33–35]. Souza et al. [36] also confirmed green manuring with legumes in precrops decreased soil pH.

The dairy manure combined with green manuring promoted supply of N and P nutrients in mudflat saline soil, which provided sufficient nutrients for the succeeding plant. The nutrients come from two sources. On the one hand, there are high N and P concentrations in the dairy manure; on the other hand, planting and tilling green manures might increase N and P accumulation in mudflat saline soil. Legume such as *Sesbania* may accumulate N through the N fixation process, thereby supplying N for the succeeding crop after

it is tilled into soil [37, 38]. Previous studies also found that green manuring with legumes and hairy vetch in precrops significantly increased N and P uptake in rice and oat [38, 39]. In the present study, alkaline N and available P content in mudflat saline soil during 2-year experimental period increased with increasing tilled biomass of green manures. Therefore, the more dairy manure applied, the more biomass of green manures tilled, and the more nutrients held in amended mudflat saline soils.

The dairy manure amendment increased first green manure growth, and green manuring for three consecutive seasons might have enhanced the growth of succeeding maize and increased grain yield of the maize. Previous studies also confirmed that grain yield of maize was significantly enhanced by green manuring [8, 40, 41]. In this study, the 30 t ha^{-1} DMA with green manuring significantly increased biomass and grain yield of maize by 79.9% and 78.6%, respectively. The direct cause of this result was green manuring in the premaize, and planting and tilling green manures improved physicochemical properties of mudflat soil. But the ultimate cause was that DMA decreased salinity and boosted initial fertility for growth of green manures.

The DMA increased Cu and Zn concentrations in grain of maize grown in the mudflat saline soil. Brock et al. [42] also confirmed that the application of dairy manure increased Cu and Zn uptake in maize. Previous studies found that the dairy manure increased Cu and Zn concentration in wheat and barley [32]. In the present study, the increment of Cu and Zn concentration in grain of maize at DMA rates was 39.4–88.3% and 17.3–35.7% compared with the unamended saline soil, which can be attributed to total Cu and Zn concentrations in dairy manure used in this study being 48.4 and 2.6-times higher than those in mudflat saline soil. Both Cu and Zn are usually added to most dairy rations as part of a mineral mix [42]. Cd, Cr, Mn, and Ni concentrations in grain of maize did not show significant changes between DMA and the control, which was due to the fact that Cd, Cr, Mn, and Ni concentrations in the dairy manure were similar to the mudflat soil, and addition of dairy manure did not elevate the concentrations of these metals although it may increase total amount of these metals. In this study, concentrations of heavy metal in grain of maize did not exceed the safety standard for food in China (GB 2762-2012). Therefore, dairy manure combined with green manuring can improve mudflat saline soil chemical properties without compromising environmental safety for heavy metals in crops grown in mudflats. The high input of dairy manure did introduce high amount of nitrogen to the mudflat soil, which may cause environmental concerns. N in dairy manure was mostly in organic form and its release process was slower in mudflat soil than that in fertile farmland. Attention should however be paid to the environmental risk in future research.

5. Conclusions

One-time application of dairy manure for infertile mudflat saline soil promoted growth of ryegrass as the first season green manure. Ryegrass after tilling can be decomposed and converted into soil organic carbon. By the cycles of the planting and tilling ryegrass, *Sesbania*, and ryegrass as green manures for three consecutive seasons, it rapidly improved the chemical properties of mudflat saline soil by decreasing soil salinity and pH and increasing soil organic carbon and available N and P. As a result, dry matter of maize plant parts increased with increasing DMA rates. Cu and Zn concentrations in grain of maize at DMA rates were higher than those in the unamended soil, whereas there were no significant differences ($p > 0.05$) of Cd, Cr, Mn, and Ni concentrations in grain of maize in the DMA soils, compared to the unamended soil. Dairy manure combined with green manuring can be applied for mudflat saline soil amendment, which is a potential win-win solution with respect to new arable land creating, waste disposal, and resource recycling.

Competing Interests

The authors declare that there is no conflict of interests regarding the publication of this paper.

Acknowledgments

This study is supported by National Natural Science Foundation of China (31101604), Fund for Agricultural Independent Innovation of Jiangsu Province [CX(15)1005], Postdoctoral Science Foundation of China (2016M601755), Fund for Ministry of Science and Technology of China (2015BAD01B03), Fund for Important Research and Development of Jiangsu Province (BE2015337), Fund for State Key Laboratory of Soil and Sustainable Agriculture (Y412201402), Fund for Ministry of Housing and Urban-Rural Department of China (2014-K6-009), Fund for Three New Agricultural Project of Jiangsu Province (SXGC[2016]277), Fund for Environmental Protection of Yangzhou City (YHK1414), and Shuangchuang Talent Plan of Jiangsu Province, China.

References

[1] F. Wang and G. Wall, "Mudflat development in Jiangsu Province, China: practices and experiences," *Ocean and Coastal Management*, vol. 53, no. 11, pp. 691–699, 2010.

[2] W. Cao and M. H. Wong, "Current status of coastal zone issues and management in China: a review," *Environment International*, vol. 33, no. 7, pp. 985–992, 2007.

[3] D. Song, X. H. Wang, X. Zhu, and X. Bao, "Modeling studies of the far-field effects of tidal flat reclamation on tidal dynamics in the East China Seas," *Estuarine, Coastal and Shelf Science*, vol. 133, pp. 147–160, 2013.

[4] K. Guo and X. Liu, "Dynamics of meltwater quality and quantity during saline ice melting and its effects on the infiltration and desalinization of coastal saline soils," *Agricultural Water Management*, vol. 139, pp. 1–6, 2014.

[5] Y. Bai, C. Gu, T. Tao, L. Wang, K. Feng, and Y. Shan, "Growth characteristics, nutrient uptake, and metal accumulation of ryegrass (*Lolium perenne* L.) in sludge-amended mudflats," *Acta Agriculturae Scandinavica Section B: Soil and Plant Science*, vol. 63, no. 4, pp. 352–359, 2013.

[6] R. J. Yao, J. S. Yang, D. H. Wu, W. P. Xie, P. Gao, and X. P. Wang, "Characterizing spatial–temporal changes of soil and

crop parameters for precision management in a coastal rainfed agroecosystem," *Agronomy Journal*, vol. 108, no. 6, p. 2462, 2016.

[7] R. J. Yao, J. S. Yang, P. Gao, J. B. Zhang, and W. H. Jin, "Determining minimum data set for soil quality assessment of typical salt-affected farmland in the coastal reclamation area," *Soil and Tillage Research*, vol. 128, pp. 137–148, 2013.

[8] T. O. Fabunmi, M. U. Agbonlahor, J. N. Odedina, and S. O. Adigbo, "Profitability of pre-season green manure practices using maize as a test crop in a derived savanna of Nigeria," *Pakistan Journal of Agricultural Sciences*, vol. 49, no. 4, pp. 593–596, 2012.

[9] M. Wiesmeier, M. Lungu, R. Hübner, and V. Cerbari, "Remediation of degraded arable steppe soils in Moldova using vetch as green manure," *Solid Earth*, vol. 6, no. 2, pp. 609–620, 2015.

[10] R. V. Valadares, R. F. Duarte, J. B. C. Menezes et al., "Soil fertility and maize yields in green manure systems in northern Minas Gerais," *Planta Daninha*, vol. 30, no. 3, pp. 505–516, 2012.

[11] L. Talgre, E. Lauringson, H. Roostalu, A. Astover, and A. Makke, "Green manure as a nutrient source for succeeding crops," *Plant, Soil and Environment*, vol. 58, no. 6, pp. 275–281, 2012.

[12] T. Jun, G. Wei, B. Griffiths, L. Xiaojing, X. Yingjun, and Z. Hua, "Maize residue application reduces negative effects of soil salinity on the growth and reproduction of the earthworm Aporrectodea trapezoides, in a soil mesocosm experiment," *Soil Biology and Biochemistry*, vol. 49, pp. 46–51, 2012.

[13] T. Zhang, T. Wang, K. S. Liu, L. Wang, K. Wang, and Y. Zhou, "Effects of different amendments for the reclamation of coastal saline soil on soil nutrient dynamics and electrical conductivity responses," *Agricultural Water Management*, vol. 159, pp. 115–122, 2015.

[14] L. Wu, M. Cheng, Z. Li et al., "Major nutrients, heavy metals and PBDEs in soils after long-term sewage sludge application," *Journal of Soils and Sediments*, vol. 12, no. 4, pp. 531–541, 2012.

[15] R. A. Blaustein, R. L. Hill, S. A. Micallef, D. R. Shelton, and Y. A. Pachepsky, "Rainfall intensity effects on removal of fecal indicator bacteria from solid dairy manure applied over grass-covered soil," *Science of the Total Environment*, vol. 539, pp. 583–591, 2016.

[16] P. J. A. Kleinman, P. Salon, A. N. Sharpley, and L. S. Saporito, "Effect of cover crops established at time of corn planting on phosphorus runoff from soils before and after dairy manure application," *Journal of Soil and Water Conservation*, vol. 60, no. 6, pp. 311–322, 2005.

[17] A. N'Dayegamiye, "Response of silage corn and wheat to dairy manure and fertilizers in long-term fertilized and manured trials," *Canadian Journal of Soil Science*, vol. 76, no. 3, pp. 357–363, 1996.

[18] F. Domingo-Olivé, À. D. Bosch-Serra, M. R. Yagüe, R. M. Poch, and J. Boixadera, "Long term application of dairy cattle manure and pig slurry to winter cereals improves soil quality," *Nutrient Cycling in Agroecosystems*, vol. 104, no. 1, pp. 39–51, 2016.

[19] D. H. Min, K. R. Islam, L. R. Vough, and R. R. Weil, "Dairy manure effects on soil quality properties and carbon sequestration in alfalfa-orchardgrass systems," *Communications in Soil Science and Plant Analysis*, vol. 34, no. 5-6, pp. 781–799, 2003.

[20] C. Mallol, "What's in a beach? Soil micromorphology of sediments from the Lower Paleolithic site of 'Ubeidiya, Israel," *Journal of Human Evolution*, vol. 51, no. 2, pp. 185–206, 2006.

[21] S. D. Bao, *Soil and Agro-Chemistry Analysis*, China Agricultural Press, Beijing, China, 3rd edition, 2000.

[22] Y. Bai, T. Tao, C. Gu, L. Wang, K. Feng, and Y. Shan, "Mudflat soil amendment by sewage sludge: soil physicochemical properties, perennial ryegrass growth, and metal uptake," *Soil Science and Plant Nutrition*, vol. 59, no. 6, pp. 942–952, 2013.

[23] H. M. Waldrip, Z. He, and T. S. Griffin, "Effects of organic dairy manure on soil phosphatase activity, available soil phosphorus, and growth of Sorghum-Sudangrass," *Soil Science*, vol. 177, no. 11, pp. 629–637, 2012.

[24] D. Espinosa, P. W. G. Sale, and C. Tang, "Changes in pasture root growth and transpiration efficiency following the incorporation of organic manures into a clay subsoil," *Plant and Soil*, vol. 348, no. 1-2, pp. 329–343, 2011.

[25] M. A. Rudisill, B. P. Bordelon, R. F. Turco, and L. A. Hoagland, "Sustaining soil quality in intensively managed high tunnel vegetable production systems: a role for green manures and chicken litter," *HortScience*, vol. 50, no. 3, pp. 461–468, 2015.

[26] W. T. Jeon, B. Choi, S. A. M. Abd El-Azeem, and Y. S. Ok, "Effect of different seeding methods on green manure biomass, soil properties and rice yield in rice-based cropping systems," *African Journal of Biotechnology*, vol. 10, no. 11, pp. 2024–2031, 2011.

[27] K. Panitnok, S. Thongpae, E. Sarobol et al., "Effects of vetiver grass and green manure management on properties of map bon, coarse-loamy variant soil," *Communications in Soil Science and Plant Analysis*, vol. 44, no. 1–4, pp. 158–165, 2013.

[28] M. H. Jorenush and A. R. Sepaskhah, "Modelling capillary rise and soil salinity for shallow saline water table under irrigated and non-irrigated conditions," *Agricultural Water Management*, vol. 61, no. 2, pp. 125–141, 2003.

[29] R.-J. Yao, J.-S. Yang, T.-J. Zhang et al., "Studies on soil water and salt balances and scenarios simulation using SaltMod in a coastal reclaimed farming area of eastern China," *Agricultural Water Management*, vol. 131, pp. 115–123, 2014.

[30] A. B. Leytem, R. S. Dungan, and A. Moore, "Nutrient availability to corn from dairy manures and fertilizer in a calcareous soil," *Soil Science*, vol. 176, no. 8, pp. 426–434, 2011.

[31] Z. Wu and J. M. Powell, "Dairy manure type, application rate, and frequency impact plants and soils," *Soil Science Society of America Journal*, vol. 71, no. 4, pp. 1306–1313, 2007.

[32] A. K. M. Arnesen and B. R. Singh, "Plant uptake and DTPA-extractability of Cd, Cu, Ni and Zn in a Norwegian alum shale soil as affected by previous-addition of dairy and pig manures and peat," *Canadian Journal of Soil Science*, vol. 78, no. 3, pp. 531–539, 1998.

[33] F. Yan, S. Schubert, and K. Mengel, "Soil pH changes during legume growth and application of plant material," *Biology and Fertility of Soils*, vol. 23, no. 3, pp. 236–242, 1996.

[34] F. Yan, S. Schubert, and K. Mengel, "Soil pH increase due to biological decarboxylation of organic anions," *Soil Biology and Biochemistry*, vol. 28, no. 4-5, pp. 617–624, 1996.

[35] S. Pocknee and M. E. Sumner, "Cation and nitrogen contents of organic matter determine its soil liming potential," *Soil Science Society of America Journal*, vol. 61, no. 1, pp. 86–92, 1997.

[36] J. L. Souza, G. P. Guimarães, and L. F. Favarato, "Development of vegetables and soil characteristics after green manuring and organic composts under levels of N," *Horticultura Brasileira*, vol. 33, no. 1, pp. 19–26, 2015.

[37] L. O. Brandsæter, H. Heggen, H. Riley, E. Stubhaug, and T. M. Henriksen, "Winter survival, biomass accumulation and N mineralization of winter annual and biennial legumes sown at various times of year in Northern Temperate Regions," *European Journal of Agronomy*, vol. 28, no. 3, pp. 437–448, 2008.

[38] Z. Shah, S. R. Ahmad, and H. U. Rahman, "Sustaining rice-wheat system through management of legumes I: effect of green manure legumes on rice yield and soil quality," *Pakistan Journal of Botany*, vol. 43, no. 3, pp. 1569–1574, 2011.

[39] A. Tarui, A. Matsumura, S. Asakura, K. Yamawaki, R. Hattori, and H. Daimon, "Evaluation of mixed cropping of oat and hairy vetch as green manure for succeeding corn production," *Plant Production Science*, vol. 16, no. 4, pp. 383–392, 2013.

[40] G. S. A. Castro and C. A. C. Crusciol, "Effects of surface application of dolomitic limestone and calcium-magnesium silicate on soybean and maize in rotation with green manure in a tropical region," *Bragantia*, vol. 74, no. 3, pp. 311–321, 2015.

[41] J.-S. Bai, W.-D. Cao, J. Xiong, N.-H. Zeng, S.-J. Gao, and S. Katsuyoshi, "Integrated application of February Orchid (*Orychophragmus violaceus*) as green manure with chemical fertilizer for improving grain yield and reducing nitrogen losses in spring maize system in northern China," *Journal of Integrative Agriculture*, vol. 14, no. 12, pp. 2490–2499, 2015.

[42] E. H. Brock, Q. M. Ketterings, and M. McBride, "Copper and zinc accumulation in poultry and dairy manure-amended fields," *Soil Science*, vol. 171, no. 5, pp. 388–399, 2006.

14

Assessment of a New Approach for Systematic Subsurface Drip Irrigation Management

Hédi Ben Ali,[1] Moncef Hammami,[2] Ahmed Saidi,[3] and Rachid Boukchina[4]

[1]*Agence de Promotion des Investissements Agricoles, 6000 Gabès, Tunisia*
[2]*Laboratory of Hydraulic, High School of Engineers of Rural Equipment, Medjez el Bab, Tunisia*
[3]*National Research Institute of Rural Engineering, Water and Forests (INRGREF), Rue Hédi EL Karray El Menzah IV,*
  BP 10, 2080 Ariana, Tunisia
[4]*Institut des Régions Arides, 6000 Gabès, Tunisia*

Correspondence should be addressed to Ahmed Saidi; saidiahmed44@gmail.com

Academic Editor: Manuel Tejada

This paper aimed to assess the reliability of a new approach that provides systematic irrigation management based on fixed water suction in the vadose zone. Trials were carried out in the experimental farm of IRA Gabès on subsurface drip irrigated (SDI) tomato plot. The SDI system was designed so that the soil water content is to be maintained within prescribed interval ascertaining the best plant growth. Irrigation management was systematically monitored by water suction evolution in the vadose zone. Recorded results showed that all-over irrigation season lateral pressure head ranged within 93.3 ± 20.0; 119.95 ± 53.35 and 106.6 ± 40.0 mb, respectively, at the upstream, middle, and downstream. The correspondent lateral pressure head distribution uniformity ranged within 97.1% and 99.6%. Soil water content varied within 0.2175 ± 0.0165; 0.206 ± 0.0195 and 0.284 ± 0.100 beneath the inlet, the behalf, and the lateral end tip. The correspondent soil water distribution uniformity was higher than 80.7% all-over irrigation season. Based on the recorded results, the proposed approach could be a helpful tool for accurate SDI systems design and best water supplies management. Nevertheless, further trials are needed to assess the approach reliability in different cropping conditions.

1. Introduction

Water scarcity is among the main problems to be faced by many societies and the world in the 21st century [1]. The use of water-efficient irrigation is one of the most practical options to reduce global water scarcity [2]. Subsurface drip irrigation (SDI) provides the opportunity to record consistently water use efficiency over traditional methods, including surface drip irrigation (DI) [3–5]. Several field trials revealed relevant profits on managing SDI for crops' production. In fact, SDI system allows the direct application of water to the rhizosphere maintaining dry the nonrooted top soil. This pattern generates numerous advantages such as minimizing soil evaporation and then evapoconcentration phenomenon [6]. Comparing evaporation from surface and subsurface drip irrigation systems, Evett et al. [7] reported that 51 and 81 mm were saved with drip laterals buried at 15 cm and 30 cm,

respectively. Patel and Rajput [8] recorded maximum onion yield (25.7 t ha^{-1}) with drip laterals buried at 10 cm, whereas Ombódi et al. [9] recorded an average yield ranging between 40.7 and 54.6 t ha^{-1} for onion in irrigated conditions.

Also, with SDI systems more uniform moisture distribution, in the vadose zone (than with drip irrigation systems), was observed, and thus drainage and surface evaporation were less with SDI [10, 11]. Automation of irrigation systems has the potential to provide maximum water use efficiency by maintaining soil moisture within an optimal interval ascertaining the best plant growth [6].

This experimental study aimed to assess the reliability of a new approach for SDI laterals' design accounting for the soil water-retention characteristics and the roots water extraction. The proposed approach provides systematic irrigation management based on fixed water suctions in the vadose zone.

2. Material and Methods

2.1. Site. Experiments were carried out in the experimental station of Arid Regions Institute of Chenchou (Gabès) whose geographical coordinates are latitude = 33.88° North, longitude = 9.79° East, and at an altitude of 59 m. Average monthly temperature ranges were between 10.4°C (January) and 28.6°C (August). Average annual rainfall is 162 mm while potential evapotranspiration (ETP) is 1430 mm/year.

Field trials were performed from May 26th up to September the 15th 2014 in tomato (Feranzi variety) plot (86.0 × 8.0 m^2). Seedlings' rows were 1.60 m distant while crop plants were 0.40 m apart. Each row crop was irrigated by a single SDI lateral buried at 15 cm depth. According to Najafi [12] and Zotarelli et al. [13], tomato crop irrigated with laterals buried at Z_d = 15 cm depth's leads to the better yields, whereas Machado and Oliveira [14] found that tomato roots were concentrated mainly within the [0–40 cm] top soil layer under DI and SDI irrigation systems.

For soil physical characterization, four representative profiles were randomly chosen (within the plot). In each profile, soil samples were collected on four layers: 0–20, 20–40, 40–60, and 60–80 cm. Analyses were focused on properties that account for soil moisture holding and water suction evolution, namely: texture, bulk density (D_a), and water content-pressure head relationship.

2.2. Method. According to Hammami et al. [6], the minimum pressure required at the upstream end of nontapered flat SDI lateral is

$$h_{Lm} = Z_d + J_L + \Delta h_{\min} + h_{op} - \Delta h_{op} \quad (1)$$

whereas the maximum pressure head (h_{LM}) required at the upstream end of the lateral is

$$h_{LM} = Z_d + J_L + \Delta h_{\min} + h_{op} + \Delta h_{op} \quad (2)$$

with h_{Lm} and h_{LM} being minimum and maximum required pressure heads (m) at the beginning of the lateral. Z_d is laterals depth of burial (m). J_L are total pressure head losses (m) along the lateral. h_{op} is optimal soil water suction (m) for crop's growth. Δh_{op} is interval of variation of the optimal soil suction (m). $\Delta h_{\min}$ is minimum differential pressure head for emitters operating.

According to Hammami et al. [6], the soil capillary capacity (C) is the highest if the second derivative of the soil moisture content with respect to the suction head is zero. Thus, using van Genuchten [15] model,

$$\theta(h) = \theta_r + \frac{(\theta_s - \theta_r)}{(1 + (\alpha |h|)^n)^m}. \quad (3)$$

The optimal suction is straight fully derived as follows:

$$h_{op} = -\frac{m^{1/n}}{\alpha}. \quad (4)$$

Nonlinear adjustment of discrete data (θ, h) allows deducing θ_r, α, and n values from the fitted expression $\theta(h)$ (14).

Substituting m, α, and n in (4) gives the correspondent h_{op} = −1.47 cm.

Gärdenäs et al. [16] reported that tomato crop tolerates (without noticeable yield decrease) a soil water pressure variation in the interval [−800, −2 cm]. Then, $\Delta h_{op} = \pm 400$ cm was considered. Therefore, for an optimal tomato crop's growth, the soil water pressure (h) should be maintained within the interval as follows:

$$h_{op} + 400 \geq h \text{ (cm)} \geq h_{op} - 400 \Longleftrightarrow$$
$$398.53 \geq h \text{ (cm)} \geq -401.47 \text{ cm.} \quad (5)$$

To avoid any soil saturation risk, we retained

$$00.00 > h \text{ (cm)} \geq -401.47. \quad (5')$$

Consequently, the correspondent optimum water content (θ_{op}) should be maintained within the interval as follows:

$$0.385 > \theta_{op} \geq 0.184 \text{ cm}^3 \text{ cm}^{-3}. \quad (6)$$

A minimum value $\Delta h_{\min}$ for the emitter operation is required. This threshold $\Delta h_{\min}$ is dependent on the structural form, dimension, and material of the emitter pathway. For any emitter model, $\Delta h_{\min}$ may be inferred from the emitter discharge-pressure head relationship provided by the manufacturer. Then, the minimum pressure $h^*_{\min}$ into emitter should respect the following condition:

$$h^*_{\min} \geq h_{op} + \Delta h_{\min}. \quad (7)$$

A trapezoidal labyrinth long-path emitter with a minimal differential operating pressure head of $\Delta h_{\min} = 500$ cm was used; then

$$h^*_{\min} \geq 498.53 \text{ cm.} \quad (8)$$

So, the required pressure in the emitters should be between $h^{\min}_{req}$ and $h^{\max}_{req}$, with

$$h^{\min}_{req} = h_{op} - \Delta h_{op} + \Delta h_{\min} \Longrightarrow$$
$$h^{\min}_{req} = 98.53 \text{ cm}$$
$$h^{\max}_{req} = h_{op} + \Delta h_{op} + \Delta h_{\min} \Longrightarrow$$
$$h^{\max}_{req} = 500.0 \text{ cm.} \quad (9)$$

Since the pressure head, in the soil around the laterals, should vary between −401.47 and 00.00 cm, emitters discharge q (l/h) should be maintained between

$$2.15 \geq q \text{ (l/h)} \geq 0.75. \quad (10)$$

Each lateral is equipped with $N = 86/0.4 = 215$ emitters; therefore its flow rate Q should comply with

$$160.5 \leq Q \text{ (l/h)} \leq 462.5. \quad (11)$$

The proper laterals' diameter used to ensure the maximum discharge ($Q_{\max}$ = 462.5 l/h) was Ø = 16 mm. Thus using Watters and Keller [17] formula, the total lateral's pressure head loss is equal to

$$J_L = 78 \text{ cm}. \tag{12}$$

Finally, the maximum inlet lateral pressure head was determined using (2) as follows:

$$h_{LM} = 641.53 \text{ cm}. \tag{13}$$

In order to maintain the lateral inlet pressure head (h_{LI}) constant (equal to or less than 641.53 cm), two interconnected reservoirs were used. Water was pumped to the first reservoir (capacity = 120.0 m^3) that supplies the second one (capacity = 1.00 m^3) which diverts water to the irrigation network. The water level inside the second reservoir was maintained constant thanks to a mechanical float. The pump was controlled by an electric float (Figure 1).

To record lateral's pressure head (h_L), suction, and the correspondents soil water content $\theta(h)$ spatial-temporal evolutions, three measurement sites were set along the lateral: at the inlet $X = 0.0$ m, at the behalf $X = L/2$, and at the end tip $X = L$. In each measurement site, the installed pieces of equipment were a U piezometer (connected on the lateral), three TDR access tubes, and 9 Watermark probes (three probes per layer buried at the distances $R = 0.0$ cm; $R = 16.0$ cm; and $R = 32.0$ cm perpendicular to the lateral (Figure 2)). Soil water content values were recorded for the following depths: $Z = 10; 15; 30; 50;$ and 70 cm. A water meter device has been installed at the laterals' inlet in order to record the delivered water volume. Simultaneously lateral flow rate was measured several times a day. Such measurements allow determining the average daily flow rate variation (from crop transplantation to harvest season). In sum, the following variables were recorded:

(i) The spatial-temporal soil water content $\theta(x, z, t)$ variation within the root zone around the lateral.

(ii) The spatial-temporal soil water suction $h(x, z, t)$ variation.

(iii) The spatial-temporal pressure head $h_L(x, t)$ variation inside the lateral.

(iv) The temporal lateral's flow rate $Q(t)$ variation.

3. Results and Discussions

3.1. Physical Soil Characteristics. Mean values of particle size proportion, bulk density (D_a), and soil water contents (at saturation θ_s, field capacity θ_c, and wilting point θ_w) for the four sampled soil layers are summarized in Table 1. These results showed that clay and silt proportions are relatively equiponderant all-over the soil profile while sand proportion decreases from the surface up to 60 cm depth. So the experimental plot is loamy sand textured soil all-over the profile but becomes as fine as it is deep. The bulk density and the soil holding capacity (roughly 100 mm/m) values confirm such texture tendency.

The θ_i and their h_i correspondent values (measured in situ) were fitted to van Genuchten [15] formula, using RETC

model (Figure 3). So the inferred analytical expression of the soil retention curve was

$$\theta = 0.096 + \frac{0.289}{\left(1 + (0.01321 \, |h|)^{4.319}\right)^{0.768}}. \tag{14}$$

3.2. Soil Moisture Distribution. Temporal soil water content θ evolution in the soil depth $Z = 10$ cm, at the inlet ($X = 0$), at the behalf ($X = L/2$), and at the lateral end tip ($X = L$), is depicted in Figure 4. All-over irrigation season, recorded θ values ranged within $0.385 > \theta \geq 0.184$ cm^3 cm^{-3} for $X = 0$ and $X = L/2$. Then, it was maintained within the predicted interval $0.385 > \theta_{op} \geq 0.184$ cm^3 cm^{-3} (6) optimal for the tomato growth, while, underneath lateral end tip ($X = L$), θ values were almost slightly lower than 0.184 cm^3 cm^{-3}. This difference could be attributed to the total pressure head losses occurring along the lateral that subsequently induces a slight emitter discharge decrease. Safi et al. [18] reported that an increase of SDI laterals' length leads to a decrease of all uniformity parameters. Also, such discrepancy could be due to measurement errors on θ and/or h values. Haverkamp et al. [19] reported that an error of only 2% of θ value could cause a relative error of 24% of soil water pressure head. The same trends of the soil moisture distribution were recorded in the soil depth $Z = 15$ cm (Figure 5), where θ remained higher than the minimum prescribed threshold $\theta(h_{op} - \Delta h_{op}) = 0, 184$ cm^3 cm^{-3}, at $X = 0$ and $X = L/2$ but still slightly lower than that threshold at the lateral end tip. Such trend confirms the above finding. In the soil depth $Z = 30$ cm, water content values remained roughly confused with the prescribed minimum threshold (at the inlet $X = 0$) at the lateral behalf ($X = L/2$) but slightly lower (at the end tip $X = L$) than such threshold 0.184 cm^3 cm^{-3} (Figure 6). However, in the deeper soil layers $Z = 50$ cm and $Z = 70$ cm, water content values remained approximately invariant lower than 0.184 cm^3 cm^{-3} all-over irrigation season and for whole lateral length (Figures 7 and 8). These results could be explained by the fact that supplied water (by the lateral) was not so enough to reach such depths. So, deep water and then nutrients losses were negligible. Thus the used approach could be useful tool to improve SDI irrigation efficiency.

The above results validate the systematic SDI irrigation management. Lazarovitch et al. [20] proved that soil hydraulic properties affect outlets flow rate in SDI irrigation system. To assess the water distribution uniformity along the laterals, we determined the coefficient of uniformity (CU) values throughout irrigation season.

$$CU = \left(1 - \frac{\sum |\theta_a - \theta_{(xi,zi)}|}{N \cdot \theta_a}\right) 100, \tag{15}$$

where θ_a is average soil water content for different depths in the three soil profiles ($X = 0$, $X = L/2$, and $X = L$) and, at a given date, $\theta_{(xi,zi)}$ is soil water content in the coordinates (xi, zi) and N is number of the sampled points.

The recorded CU values are always higher than 80.7% and the mean value was 84,3%. These results confirm those of Ben

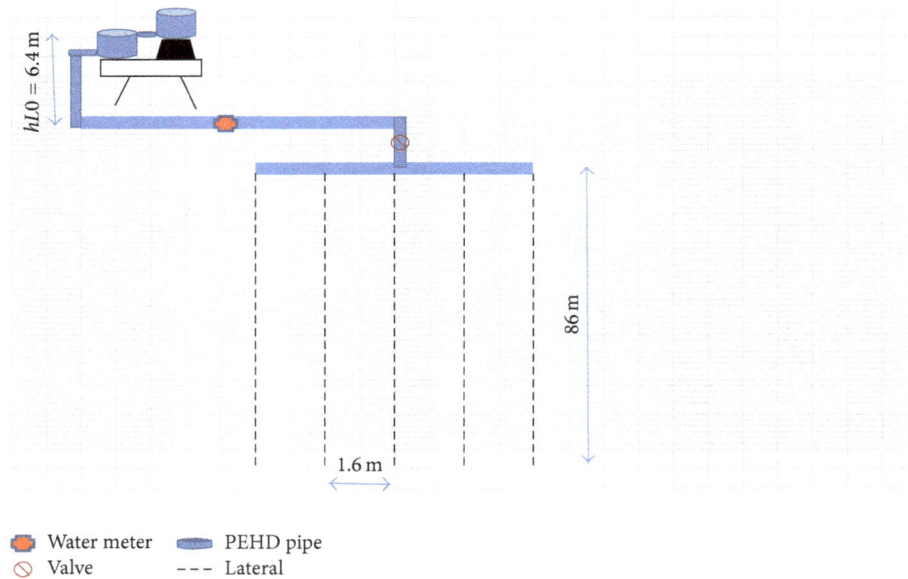

FIGURE 1: Experimental layout scheme.

TABLE 1: Soil physical characteristics.

Soil layer (cm)	Sand (%)	Silt (%)	Clay (%)	D_a (g/cm^3)	θ_c (%)	θ_w (%)	θ_s (%)
0–20	22	73	5	1.56	19.3	10.5	38.5
20–40	19	77	4	1.62	21.4	10.1	36.3
40–60	11	85	4	1.62	18.7	8.8	37.4
60–80	17	80	3	1.53	17.2	7.4	42.2

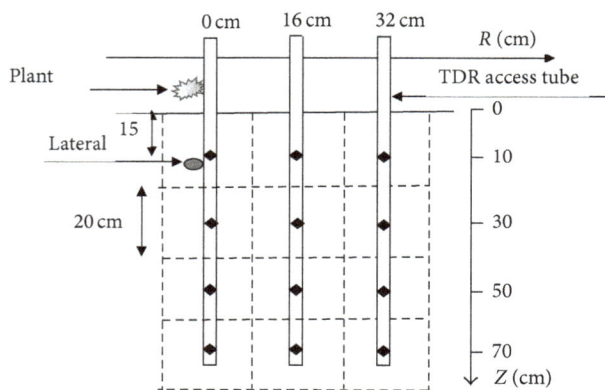

FIGURE 2: Profile of a measurement site.

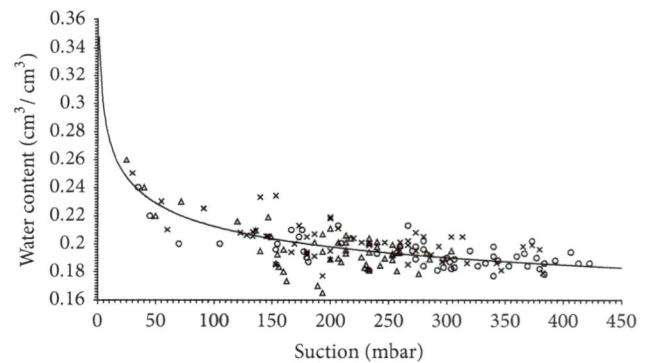

FIGURE 3: Soil retention curve $\theta(h)$ fitted (solid line) against experimental data determined at $Z = 10$ cm ($\times$), $Z = 30$ cm ($\triangle$), and $Z = 50$ cm (o) depths.

Ali et al. [3] who reported that soil water content underneath SDI system was always higher and especially varied within narrower interval than under drip irrigation system. Gil et al. [21] recorded a lower variability of buried emitters' discharges compared to on surface ones.

3.3. *Soil Suction Distribution.* Temporal soil suction evolution in the depth $Z = 15$ cm for the three sites $X = 0$;

$X = L/2$; and $X = L$ along the lateral is shown in Figure 9. Throughout irrigation season, the soil pressure (h) varied within the following intervals: $[-73.3 \geq h\,(\mathrm{mb}) \geq -113.3]$, $[-66.6 \geq h\,(\mathrm{mb}) \geq -173.3]$, and $[-66.66 \geq h\,(\mathrm{mb}) \geq -146.7]$, respectively, at the abscissas ($X = 0$), ($X = L/2$), and $X = L$. Thus it was ranged within the optimal predicted values (5'). Yet, neither saturation risks nor deep percolation

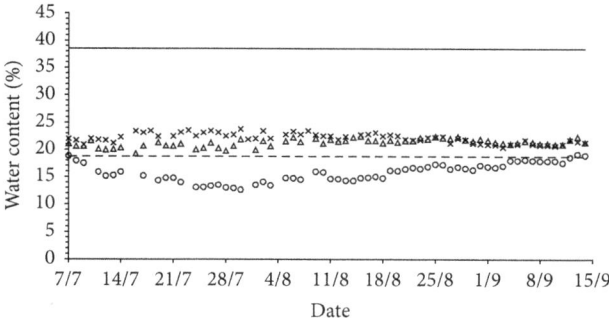

FIGURE 4: Temporal soil water content variation in the soil depth $Z = 10$ cm: at the inlet $X = 0$ ($\times$), at the behalf $X = L/2$ ($\triangle$), and at the lateral end tip $X = L$ (o) against $(\theta_{op} + \Delta\theta_{op})$ solid and $(\theta_{op} - \Delta\theta_{op})$ dashed lines.

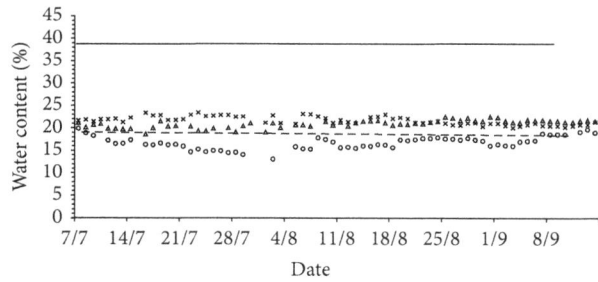

FIGURE 5: Temporal soil water content variation in the soil depth $Z = 15$ cm: at the inlet $X = 0$ ($\times$), at the behalf $X = L/2$ ($\triangle$), and at the lateral end tip $X = L$ (o) against $(\theta_{op} + \Delta\theta_{op})$ solid and $(\theta_{op} - \Delta\theta_{op})$ dashed lines.

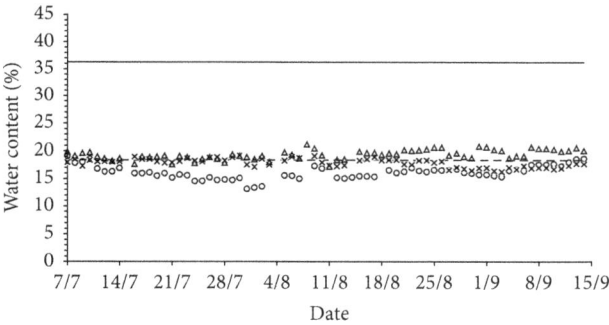

FIGURE 6: Temporal soil water content variation in the soil depth $Z = 30$ cm: at the inlet $X = 0$ ($\times$), at the behalf $X = L/2$ ($\triangle$), and at the lateral end tip $X = L$ (o) against $(\theta_{op} + \Delta\theta_{op})$ solid and $(\theta_{op} - \Delta\theta_{op})$ dashed lines.

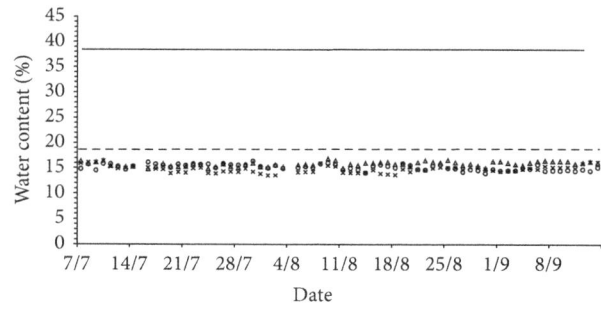

FIGURE 7: Temporal soil water content variation in the soil depth $Z = 50$ cm: at the inlet $X = 0$ ($\times$), at the behalf $X = L/2$ ($\triangle$), and at the lateral end tip $X = L$ (o) against $(\theta_{op} + \Delta\theta_{op})$ solid and $(\theta_{op} - \Delta\theta_{op})$ dashed lines.

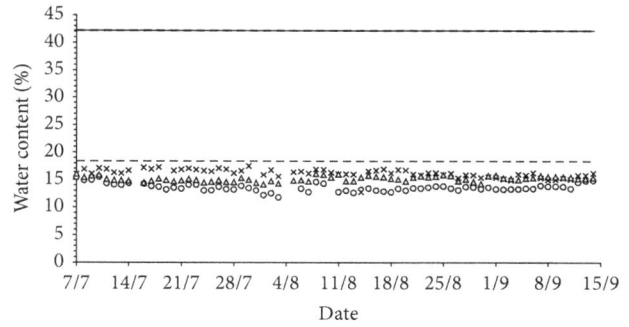

FIGURE 8: Temporal soil water content variation in the soil depth $Z = 70$ cm: at the inlet $X = 0$ ($\times$), at the behalf $X = L/2$ ($\triangle$), and at the lateral end tip $X = L$ (o) against $(\theta_{op} + \Delta\theta_{op})$ solid and $(\theta_{op} - \Delta\theta_{op})$ dashed lines.

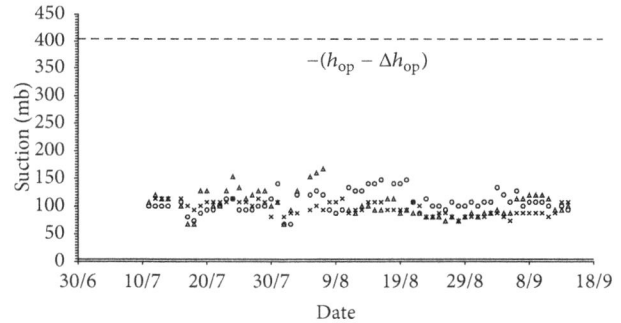

FIGURE 9: Temporal soil suction variation around the inlet $X = 0$ ($\times$), the behalf $X = L/2$ ($\triangle$), and the lateral end tip $X = L$ (o) compared to the minimum (dashed line) and the maximum (solid line) required values.

water losses were recorded. Yao et al. [22] reported that the back pressure risk (or over pressure) occurring underneath subsurface lateral could be addressed by rigorous network design.

3.4. Lateral Pressure Head and Flow Rate. Because the pressure head H in the supplying reservoir was maintained constant equal to 641.53 cm, the pressure head at the lateral inlet ($X = 0$) remained also constant ($H \approx 640$ cm). However, H values inside the behalf and at the lateral end

tip were slightly lowered (ranged between 600 and 640 cm) (Figure 10). Such slight variation could be attributed to the linear and nonlinear head losses along the lateral. Though the lateral inlet pressure head was maintained constant, the correspondent flow rate Q_L was noticeably variable within $236 \geq Q_L(l/h) \geq 184$ but ranged within the fixed interval (11). Such variation could be explained by the soil (around the lateral) suction variation due to the soil water redistribution enhanced essentially by roots' water uptake. It should be

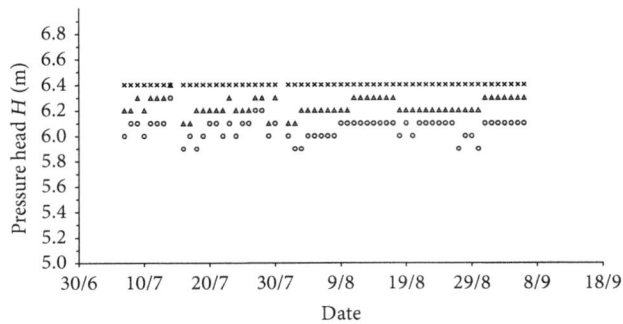

FIGURE 10: Pressure head values recorded at the inlet $X = 0$ (×), the behalf $X = L/2$ (△), and at the lateral end tip $X = L$ (o).

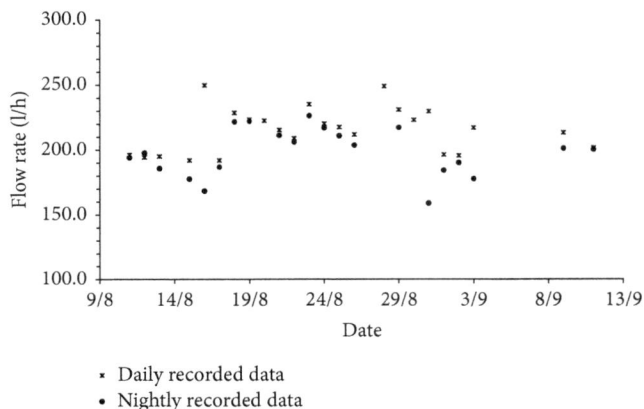

- Daily recorded data
- Nightly recorded data

FIGURE 11: Lateral inlet discharge variation.

stressed that daily lateral discharge was always higher than nightly one (Figure 11). This slight difference (between daily and nightly discharges) highlighted the higher roots' water uptake enhanced by intensive physiologic activities by day times.

4. Conclusion

The objective of this work aimed to check the reliability of a new approach of SDI laterals design for a systematic irrigation management monitored by soil suction variation close to the outlets. Recorded results showed that, without human intervention for irrigation management, water content, in the soil layer 0–40 cm, remained within the interval $]\theta(h_{opt} + \Delta h_{opt}); \theta(h_{opt} - \Delta h_{opt})]$, corresponding to the optimal humidity interval for tomato growth. Soil water content in the deep layers ($Z = 50$ cm and $Z = 70$ cm) remained roughly constant but lower than $\theta(h_{opt} - \Delta h_{opt})$. So neither saturation risks nor water and nutrients losses by deep percolation were observed within the vadose zone. In addition, irrigation water uniformity along the lateral was almost higher than 80.7%. So the design procedure illustrated in this paper provides the appropriate emitters discharge and the inlet lateral pressure head that fit the plant roots water uptake. Even though, soil water content recorded at the lateral end tip remained lower than the minimum optimal threshold throughout the entire

cropping cycle. Therefore, the proposed approach could be an efficient tool for rigorous SDI lateral design. But further field trials are needed to effectively confirm such finding.

Competing Interests

The authors declare that they have no competing interests.

References

[1] FAO, *Coping with Water Scarcity. Challenge of the Twenty-First Century*, UN Water, 2007, http://www.fao.org/nr/water/docs/escarcity.pdf.

[2] L. Levidow, D. Zaccaria, R. Maia, E. Vivas, M. Todorovic, and A. Scardigno, "Improving water-efficient irrigation: prospects and difficulties of innovative practices," *Agricultural Water Management*, vol. 146, pp. 84–94, 2014.

[3] H. Ben Ali, M. Hammami, R. Boukhchina, and A. Saidi, "Comparative study between surface and subsurface drip irrigation systems: case of potato crop," *International Journal of Innovation and Applied Studies*, vol. 6, no. 4, pp. 860–870, 2014.

[4] B. Douh and A. Boujelben, "Diagnosis of subsurface drip irrigation practices in Tunisia: impacts on soil water storage variation, corn yield and water use efficiency," *Larhyss Journal*, no. 10, pp. 115–126, 2012 (French).

[5] N. Patel and T. B. S. Rajput, "Effect of drip tape placement depth and irrigation level on yield of potato," *Agricultural Water Management*, vol. 88, no. 1–3, pp. 209–223, 2007.

[6] M. Hammami, K. Zayani, and H. Ben Ali, "Required lateral inlet pressure head for automated subsurface drip irrigation management," *International Journal of Agronomy*, vol. 2013, Article ID 162354, 6 pages, 2013.

[7] S. R. Evett, T. A. Howell, and A. D. Schneider, "Energy and water balances for surface and subsurface drip irrigated corn," in *Proceedings of the 5th International Micro Irrigation Congress*, pp. 135–140, Orlando, Fla, USA, 1995.

[8] N. Patel and T. B. S. Rajput, "Dynamics and modeling of soil water under subsurface drip irrigated onion," *Agricultural Water Management*, vol. 95, no. 12, pp. 1335–1349, 2008.

[9] A. Ombódi, N. Koczka, A. Lugasi, H. G. Daood, M. Berki, and L. Helyes, "Nutritive constituents of onion grown from sets as affected by water supply," *HortScience*, vol. 48, no. 12, pp. 1543–1547, 2013.

[10] G. S. Ghali and Z. J. Svehlik, "Soil-water dynamics and optimum operating regime in trickle-irrigated fields," *Agricultural Water Management*, vol. 13, no. 2–4, pp. 127–143, 1988.

[11] C. J. Phene, K. R. Davis, R. B. Hutmacher, and R. L. McCormick, "Advantages of subsurface irrigation for processing tomato," *Acta Horticulturae*, vol. 200, pp. 101–114, 1987.

[12] P. Najafi, "Effects of using subsurface drip irrigation and treated municipal waste water in irrigation of tomato," *Pakistan Journal of Biological Sciences*, vol. 9, no. 14, pp. 2672–2676, 2006.

[13] L. Zotarelli, J. M. Scholberg, M. D. Dukes, R. Muñoz-Carpena, and J. Icerman, "Tomato yield, biomass accumulation, root distribution and irrigation water use efficiency on a sandy soil, as affected by nitrogen rate and irrigation scheduling," *Agricultural Water Management*, vol. 96, no. 1, pp. 23–34, 2009.

[14] R. M. A. Machado and G. M. R. Oliveira, "Tomato root distribution, yield and fruit quality under different subsurface drip irrigation regimes and depths," *Irrigation Science*, vol. 24, no. 1, pp. 15–24, 2005.

[15] M. T. van Genuchten, "Closed-form equation for predicting the hydraulic conductivity of unsaturated soils," *Soil Science Society of America Journal*, vol. 44, no. 5, pp. 892–898, 1980.

[16] A. I. Gärdenäs, J. W. Hopmans, B. R. Hanson, and J. Šimůnek, "Two-dimensional modeling of nitrate leaching for various fertigation scenarios under micro-irrigation," *Agricultural Water Management*, vol. 74, no. 3, pp. 219–242, 2005.

[17] G. Z. Watters and J. Keller, "Trickle irrigation tubing hydraulics," Tech. Ref. 78-2015, ASCE, Reston, Va, USA, 1978.

[18] B. Safi, M. R. Neyshabouri, A. H. Nazemi, S. Massiha, and S. M. Mirlatifi, "Water application uniformity of a subsurface drip irrigation system at various operating pressures and tape lengths," *Turkish Journal of Agriculture and Forestry*, vol. 31, no. 5, pp. 275–285, 2007.

[19] R. Haverkamp, M. Vauclin, J. Touma, P. J. Wierenga, and G. Vachaud, "Comparison of numerical simulation models for one-dimensional infiltration," *Soil Science Society of America Journal*, vol. 41, no. 2, pp. 285–294, 1977.

[20] N. Lazarovitch, U. Shani, T. L. Thompson, and A. W. Warrick, "Soil hydraulic properties affecting discharge uniformity of gravity-fed subsurface drip irrigation systems," *Journal of Irrigation and Drainage Engineering*, vol. 132, no. 6, pp. 531–536, 2006.

[21] M. Gil, L. Rodríguez-Sinobas, L. Juana, R. Sánchez, and A. Losada, "Emitter discharge variability of subsurface drip irrigation in uniform soils: effect on water-application uniformity," *Irrigation Science*, vol. 26, no. 6, pp. 451–458, 2008.

[22] W. W. Yao, X. Y. Ma, J. Li, and M. Parkes, "Simulation of point source wetting pattern of subsurface drip irrigation," *Irrigation Science*, vol. 29, no. 4, pp. 331–339, 2011.

15

Environmental Influences on Growth and Reproduction of Invasive *Commelina benghalensis*

Mandeep K. Riar,[1] **Danesha S. Carley,**[1] **Chenxi Zhang,**[1] **Michelle S. Schroeder-Moreno,**[1] **David L. Jordan,**[1] **Theodore M. Webster,**[2] **and Thomas W. Rufty**[1]

[1]*Department of Crop Science, North Carolina State University, Raleigh, NC 27695, USA*
[2]*Crop Protection and Management Research Unit, USDA-ARS, Tifton, GA 31793, USA*

Correspondence should be addressed to David L. Jordan; david_jordan@ncsu.edu

Academic Editor: Rodomiro Ortiz

Commelina benghalensis (Benghal dayflower) is a noxious weed that is invading agricultural systems in the southeastern United States. We investigated the influences of nutrition, light, and photoperiod on growth and reproductive output of *C. benghalensis*. In the first experimental series, plants were grown under high or low soil nutrition combined with either full light or simulated shade. Lowered nutrition strongly inhibited vegetative growth and aboveground spathe production. Similar but smaller effects were exerted by a 50% reduction in light, simulating conditions within a developing canopy. In the second series of experiments, *C. benghalensis* plants were exposed to different photoperiod conditions that produced short- and long-day plants growing in similar photosynthetic periods. A short-day photoperiod decreased time to flowering by several days and led to a 40 to 60% reduction in vegetative growth, but reproduction above and below ground was unchanged. Collectively, the results indicate that (1) fertility management in highly weathered soils may strongly constrain competitiveness of *C. benghalensis*; (2) shorter photoperiods will limit vegetative competitiveness later in the growing seasons of most crops; and (3) the high degree of reproductive plasticity and output possessed by *C. benghalensis* will likely cause continual persistence problems in agricultural fields.

1. Introduction

Commelina benghalensis L. is among the world's worst weeds in agricultural systems, with infestations occurring in 25 crops in 29 countries. In the US, it became established in Florida in the early 1930s [1] and is now moving northward. The prevalence in Georgia has led to challenges with *Commelina benghalensis* control in cotton (*Gossypium hirsutum* L.) and peanut (*Arachis hypogaea* L.) production [2], as infestations have commonly caused 60% to 100% yield reductions [3, 4]. And there are observations indicating that *C. benghalensis* may be spreading further northward as far as North Carolina [5].

A key to *C. benghalensis* invasiveness is its reproductive flexibility. In its native geographical areas, that is, tropical Asia, Africa, and the Pacific Islands, *C. benghalensis* grows as a perennial, but it can survive as an annual in temperate regions [6]. It is fast growing and a prolific seed producer [7]. Both aerial and subterranean seeds are produced in dimorphic flowers [8], and seeds have variable dormancy and germination characteristics [9, 10]. Furthermore, *C. benghalensis* has the ability to regenerate from stem fragments [11]. These characteristics, plus a high degree of tolerance to glyphosate [12], make *C. benghalensis* exceptionally difficult to control in agronomic systems when it becomes established.

We are investigating environmental effects on *C. Benghalensis* growth and development. The purpose is to gain insights into factors that influence its competitiveness and persistence, which are keys for assessment of the risk of invasions and formation of long-term control strategies. In recent experiments, it was found that seeds of *C. benghalensis* can persist in soil for up to four years in areas extending from Florida and Georgia to North Carolina [13]. This implied that the management programs must prevent seed production for that period of time to effectively reduce seed banks. In further agroecological studies, it was found that control of *C. benghalensis* will be especially problematic in sustainable farming systems not using herbicides [14]. The viability of *C. benghalensis* seed acquired during grazing or consumption of fresh hay is not reduced during animal digestion and generation of manure, so there is no restraint on seed dispersal. Also, vegetative regeneration was near its maximum in the temperature range typical in summers in the southeastern US, severely limiting the potential effectiveness of cultivation.

In the study described in this paper, we examine *C. benghalensis* responses to several environmental variables that can be related to prediction of field behavior. Two of the variables examined, light and photoperiod, required that experiments be conducted in controlled-environment growth chambers. Specific questions were being addressed. One was "how are growth and development of *C. benghalensis* altered by changes in nutrition and light?" Crops are fertilized at different levels, so the magnitude of the fertility response would help predict cropping systems where *C. benghalensis* would be more or less aggressive. Exposure to low light will reveal the extent that growth and development might be expected in crop understories or small gaps in crop populations. Another question was "how is *C. benghalensis* affected by altered photoperiod?" With current warming patterns [15, 16], cropping seasons can be extended, unless a counterbalancing effect is exerted by shorter day lengths. To assess responses in a worst case scenario, the environmental treatments were imposed under relatively high temperatures of 30–35°C, a temperature range common in the southeastern states in the US, when crop interference by weeds most often occurs, and a range where *C. benghalensis* exists [17].

2. Materials and Methods

2.1. Nutrition and Shade. Experiments were conducted in walk-in environmental chambers at the Southeastern Plant Environment Laboratory, Raleigh, NC. Large aerial seeds of *C. benghalensis* were germinated in 6 L plastic pots (25 cm diameter) containing Norfolk sandy loam soil (kaolinitic, thermic Typic Kandiudults, bulk density 1.2 g cm^{-3} and pH 6.1). Prior to sowing, seeds were disinfected by soaking in 5% bleach solution (0.25% NaOCl) for 5 min, rinsed with water, and then scarified with a blade to break physical dormancy and enhance germination [18]. Seeds were germinated at a constant day/night temperature of 30/30°C, with a 9 h photoperiod. Treatments were imposed just after seedlings were thinned to two per pot at the one-leaf stage (approximately 10 days after seeding). The treatments examined *C. benghalensis* responses at two nutrient levels and under two light intensities. Lighting was provided at either full (600 μmol m^{-2} s^{-1}) or reduced (324 μmol m^{-2} s^{-1}) PPFD, provided by a combination of incandescent and fluorescent lamps. Reduced lighting was achieved by covering the area with a shade cloth. PPFD was measured with a LI-191 Line Quantum Sensor (LI-COR Biosciences, Lincoln, NE). The daily light and dark periods were maintained from 08:00 to 17:00 h and 17:00 to 08:00 h, respectively. The aerial

temperature was held constant at 30°C. Soil temperatures were monitored continuously at a 2 cm depth in the pots using temperature probes (WatchDog A-Series Data Loggers, Spectrum Technologies). Soil temperatures were within a ±3°C range of aerial temperatures under all treatments.

The experimental design was a split-plot with light intensity as the main plot and nutrient level as the subplot. Each light treatment was subdivided into two nutrient treatments. Pots with high nutrition received 200 mL of complete Hoagland nutrient solution [19] every day, and pots with low nutrient level received 200 mL of the same solution once a week, with deionized water added on the other days. Each pot was flushed with deionised water prior to nutrient additions to minimize residual nutrient accumulation. Three pots from each treatment combination were randomly selected and harvested at 14, 28, 35, 42, 49, and 56 d after treatments started. At harvests, plants were separated into aerial and subterranean tissues. The tissues were further separated into aerial vegetative (shoots and leaves), aerial reproductive (aerial spathes), subterranean vegetative (roots), and subterranean reproductive (rhizomes and spathes). All tissues were dried to a constant mass at 60°C in a drying oven and weighed afterwards. To avoid seed dissemination, aerial fruits were collected prior to dehiscence throughout the treatment period. Plant height, number of leaves, and leaf area were also measured at time of harvest. A standard Li-Cor model LI-3100C Area Meter (LI-COR Biosciences, Lincoln, NE) was used to measure leaf area per plant. There were 3 replications for each treatment and the experiment was conducted twice. Data were analyzed using the GLIMMIX procedure in SAS version 9.3 (SAS Institute, Inc., Cary, NC). Mean separation was performed using Tukey's honest significant difference (HSD) at $\alpha = 0.05$.

2.2. Photoperiod. Photoperiod experiments also were conducted in growth chambers at the Southeastern Plant Environment Laboratory. Seed treatment, germination, seedling establishment conditions, and soil were the same as described previously. In this circumstance, plants were exposed to five temperature-photoperiod regimes established in five reach-in environmental chambers. Twelve pots were kept in each chamber.

The five day/night temperature-photoperiod regimes included 30/22°C with long-day (12 h), 30/22°C with short-day (9 h), 35/28°C with long-day, 35/28°C with short-day, and a constant day/night temperature at 30°C with short-day. Each growth chamber was programmed with one unique combination of day/night temperature and photoperiod. In the short-day regime, day and night periods were maintained from 08:00 to 17:00 h and from 17:00 to 08:00 h, respectively. Illumination during day hours was provided by a combination of incandescent and fluorescent lamps with a PPFD of 600 μmol m^{-2} s^{-1}. The long-day regime was achieved by imposing the 3 h night interruption with incandescent lighting (from 00:00 to 03:00 h by incandescent lamps generating a nonphotosynthetic PPFD of 40 μmol m^{-2} s^{-1}). By disrupting continuous night hours, the night interruption leads to similar flowering transition effects as those occurring with actual long-day hours [20]. But, with similar photosynthetic

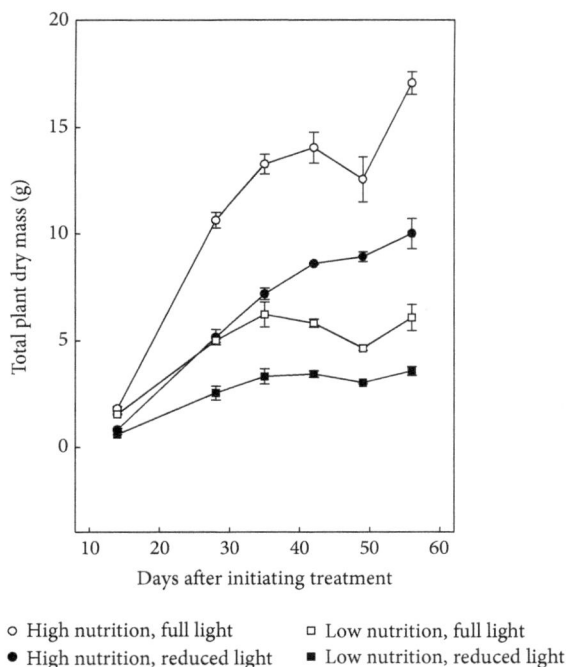

FIGURE 1: Effects of nutrition level and light intensity on total dry mass of *C. benghalensis*. Data points are means of six replicates, and bars indicate standard errors of the mean.

period lengths, the conditions allow the physiological effects exerted by vegetative/reproductive shifts to be evaluated alone. The night interruption technique has been used in many experiments over the years in controlled environment to suppress flowering of crop plants [19].

Plants were watered twice daily with 150 mL deionized water and once every other day with 200 mL complete Hoagland nutrient solution [19]. The plants thus were growing under high nutrition similar to that in the previous nutrition experiments. Three pots from each chamber were randomly selected and harvested at 14, 28, 42, and 56 d after treatment. Tissues were separated and processed as described previously. The experiment was repeated twice and data were combined and analyzed using the GLM procedure in SAS 9.3. (SAS Institute, Inc., Cary, NC).

3. Results and Discussion

3.1. Responses to Nutrition and Reduced Light. C. benghalensis whole-plant mass accumulation increased greatly with a higher level of nutrition (Figure 1), especially under full light. Increased growth included increases in root mass and plant height (Figures 2(a) and 2(b)). Impacts on leaf morphology were particularly pronounced, as the number of leaves and the leaf area per plant were considerably greater in plants receiving higher nutrition, regardless of light intensity (Figures 2(c) and 2(d)). This response is typical for fast-growing plants under high nutrition, where shoot growth is often stimulated more than root growth, resulting in an increase in the shoot to root growth ratio [21, 22].

Reproduction of *C. benghalensis* was also greater at high nutrition. Time to flowering was not changed (data not shown), but the number of aerial spathes was constantly greater in plants grown in high nutrition after the first sampling date, and by the end of the 56 day experiment, it was 3 times greater than that with low nutrition (Table 1). In contrast, the statistical analyses indicated that production of subterranean spathes varied little between the two nutrition levels.

The fertility responses are especially relevant for field responses in the southeastern US because soils are highly weathered with inherently low nutrition [23], and fertility requirements can be very different among cropping systems. Based on studies done with a number of higher plant species, it has been proposed that weed invasiveness is strongly linked with increased growth rates under improved resource conditions, but that invasive species may not outperform native or noninvasive species when resources are limited [24–27]. The strong response of *C. benghalensis* to increased nutrition in our experiments suggests that it would be much more aggressive in cropping systems with high rates of fertilization like those with corn (130–170 kg N/ha) and much less so in crops grown with lower fertilizer additions like N_2-fixing soybean that often receives low amounts of fertilizer (<33 kg N/ha) if any at all.

There are circumstances where weed species might acquire nitrogen from alternative sources. Experiments using ^{15}N natural abundance, for example, found that weeds could obtain large amounts of nitrogen transferred from N_2-fixing soybean [28]. However, because transfer occurred through and was dependent on soil-borne mycorrhizae, this type of alternative nitrogen acquisition would not be available to *C. benghalensis*. Our recent analyses have indicated that it is not a mycorrhizal host species (Riar, unpublished observation). Therefore, growth and competitiveness of *C. benghalensis* would primarily be influenced by fertilization in the crop system.

Shading was used in this study to examine *C. benghalensis* growth and reproductive output under reduced light, in an attempt to simulate conditions that exist when growing concurrently with a developing leaf canopy. Shading greatly reduced total biomass of *C. benghalensis* (Figure 1) when plants received the same nutrition level. Similarly, root mass was consistently less in shaded plants, especially under the high nutrition treatment (Figure 2(a)). Low light conditions tended to increase plant height, but differences were not statistically significant (Figure 2(b)). Shading had little effect on total leaf number and leaf area per plant (Figures 2(c) and 2(d)) but because leaf canopy mass was decreased substantially, shaded plants had much thinner leaves. This evidently reflects a morphological compensation response, where decreased amounts of available photosynthate are prioritized to maximize photosynthetic surface. In past experiments by others, greater specific leaf area and leaf area ratio were observed under reduced light conditions with jimsonweed (*Datura stramonium* L.), velvetleaf (*Abutilon theophrasti* Medik.), and soybean [29]. Thinner leaves and a less dense canopy evidently are an important adaptation of

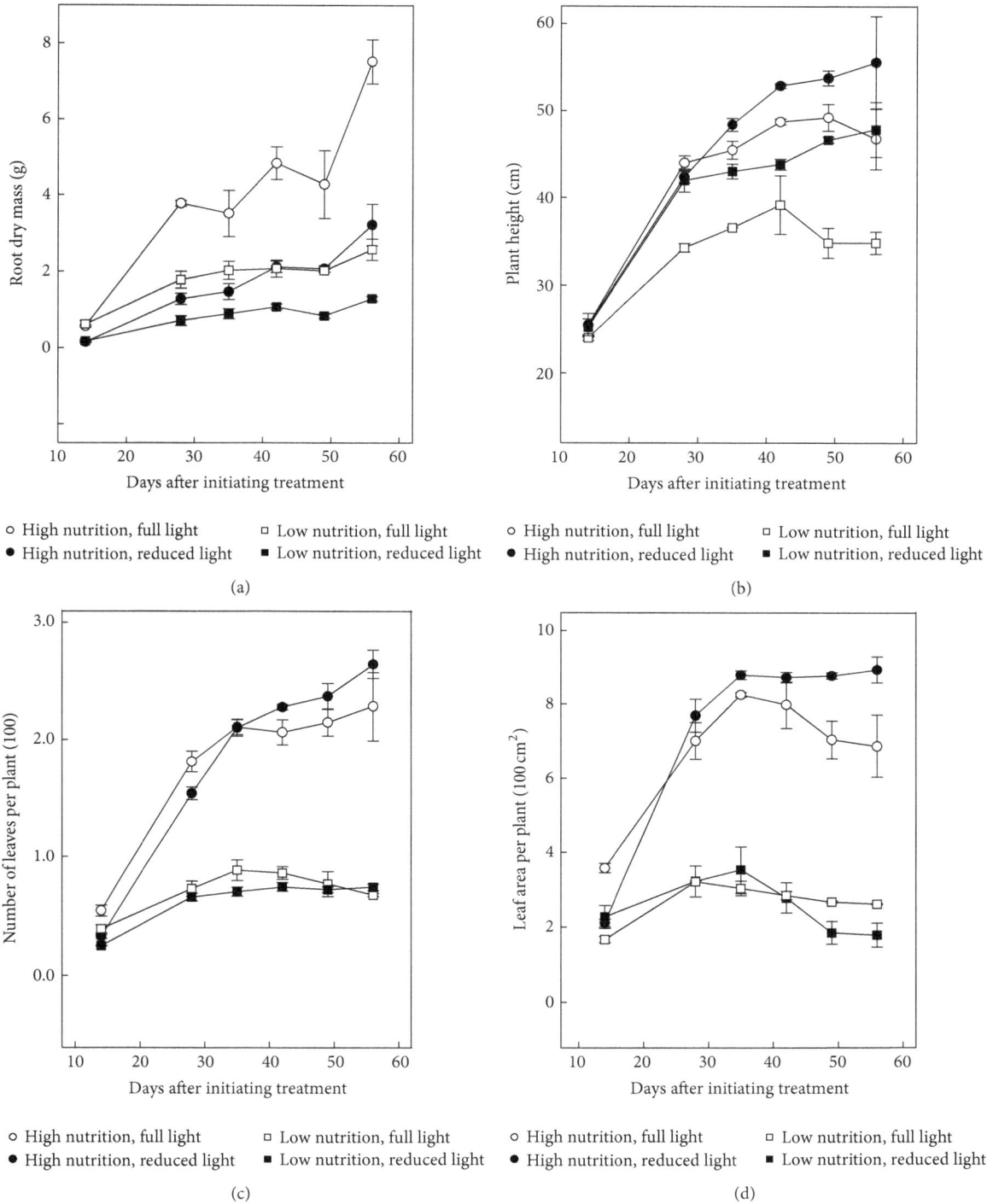

○ High nutrition, full light □ Low nutrition, full light
● High nutrition, reduced light ■ Low nutrition, reduced light

(a)

○ High nutrition, full light □ Low nutrition, full light
● High nutrition, reduced light ■ Low nutrition, reduced light

(b)

○ High nutrition, full light □ Low nutrition, full light
● High nutrition, reduced light ■ Low nutrition, reduced light

(c)

○ High nutrition, full light □ Low nutrition, full light
● High nutrition, reduced light ■ Low nutrition, reduced light

(d)

FIGURE 2: Effects of nutrition level and light intensity on (a) root dry mass, (b) plant height, (c) number of leaves per plant, and (d) leaf area per plant of *C. benghalensis*. Data points are means of six replicates, and bars indicate standard errors of the mean.

plants growing under shaded conditions that permits light penetration to lower leaves [30, 31].

As was the case with altered nutrition, shading had no impact on initiation of flowering (data not shown), but there were significant decreases in aerial spathe production (Table 1). Subterranean spathe production tended to be suppressed slightly by shading, but most of the observed differences were not statistically significant (Table 1).

The degrees of adjustment in spathe production above and below ground with plants that received low nutrition and those under shading indicate that *C. benghalensis* shifts to a "survival" strategy when resources are limited [32]. Even

TABLE 1: Effects of nutrient supply and light on *C. benghalensis* aerial and subterranean spathe production.

| Treatment | | Aerial spathes | | | | | Subterranean spathes | | | | |
| | | Days after treatment | | | | | Days after treatment | | | | |
Nutrient[a]	Light[b]	28	35	42	49	56	28	35	42	49	56
		Number of spathes plant^{-1}									
High	Full	132[c]	251	286	379	558	16	23	34	55	80
High	Shade	68	150	212	284	398	7	13	26	37	54
Low	Full	65	115	126	124	150	11	20	30	35	52
Low	Shade	38	66	92	90	124	5	13	22	29	41
HSD$_{0.05}$[d]		81	56	61	75	35	NS	NS	NS	NS	30

[a] Plants treated with high nutrient level received complete Hoagland's nutrient solution on a daily basis; plants treated with low nutrient level received complete Hoagland's nutrient solution once a week.
[b] PPFD under full light was at 600 μmol m^{-2} s^{-1}; PPFD under shade was at 324 μmol m^{-2} s^{-1}.
[c] Values are means of six replicates.
[d] Tukey's honest significant difference at $\alpha = 0.05$.

with severely reduced growth and restriction of aboveground spathe production, belowground reproduction tended to be maintained. On an individual plant basis, this increases the likelihood of genetic persistence, mainly because of less predation than occurs above ground. From an agronomic viewpoint, genetic persistence equates with a high likelihood of persistence in field soils. Difficulties with eradication of *C. benghalensis* in agricultural fields are clearly implied.

3.2. Response to Altered Photoperiod. Little has been published about *C. benghalensis*'s photoperiod sensitivity, particularly in the high temperature range where its growth and potential interference with crops are the greatest. High temperature interactions with photoperiod are becoming more important with global warming, as extension of high temperatures into months with shorter photoperiods could result in growth and reproductive characteristics unlike those in summer months. These experiments were specifically designed (a) to examine whether flowering was photoperiod sensitive and (b) to determine the extent that vegetative growth and reproductive output were altered by short days. All of the photoperiod treatments were imposed with plants growing under high nutrition.

Flowering of *C. benghalensis* was altered in short days (Figure 3). With the shortened light period of 9 hours and at the high temperature of 35/28°C, flowering occurred at day 27 whereas with long-day plants (with the night interruption) flowering occurred 3 days later at day 30. With the 9-hour light period and 30/22°C temperature, the short-day plants flowered at day 33 compared to day 40 with the long-day plants. Plants growing in the 30/30°C temperature also flowered sooner in short days, in this case on day 29, when compared to short-day plants at 30/26°C. The difference in short-day and long-day flowering with *C. benghalensis* contrasts with the much stronger suppression of flowering in soybean (*Glycine max* L.) and tobacco (*Nicotiana tabacum* L.) under similar conditions with a night interruption [33, 34].

Even though short-day plants flowered only a few days earlier than long-day plants, whole plant growth was greatly reduced for plants exposed to short days (Figure 4). As might have been predicted from earlier studies examining *C.*

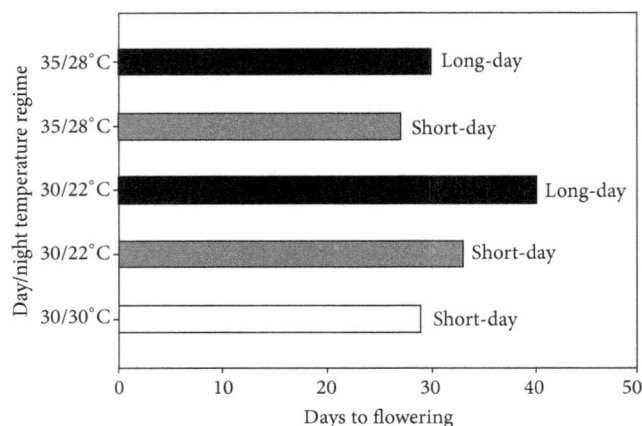

FIGURE 3: Photoperiod effects on time of flowering of *C. benghalensis* plants. Long-day plants (LD) were exposed to a 9 h light period and a 3 h night interruption with nonphotosynthetic light to suppress flowering. Short-day plants (SD) were exposed only to the 9 h light period.

benghalensis response to high temperature [17, 35], growth processes were enhanced at the high 35/28°C temperature. However, mass of short-day plants, when compared to long-day plants, was decreased by about 40% at 56 d in 35/28°C and growth was decreased 60% by short days in the slower growing plants at 30/22°C. The decreases in plant mass were accompanied by decreases in root mass (Figure 5(a)), plant height (Figure 5(b)), and total leaf area (Figure 5(d)) compared to plants in the long-day photoperiod. Effects on the number of leaves were not as consistent, with reductions in leaf number occurring in short-day plants at 30/22°C but not at 35/28°C by the final 56-day harvest (Figure 5(c)). Taken as a whole, our results are at odds with a previous study where it was concluded that photoperiod would not greatly affect growth of *C. benghalensis* [36].

Reproductive performance in the short-day plants was statistically similar to that in the long-day plants at in 35/28°C and 30/22°C temperatures. Thus, reproductive data for the short- and long-day plants were combined (Table 2).

TABLE 2: Effects of three temperature regimes on *C. benghalensis* aerial and subterranean spathe production. Data for the photoperiod treatments were similar, statistically, so they were combined.

Temperature regime	Aerial spathes Days after treatment			Subterranean spathes Days after treatment		
	28	42	56	28	42	56
°C	Number of spathes plant^{-1}					
35/28[a]	18[b]	68	87	4	30	31
30/22	22	94	120	1	15	22
30/30	42	84	98	5	25	37
HSD$_{0.05}$[c]	4	15	20	1	5	5

[a]Numbers indicate day/night temperatures.
[b]Values are means of 24 replicates. Photoperiod effects were nonsignificant at the 35/28 and 30/22°C temperatures, so data were pooled for each temperature regime.
[c]Tukey's honest significant difference at $\alpha = 0.05$.

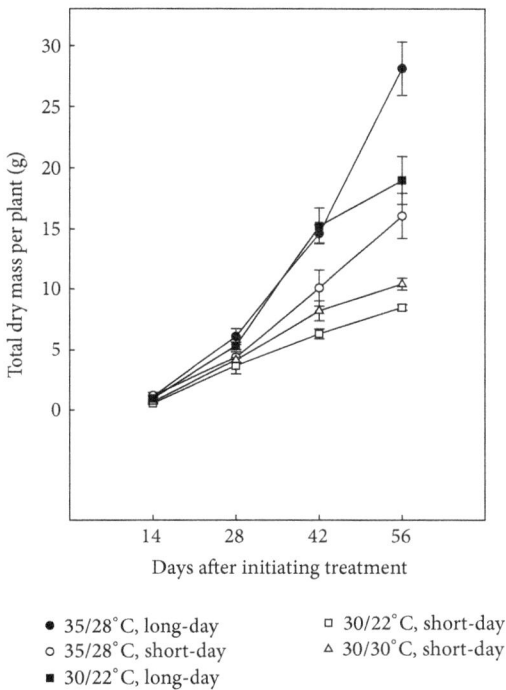

FIGURE 4: Effects of day/night temperature regime and altered photoperiod on total dry mass of *C. benghalensis*. The long-day treatment was imposed by a 3 h nonphotosynthetic night interruption in the middle of dark periods. LD: long-day; SD: short-day. Data points are means of six replicates, and bars indicate standard errors of the mean.

Some differences in the number of reproductive structures occurred among temperatures. At the higher temperature of 35/28°C, the higher vegetative mass (Figure 4) was associated with lower production of aerial spathes and greater production of belowground spathes.

As indicated above with low fertility or shaded plants, it is quite evident that *C. benghalensis* retains an ability to sustain a high level of reproductive output in high temperature ranges, in this case even when whole plant growth is constrained by the physiological response to a shorter photoperiod. This resilience in reproduction is obviously of concern when it is considered that one of the biggest challenges in weed management is to limit seed production in agronomic fields, reducing the long-term soil seed bank. Studies have shown that as many as 8,000 to 12,000 seeds m^{-2} can be produced by *C. benghalensis* [7], and the seeds can remain viable for at least 3 to 4 years [13].

4. Conclusions

Much is unknown about how the invasive weed *C. benghalensis* interacts with the environment. These experiments with different nutrition, low light, and photoperiod treatments offer new insights into environmental factors that enhance or suppress the invasiveness of *C. benghalensis* in agricultural systems. One insight is that *C. benghalensis* competitiveness and reproduction are strongly increased by the high nutrition typically used with grain crops. A valid containment strategy with large infestations of *C. benghalensis* would be to grow a series of crops with low or no added nitrogen fertilizer, like soybean or an N$_2$-fixing pasture. It should be emphasized that substantial growth and especially reproductive output still occurred at lower nutrition in our experiments, so it is anticipated that a degree of weed pressure will persist.

A second insight is that vegetative growth is suppressed under short days, even when flowering shifts only a few days. Thus, *C. benghalensis* competitiveness is likely to be reduced when germination and growth occur during shorter photoperiods outside of normal crop growing seasons, even though high temperatures may become more prevalent. A third important observation is that, regardless of the type of environment it faces, *C. benghalensis* exhibits a substantial reproductive capability, and this is true even when vegetative growth is suppressed by nutrition, low light, or a photoperiod shift. Underground reproduction appears to be particularly resilient. The persistent reproductive output underscores the importance of intensive management strategies to prevent introduction of *C. benghalensis* into agricultural fields.

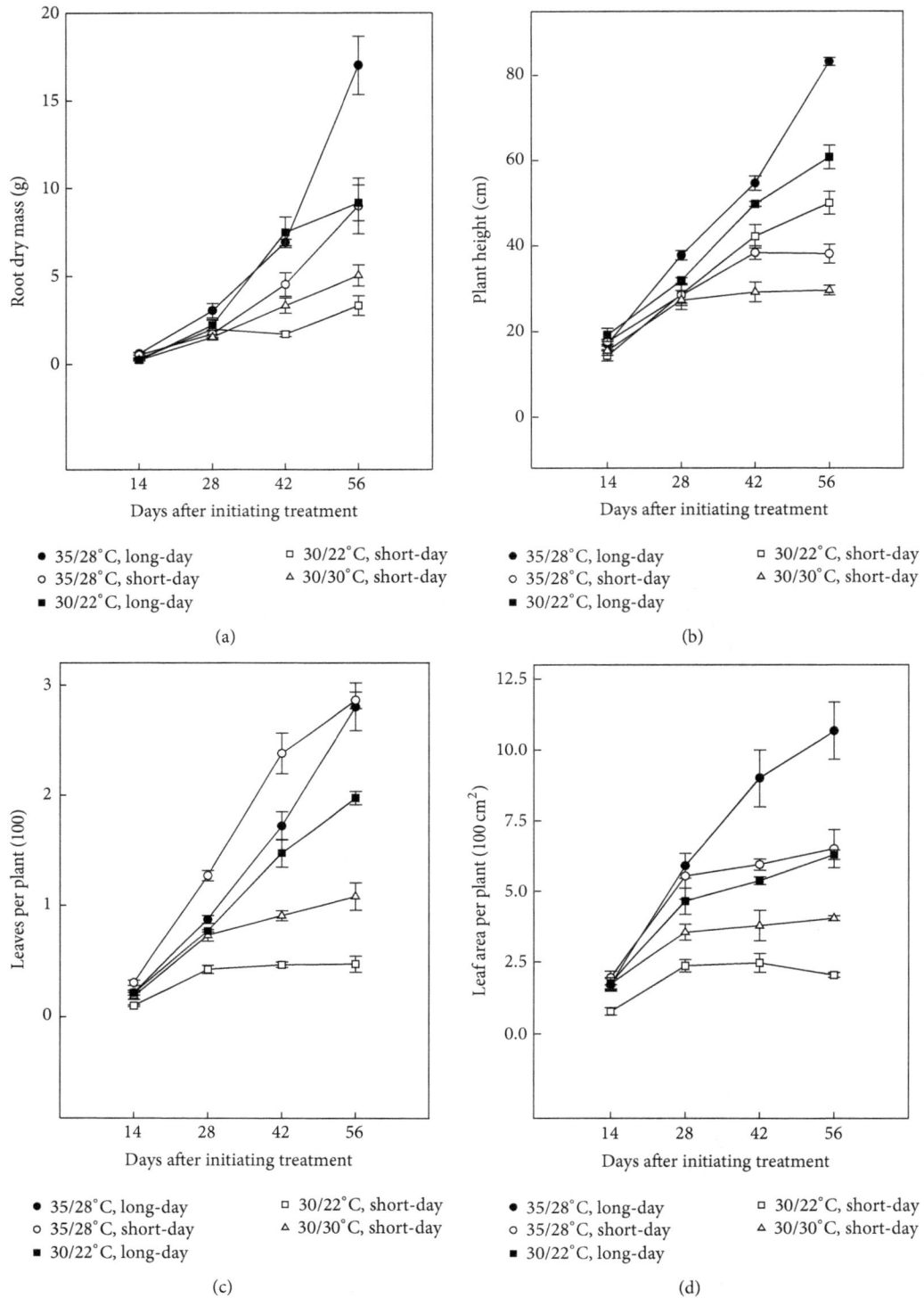

FIGURE 5: Effects of day/night temperature regime and altered photoperiod on (a) root dry mass, (b) plant height, (c) number of leaves per plant, and (d) leaf area per plant of *C. benghalensis*. The long-day treatment was imposed by a 3 h nonphotosynthetic night interruption in the middle of dark periods. LD: long-day; SD: short-day. Data points are means of six replicates, and bars indicate standard errors.

Conflict of Interests

The authors declare that there is no conflict of interests regarding the publication of this paper.

Acknowledgments

This research was funded by the US Department of Agriculture—Cooperative State Research, Education, and Extension Service—National Research Initiative (USDA-CSREES-NRI, Grant no. 2008-35320-18689) Program. The authors express appreciation to NCSU Phytotron staff for their help. Thanks are also due to Shannon Sermons and Laura Vance for constructive suggestions on the paper.

References

[1] R. B. Faden, "The misconstrued and rare species of Commelina (Commelinaceae) in the eastern United States," *Annals of the Missouri Botanical Garden*, vol. 80, no. 1, pp. 208–218, 1993.

[2] T. M. Webster and L. M. Sosnoskie, "Loss of glyphosate efficacy: a changing weed spectrum in Georgia cotton," *Weed Science*, vol. 58, no. 1, pp. 73–79, 2010.

[3] T. M. Webster, W. H. Faircloth, J. T. Flanders, E. P. Prostko, and T. L. Grey, "The critical period of Bengal dayflower (Commelina bengalensis) control in peanut," *Weed Science*, vol. 55, no. 4, pp. 359–364, 2007.

[4] T. M. Webster, T. L. Grey, J. T. Flanders, and A. S. Culpepper, "Cotton planting date affects the critical period of Benghal dayflower (Commelina benghalensis) control," *Weed Science*, vol. 57, no. 1, pp. 81–86, 2009.

[5] T. M. Webster, M. G. Burton, A. S. Culpepper, A. C. York, and E. P. Prostko, "Tropical spiderwort (Commelina benghalensis): a tropical invader threatens agroecosystems of the Southern United States," *Weed Technology*, vol. 19, no. 3, pp. 501–508, 2005.

[6] L. G. Holm, D. L. Plucknett, J. V. Pancho, and J. P. Herberger, *The World's Worst Weeds: Distribution and Biology*, University Press of Hawaii, Honolulu, Hawaii, USA, 1977.

[7] S. R. Walker and J. P. Evenson, "Biology of Commelina benghalensis L. in south-eastern Queensland. 1. Growth, development and seed production," *Weed Research*, vol. 25, no. 4, pp. 239–244, 1985.

[8] P. Maheshwari and J. K. Maheshwari, "Floral dimorphism in Commelina forskalaei Vahl. and Benghal dayflower L.," *Phytomorphology*, vol. 5, pp. 413–422, 1955.

[9] S. Y. Kim, S. K. De Datta, and B. L. Mercado, "The effect of chemical and heat treatments on germination of Commelina benghalensis L. aerial seeds," *Weed Research*, vol. 30, no. 2, pp. 109–116, 1990.

[10] S. R. Walker and J. P. Evenson, "Biology of Commelina benghalensis L. in South-eastern Queensland. 2. Seed dormancy, germination and emergence," *Weed Research*, vol. 25, no. 4, pp. 245–250, 1985.

[11] G. D. Budd, P. E. L. Thomas, and J. C. S. Allison, "Vegetative regeneration, depth of germination and seed dormancy in Commelina benghalensis L.," *Rhodesian Journal of Agricultural Research*, vol. 17, no. 2, pp. 151–153, 1979.

[12] A. S. Culpepper, J. T. Flanders, A. C. York, and T. M. Webster, "Tropical spiderwort (Commelina benghalensis) control in glyphosate-resistant cotton (Gossypium hirsutum)," *Weed Technology*, vol. 18, pp. 432–436, 2004.

[13] M. K. Riar, T. M. Webster, B. J. Brecke et al., "Benghal dayflower (Commelina benghalensis) seed viability in soil," *Weed Science*, vol. 60, no. 4, pp. 589–592, 2012.

[14] M. K. Riar, J. F. Spears, J. C. Burns, D. L. Jordan, C. Zhang, and T. W. Rufty, "Persistence of Benghal dayflower (Commelina benghalensis) in sustainable agronomic systems: potential impacts of hay bale storage, animal digestion, and cultivation," *Agroecology and Sustainable Food Systems*, vol. 38, no. 3, pp. 283–298, 2014.

[15] J. Hansen, *Storms of My Grandchildren*, Bloomsbury Publishing, New York, NY, USA, 2009.

[16] J. Hansen, M. Sato, G. Russell, and P. Kharecha, "Climate sensitivity, sea level and atmospheric carbon dioxide," *Philosophical Transactions of the Royal Society A*, vol. 371, no. 2001, Article ID 20120294, 2013.

[17] S. M. Sermons, M. G. Burton, and T. W. Rufty, "Temperature response of Benghal dayflower (Commelina benghalensis): implications for geographic range," *Weed Science*, vol. 56, no. 5, pp. 707–713, 2008.

[18] R. H. Goddard, T. M. Webster, R. Carter, and T. L. Grey, "Resistance of Benghal dayflower (Commelina benghalensis) seeds to harsh environments and the implications for dispersal by mourning Doves (Zenaida macroura) in Georgia, U.S.A," *Weed Science*, vol. 57, no. 6, pp. 603–612, 2009.

[19] C. H. Saravitz, R. J. Downs, and J. F. Thomas, "Phytotron procedural manual for controlled-environment research at the Southeastern Plant Environment Laboratory," Technical Bulletin 244 (Revised), North Carolina State University, North Carolina Agricultural Research Service, Raleigh, NC, USA, 2009.

[20] J. Hanke, K. M. Hartmann, and H. Mohr, "The effects of night breaks on flowering of Sinapis alba L.," *Planta*, vol. 86, no. 3, pp. 235–249, 1969.

[21] F. S. Chapin, "Integrated responses of plants to stress," *BioScience*, vol. 41, no. 1, pp. 29–36, 1991.

[22] T. W. Rufty, "Probing the carbon and nitrogen interaction: a whole plant perspective," in *A Molecular Approach to Primary Metabolism in Higher Plants*, C. H. Foyer and P. Quick, Eds., pp. 255–273, Taylor & Francis, London, UK, 1997.

[23] S. W. Buol, R. J. Southard, R. C. Graham, and P. A. McDaniel, *Soil Genesis and Classification*, Wiley-Blackwell, London, UK, 6th edition, 2011.

[24] J. Maillet and C. Lopez-Garcia, "What criteria are relevant for predicting the invasive capacity of a new agricultural weed? The case of invasive American species in France," *Weed Research*, vol. 40, no. 1, pp. 11–26, 2000.

[25] A. Kolb and P. Alpert, "Effects of nitrogen and salinity on growth and competition between a native grass and an invasive congener," *Biological Invasions*, vol. 5, no. 3, pp. 229–238, 2003.

[26] J. H. Burns, "A comparison of invasive and non-invasive dayflowers (Commelinaceae) across experimental nutrient and water gradients," *Diversity and Distributions*, vol. 10, no. 5-6, pp. 387–397, 2004.

[27] J. H. Burns and A. A. Winn, "A comparison of plastic responses to competition by invasive and non-invasive congeners in the Commelinaceae," *Biological Invasions*, vol. 8, no. 4, pp. 797–807, 2006.

[28] K. A. Moyer-Henry, J. W. Burton, D. W. Israel, and T. W. Rufty, "Nitrogen transfer between plants: a ^{15}N natural abundance study with crop and weed species," *Plant and Soil*, vol. 282, no. 1-2, pp. 7–20, 2006.

[29] E. E. Regnier, M. E. Salvucci, and E. W. Stoller, "Photosynthesis and growth responses to irradiance in soybean (*Glycine max*) and three broadleaf weeds," *Weed Science*, vol. 36, no. 4, pp. 487–496, 2008.

[30] O. Björkman, "Responses to different quantum flux densities," in *Physiological Plant Ecology I*, O. L. Lange, P. S. Nobel, C. B. Osmond, and H. Ziegler, Eds., vol. 12/A of *Encyclopedia of Plant Physiology*, pp. 57–107, Springer, Berlin, Germany, 1981.

[31] E. W. Stoller and T. W. Myers, "Response of soybeans (*Glycine max*) and four broadleaf weeds to reduced irradiance," *Weed Science*, vol. 37, no. 4, pp. 570–574, 1989.

[32] G. P. Cheplick and J. A. Quinn, "Amphicarpum purshii and the 'pessimistic strategy' in amphicarpic annuals with subterranean fruit," *Oecologia*, vol. 52, no. 3, pp. 327–332, 1982.

[33] J. F. Thomas, C. E. Anderson, C. D. Jr. Raper, and R. J. Downs, "Time of floral initiation in tobacco as a function of temperature and photoperiod," *Canadian Journal of Botany*, vol. 53, no. 14, pp. 1400–1410, 1975.

[34] J. F. Thomas and C. D. Raper, "Morphological response of soybeans as governed by photoperiod, temperature, and age at treatment," *Botanical Gazette*, vol. 138, no. 3, pp. 321–328, 1977.

[35] M. H. Sabila, T. L. Grey, T. M. Webster, W. K. Vencill, and D. G. Shilling, "Evaluation of factors that influence benghal dayflower (*Commelina benghalensis*) seed germination and emergence," *Weed Science*, vol. 60, no. 1, pp. 75–80, 2012.

[36] C. B. Gonzalez and C. R. B. Haddad, "Light and temperature effects on flowering and seed germination of *Commelina benghalensis* L.," *Archives of Biological Tecnology*, vol. 38, pp. 651–659, 1995.

16

The Effects of Biochar and Its Combination with Compost on Lettuce (*Lactuca sativa* L.) Growth, Soil Properties, and Soil Microbial Activity and Abundance

Dalila Trupiano,[1] **Claudia Cocozza,**[1] **Silvia Baronti,**[2] **Carla Amendola,**[1]
Francesco Primo Vaccari,[2] **Giuseppe Lustrato,**[1] **Sara Di Lonardo,**[2] **Francesca Fantasma,**[1]
Roberto Tognetti,[1] **and Gabriella Stefania Scippa**[1]

[1]*Dipartimento di Bioscienze e Territorio, Università degli Studi del Molise, 86090 Pesche, Italy*
[2]*Istituto di Biometeorologia-Consiglio Nazionale delle Ricerche (IBIMET-CNR), 50145 Firenze, Italy*

Correspondence should be addressed to Dalila Trupiano; dalila.trupiano@unimol.it

Academic Editor: Ibrokhim Y. Abdurakhmonov

Impacts of biochar application in combination with organic fertilizer, such as compost, are not fully understood. In this study, we tested the effects of biochar amendment, compost addition, and their combination on lettuce plants grown in a soil poor in nutrients; soil microbiological, chemical, and physical characteristics were analyzed, together with plant growth and physiology. An initial screening was also done to evaluate the effect of biochar and compost toxicity, using cress plants and earthworms. Results showed that compost amendment had clear and positive effects on plant growth and yield and on soil chemical characteristics. However, we demonstrated that also the biochar alone stimulated lettuce leaves number and total biomass, improving soil total nitrogen and phosphorus contents, as well as total carbon, and enhancing related microbial communities. Nevertheless, combining biochar and compost, no positive synergic and summative effects were observed. Our results thus demonstrate that in a soil poor in nutrients the biochar alone could be effectively used to enhance soil fertility and plant growth and biomass yield. However, we can speculate that the combination of compost and biochar may enhance and sustain soil biophysical and chemical characteristics and improve crop productivity over time.

1. Introduction

Soil fertility degradation, caused by erosion and depletion or imbalance of organic matter/nutrients, is affecting world agricultural productivity [1]. Inorganic fertilizers have played a significant role in increasing crop production since the "green revolution" [2]; however, they are not a sustainable solution for maintenance of crop yields [3]. Long-term overuse of mineral fertilizers may accelerate soil acidification, affecting both the soil biota and biogeochemical processes, thus posing an environmental risk and decreasing crop production [4]. Organic amendments, such as compost and biochar, could therefore be useful tools to sustainably maintain or increase soil organic matter, preserving and improving soil fertility and crop yield.

Biochar is a carbon-rich material obtained from thermochemical conversion (slow, intermediate, and fast pyrolysis or gasification) of biomass in an oxygen-limited environment. It can be produced from a range of feedstock, including forest and agriculture residues, such as straw, nut shells, rice hulls, wood chips/pellets, tree bark, and switch grass [5]. Biochar has been described as a possible tool for soil fertility improvement, potential toxic element adsorption, and climate change mitigation [6–8].

Indeed, several studies have shown that biochar application to soil can (i) improve soil physical and chemical properties [9, 10], (ii) enhance plant nutrient availability and correlated growth and yield [11, 12], (iii) increase microbial population and activities [13–15], and (iv) reduce greenhouse gas emissions through C sequestration [16].

TABLE 1: *Biochar and compost characteristics.* The complete biochar and compost characterization and related methods are detailed in Amendola et al. [31] and Alfano et al. [30], respectively. All concentrations refer to dry matter and represent the means of three replicates ± standard error.

	Biochar	Compost
pH	9.7 ± 0.1	7.5 ± 0.1
Alkalinity (% CaCO$_3$)	18.2 ± 0.7	6.5 ± 0.4
EC (dS/m)	7.5 ± 0.4	4.9 ± 0.3
Moisture (g/kg)	62.4 ± 1.3	3.4 ± 0.9
CEC (cmol/kg)	21.3 ± 0.2	21.0 ± 0.2
P$_{tot}$ (g/kg)	12.2 ± 3.0	5.5 ± 0.6
N$_{tot}$ (g/kg)	9.1 ± 0.2	12.0 ± 4.0
C$_{tot}$ (g/kg)	778.1 ± 0.0	337.2 ± 0.3
C/N	125.5	28.1
Cultivable aerobic bacteria (log CFU/g)	Absent	7.6 ± 0.3
Eumycetes (log CFU/g)	Absent	4.9 ± 0.1
Actinomycetes (log CFU/g)	Absent	5.18 ± 0.2
Coliform bacteria (log CFU/g)	Absent	Absent
E. coli (log CFU/g)	Absent	Absent
Salmonellae spp. (log CFU/g)	Absent	Absent

The beneficial effects of biochar on plant productivity and soil microbial population are related to the improvement of specific surface area, cation exchange capacity, bulk density, pH, water, and nutrients within the soil matrix [17]. Beside the generally positive plant growth responses to biochar amendment, especially in acidic coarse texture soil, negligible or negative effects also occur due to types of feedstock and pyrolysis process, biochar application rate, plant species, and soil characteristics [18, 19]. Furthermore, in most cases, biochar does not provide high amounts of nutrients [20, 21].

Some recent studies have indicated that combined applications of biochar with organic or inorganic fertilizers could lead to enhanced soil physical, chemical, and biological properties, as well as plant growth. In particular, several composted materials represent a sustainable source of available nutrients that could enhance plant growth, ameliorating soil physicochemical characteristics and microbiological properties [22–26]. Liu et al. [26] showed that the combined application of compost and biochar had a positive synergistic effect on soil nutrient contents and water-holding capacity under field conditions. In addition, the combination of biochar with compost has proved to be suitable, allowing the reduction of fertilizer inputs, stabilizing the soil structure, and improving its nutrient content and water retention capacity [27, 28]. Again, these studies underline that compost and biochar combination could enhance compost properties, leading to a higher added value and a much better carbon sequestration potential due to the long-term stability of biochar [24, 25].

However, the literature shows that compost effects, as also reported above for biochar, can differ on soil biophysical-chemical properties and plant growth and yield on the basis of feedstock types, methods of producing, and application [29]. The objectives of this study were thus to determine the effect of biochar application alone, obtained from orchard pruning biomass by slow pyrolysis (550°C), or combined with compost, obtained from olive mill residues, on (i) plant growth, physiology, and yield, (ii) soil chemical characteristics, and (iii) soil microbiological abundance and enzyme activities. For this purpose, a short-term potting experiment was performed, using *Lactuca sativa* L. as reference plants and a soil poor in nutrients as growing substrate, to test the following hypotheses: biochar addition together with compost improves (1) soil chemical and (2) microbiological properties and enhances (3) plant growth and physiology more than compost and biochar alone.

2. Materials and Methods

2.1. Biochar, Compost, and Toxicity Test. The biochar used was a commercial charcoal (provided by Romagna Carbone s.n.c., Italy), obtained from orchard pruning biomass through a slow pyrolysis process at a temperature of 500°C in a transportable ring kiln 2.2 m in diameter that holds around 2 t of feedstock.

An olive waste compost was used, prepared in a specific experimental composting process, following Alfano et al. [30]. Briefly, compost was prepared mixing humid olive husks, from a two-phase extraction plant, with olive leaves (8% w/w); one-year-old, humid, composted husks (25% w/w) were then added to this mixture. Biochar and compost characteristics are summarized in Table 1 and analytical methods are provided in Amendola et al. [31] and Alfano et al. [30].

Biochar and compost phytotoxicity was assessed through the germination index (GI$_\%$) of cress plants (*Lepidium sativum* L.) [26]. Three different solutions were used to evaluate the biochar and compost toxicity on seed germination: sterile deionized water as control solution and solutions containing 50% and 75% extract of biochar or compost. Solutions were added to Petri dishes containing 10 sterile seeds of *L. sativum*. Germination percentage and plant root length were recorded after incubation for 42 h at 27°C in the dark (according to Vitullo et al. [32]). Seeds were scored as germinated if the radicle exceeded the length of the longest seed coat dimension. The seed germination percentage was assessed according to the formula: $GI_\% = (G_s L_s)/(G_c L_c) \times 100$, where G_s and L_s are seed germination and root elongation (mm) for the samples and G_c and L_c the corresponding values for controls. The test was repeated in triplicate. The $GI_\%$ was obtained by means of $GI_{50\%}$ and $GI_{75\%}$ values. Potential toxicity of biochar was also tested on earthworms (*Lumbricus terrestris* L). For the earthworm avoidance test, equal amounts of unfertilized soil with and without biochar (65 g kg^{-1} of dry soil) were placed in two halves of a pot (50 × 30 cm). Forty earthworms were released between the two substrates. After 48 hours, the pot was examined to determine the soil selected by earthworms [33].

2.2. Experimental Design. One-month-old lettuce (*Lactuca sativa* L. var. *longifolia*) seedlings (Vivaio Mignogna, Ripamolisana, Ripamolisana, Molise, Italy) were transplanted into plastic pots (3.5 l) prepared with four different substrates: (i) unfertilized soil (PS); (ii) unfertilized soil plus compost (PSC); (iii) unfertilized soil plus biochar (PSB); and (iv) unfertilized soil plus compost and biochar (PSCB). Plants were then grown until maturity (9 weeks) in a screened greenhouse (University of Molise, Pesche, Italy; Lat 41°37′00″N; Long. 14°17′00″E; 510 m a.s.l.) under a controlled water regime, temperatures ranging between 12 and 25°C, and natural day length corresponding to spring-summer season (May–July). Soil was collected from an uncultivated pasture area, located in Pesche, with a floral composition predominantly of graminoid grasses, not under a rotation system and that includes hedges. This area is mainly used for grazing, but the fodder is harvested mechanically. The preplanting physicochemical properties of the soil are given in Supporting Information (Table S1 in Supplementary Material available online at https://doi.org/10.1155/2017/3158207). Briefly, the soil was moderately subalkaline (0–30 cm) with a clay texture according to United States Department of Agriculture classification [34]; it was characterized by low electrical conductivity (EC), cation exchange capacity (CEC), and nitrogen and carbon content; it is unlikely that this soil already contained charcoal as there had been no tradition of crop residue or other burning on the land. For the experiment, soil was air-dried for 72 h, weighed and finely crushed, and then mixed thoroughly before packing lightly in the pots on top of 100 g of pebbles placed on the base to improve drainage. The weight of each filled pot was 3000 g.

Biochar and compost application rates were set up on the basis of previous lettuce pot and field researches [12, 35–39]. In detail, data reported by Carter et al. [35] showed that 50 and 150 g kg^{-1} rice-husk biochar application rate led to a highly positive effect on lettuce growth in compost fertilized and unfertilized soils, respectively. Based on this finding, in the present study, biochar and compost were supplied at an application rate of 65 g and 50 g per kg of dry soil, respectively. After mixing, the pots were filled in order to ensure the same soil bulk density. There were ten pots (one plant per pot) for each treatment arranged in a complete randomized block design and rotated each day to a different position within

the block for the duration of the trial. The pots were fully irrigated to prevent water stress (twice a day, as required) and a suspended shade cover net was used to reduce exposure to sunlight.

2.3. Soil Analyses. Soils were sampled at the end of the experiment and air-dried for 72 h. The moisture content was calculated according to the Black method [40] as the difference in sample weight before and after oven drying to constant weight at 105°C. The pH was measured by potentiometry (pH meter Eutech Instruments) in H_2O and 0.01 M $CaCl_2$ using a 1 : 2.5 soil weight : extract-volume ratio. Alkalinity of samples with a pH value greater than 7.0 was determined by titrimetry according to the Higginson and Rayment method [41, 42]. Electrical conductivity (EC) was determined by a conductivity meter (Cond 510, XS Instruments) on a 1 : 5 soil : water suspension [41, 42]. Ash content was determined by igniting an oven-dried sample in a muffle furnace at 440 ± 40°C, according to the American Society for Testing and Materials [43]. Cation exchange capacity (CEC) was assessed according to Mehlich [44] using the $BaCl_2$. For total nitrogen (N_{tot}) determination a modified Kjeldahl procedure was used with Devarda's alloy pretreatment, important to recover both NO_3^--N and NO_2^--N [45]. Total phosphorus (P_{tot}) was detected by spectrophotometry (UV-1601 Shimadzu) according to the test method described by Bowman [46]. Available phosphorus (P_{av}) was extracted by a $NaHCO_3$ solution at pH 8.5 and evaluated by spectrophotometry according to the Olsen test method [47]. Total carbon (C_{tot}) content was determined using a CHN autoanalyzer (CHN 1500, Carlo Erba) [48].

2.4. Plant Analysis. Plant morphological analyses were performed weekly by measuring the main leaf parameters: number (LN); area (LA); length (LL); width (LW); and perimeter (LP). The Image J 1.41 (https://rsb.info.nih.gov/ij/) software was used for analysis. In addition, at the end of the experiment, leaf and root biomass allocation was determined before (fresh weight, FW) and after (dry weight, DW) two days of drying in an oven at 60°C. The measurements were performed on six plants.

Leaf water potential (ψ) was measured using a pressure chamber (PMS, Instrumentation Co., Corvallis, OR, USA). Leaf gas exchange measurements were performed using a portable gas exchange system (CIRAS-1, PP Systems, Hertz, Hitchin, UK). Leaf gas exchanges and ψ were measured on the fourth fully expanded and sun exposed leaf on a cloud-free day (after 3 months of growth). These measurements were taken on five plants (one leaf per plant) per treatment around midday (between 11.30 a.m. and 1.30 p.m.).

Chlorophyll content was also measured. For this, chlorophylls a (Chl$a$) and b (Chl$b$) were extracted from three randomly sampled leaf discs (10 mm) with N,N dimethylformamide (DMF). Extraction was performed for 48 h at 4°C in the dark at a ratio of 1 : 20 (plant material : solvent, w : v) [49]. The extinction coefficients proposed by Inskeep and Bloom [50] were used for the quantification by spectrophotometric analysis. The following formulas were used: Chl $a = 12.70_{A664.5} - 2.79_{A647}$; Chl $b = 20.70_{A647} - 4.62_{A664.5}$;

$Chl = 17.90_{A647} + 8.08_{A664.5}$, where A is absorbance in 1.00 cm cuvettes and Chl is mg per l. The leaf area Chl content was then calculated ($mg\,cm^{-2}$) together with the Chl a/Chl b ratio by dividing the Chl a content by the Chl b content.

2.5. Microbiological Analyses. Cellular cultivability was assessed by plate-counting. Cultivable aerobic bacteria in soil samples were analyzed on standard plate count agar (Difco Bekton Dikson, Milan, Italy), at 28 and 55°C for 48 h; actinomycetes on actinomyces agar (Difco), at 28°C and 55°C for 48–72 h; eumycetes on malt agar (Difco) + rose bengal $33\,mg\,l^{-1}$ and tetracycline $100\,ml\,l^{-1}$, at 28°C for 72 h, following the method described by Alfano et al. [30]. Cellular counts were done in triplicate by performing quantitative determinations on the basis of colony forming units (CFU) in agarized media, according to Alfano et al. [30]. All results are expressed as log CFU/g DW, after drying aliquots of the samples at 105°C for 48 h.

Soil enzymatic activities were determined by the API-ZYM system (bio-Mérieux Italia, Rome, Italy). With the API-ZYM system, semiquantitative evaluation of the activities of 19 hydrolytic enzymes [alkaline phosphatase, esterase (C4), esterase-lipase (C8), lipase (C14), leucine arylamidase, valine arylamidase, cystine arylamidase, trypsin, α-chymotrypsin, acid phosphatase, phosphoamidase, α-galactosidase, β-galactosidase, β-glucuronidase, α-glucosidase, β-glucosidase, N-acetyl-β-glucosaminidase, α-mannosidase, and α-fucosidase] was determined [51, 52]. Using a sterile Pasteur pipette, each gallery was inoculated with two drops of 10^{-1} or 10^{-2} suspensions of 20 g of soil in 180 mL of sterile saline solution (0.9% NaCl, w/v). The color that developed in each enzymatic reaction was assigned according to the color chart (range 0–5) supplied with the system, and this, in accordance with reported procedures, provided the conversion evaluation of the hydrolyzed substrates in nanomoles [51]. The samples were analyzed in triplicate and the average data were used. Results were expressed as enzyme relative activity/g of DW of substrate; data were corrected by the dilution factor.

2.6. Statistical Analysis. Analysis of variance (ANOVA) was applied in order to evaluate the effect of each treatment on soil proprieties, chlorophyll, ecophysiology, and microbiological data (one-way ANOVA) and the effect of day, treatments, and their interaction (two-way ANOVA) for plant morphological data. To assess the differences in the measured parameters among treatments, a postmeans comparison was performed using the Fisher least significant difference (LSD) test at the 0.05 significance level. Statistical analysis was conducted with OriginPro version 8.5.1 (OriginLab, Northampton, MA, USA).

3. Results

3.1. Biochar and Compost Characteristics. Biochar and compost used for the experiment were previously analyzed by Amendola et al. [31] and Alfano et al. [30], respectively. Main biochar and compost characteristics are summarized in Table 1. Briefly, as in the majority of biochar, the pH value

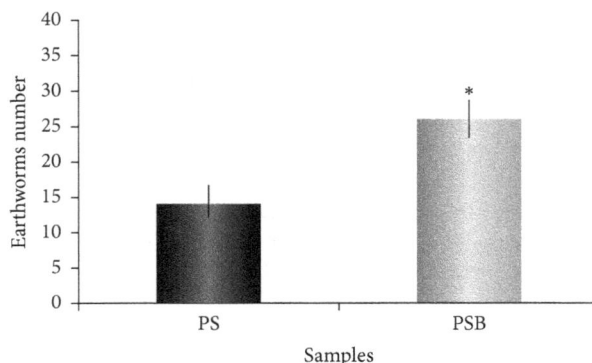

FIGURE 1: *Earthworms avoidance test.* Number of earthworms (*Lumbricus terrestris* L.) recovered in unfertilized (PS) versus unfertilized plus biochar (PSB) soil compartments after 48 h. Vertical bars represent the standard error of the mean of three replicates of 40 earthworms each. Asterisks indicate significant differences ($p \leq 0.05$).

was within the range of alkalinity. The C_{tot} and N_{tot} contents of biochar were 77.8% and 0.9%, respectively, according to Class 1 of the Guidelines for Certification of the International Biochar Initiative (IBI, http://www.european-biochar .org/en/ebc-ibi). The compost was mature, showing stable chemical and microbiological characteristics, with the potential to be used as an agricultural substrate or soil conditioner.

The results showed that biochar was not toxic for soil biotic communities, *L. sativum* seeds, or earthworms. Indeed, the phytotoxicity test with *Lepidium* showed no effects of biochar and compost on the germination index (82 ± 4 and 95 ± 4%, resp.) compared to the control (water; 100 ± 2%) (data not shown), while the avoidance test showed that *L. terrestris* preferred the biochar-amended soil (Figure 1).

3.2. Soil Characterization. Soil chemical analysis showed that the addition of biochar induced a significant increase of pH values from 6.9 (PS) to 8.0 and from 7.5 (PSC) to 7.7 in PSB and PSCB, respectively. On the contrary, the alkalinity value did not change in PSB and PSCB compared to PS and PSC, respectively (Table 2).

An increase of total N content from 0.8 (PS) to 1.2 was observed in PSB. Conversely, the EC increased about 1.4 and 1.1 fold in PSB and PSCB, respectively, while moisture decreased 0.8-fold in PSCB. Ash content slightly decreased in PSB and PSCB. The P_{tot} was increased about 1.5-fold in PSB, whereas no alteration was observed in PSBC. On the contrary, the P_{av} was 2-fold greater in PSCB than PSC. The C_{tot} content was significantly increased in PSB (5-fold) and PSCB (2-fold) compared to the relative controls (PS and PSC). The CEC value was also slightly increased in PSB and PSCB compared to PS and PSC, respectively. All the above parameters were increased in PSCB compared to PSB, except for pH, moisture, and ash that decreased.

3.3. Plant Growth. Significant differences in growth parameters were recorded between treatments (Figure 2). In detail, lettuce plants showed higher leaf number in PSB than PS

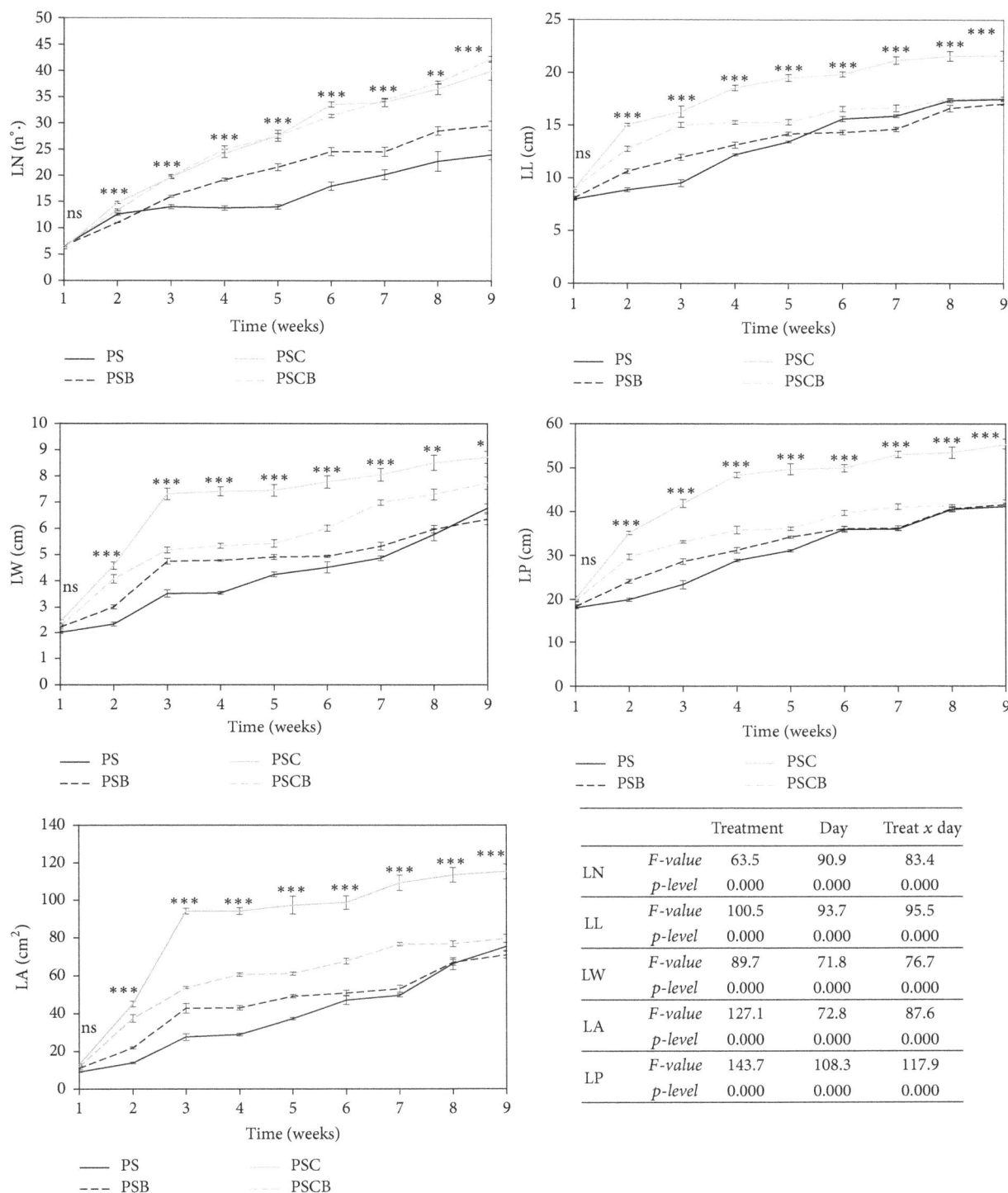

FIGURE 2: *Morphological analysis.* The main leaf parameters were analyzed: LN = leaf number; LL = leaf length; LW = leaf width; LA = leaf area; LP = leaf perimeter. Data represent the mean (n = 6) ± standard error. Mean values marked with asterisks are statistically different at ***$p \leq 0.0001$, at **$p \leq 0.005$, and at *$p \leq 0.01$. Two-way ANOVA was applied to weigh the effects of day, treatment, and their interactions on plant growth parameters (p and F level values are reported). PS = lettuce plants grown in unfertilized soil; PSB = lettuce plants grown in unfertilized soil plus biochar; PSC = lettuce plants grown in unfertilized soil plus compost; and PSCB = lettuce plants grown in unfertilized soil plus compost and biochar.

TABLE 2: *Soil chemical properties*. Data represent the mean (n = 4) ± standard error. Mean values marked with the same letter are not statistically different. One-way ANOVA was applied to weigh the effects of different treatments ($p < 0.05$).

	PS	PSB	PSC	PSCB
pH	6.9 ± 0.2^d	8.0 ± 0.0^a	7.5 ± 0.0^c	7.7 ± 0.0^b
Alkalinity (% $CaCO_3$)	7.7 ± 0.4^a	5.9 ± 0.5^a	6.4 ± 0.3^a	7.1 ± 0.4^a
EC (dS/m)	0.71 ± 0.0^c	1.0 ± 0.0^b	1.3 ± 0.1^b	1.5 ± 0.0^a
Moisture (g/kg)	63.0 ± 1.6^a	58.8 ± 0.3^b	61.8 ± 0.9^a	47.0 ± 2.1^c
Ash (%)	93.8 ± 0.2^a	90.0 ± 0.5^b	90.8 ± 0.7^b	86.0 ± 0.3^c
N_{tot} (g/kg)	0.8 ± 0.0^c	1.2 ± 0.0^b	1.9 ± 0.2^a	2.2 ± 0.0^a
P_{tot} (mg/kg)	199.9 ± 9.9^c	376.6 ± 52.3^b	455.6 ± 33.3^{ab}	471.0 ± 1.3^a
P_{av} (mg/kg)	$<12.0 \pm 0.0^c$	$<12.0 \pm 0.0^c$	24.9 ± 3.9^b	58.3 ± 4.5^a
C_{tot} (g/kg)	12.8 ± 0.5^c	59.1 ± 0.0^a	29.1 ± 0.8^b	65.9 ± 2.2^a
CEC (cmol/kg)	39.3 ± 0.0^d	40.4 ± 0.0^b	39.9 ± 0.0^c	43.6 ± 0.0^a

PS: unfertilized soil.
PSB: unfertilized soil plus biochar.
PSC: compost fertilized soil.
PSCB: compost fertilized soil plus biochar.

while leaf length, width, area, and perimeter were unchanged. Plants grown in PSCB showed lower leaf length, width, area, and perimeter compared to PSC while leaf number was unchanged. Furthermore, all leaf parameters were increased in PSC compared to PS and also in PSCB compared to PSB, although they were almost unchanged at the end of treatment (after 9 weeks).

Total plant dry weight, considering leaf and root biomass, increased in PSB compared to PS; on the contrary, leaf biomass decreased in PSCB and root biomass was unchanged compared to PSC and PSB (Figure 3). Chlorophyll content did not show significant changes between treatments (data not shown).

The leaf water potential was more negative in plants grown in PS than those grown in other substrates, followed by PSB and PSC and PSCB (Figure 4).

Stomatal conductance showed no significant differences between control and treated plants (Figure 5), while transpiration rate was slightly increased in PSCB and unchanged in PSC and PSB compared to PS (Figure 5). Assimilation rate was not altered by treatment with biochar (PSB) and biochar-compost (PSCB). In particular, plants showed the highest assimilation rate and water use efficiency in PSC (Figure 5).

3.4. Cultivable Microorganisms.

The analysis of cultivable microorganisms showed that in PSB the cultivable aerobic bacteria, actinomycetes, and eumycetes decreased ($p \leq 0.05$) compared to PS while they were unchanged in PSBC compared to PSC (Table 3). Furthermore, the abundance of cultivable aerobic bacteria and eumycetes was higher in PSC than PS while actinomycetes were unchanged; all cultivable microorganisms increased in PSCB compared to PSB (Table 3).

3.5. Soil Enzyme Activities.

Enzymatic profile analysis revealed that all enzymatic activities were increased in PSC compared to PS, except lipase-esterase and esterase that were unchanged. Biochar treatment alone or in combination with

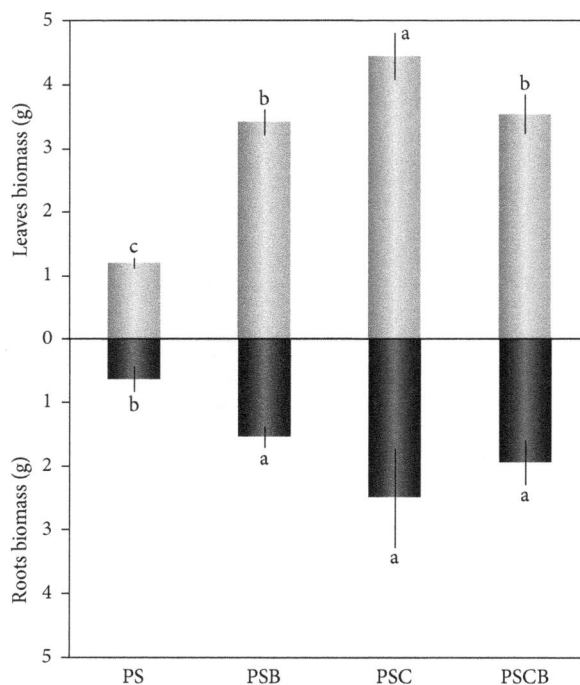

FIGURE 3: *Effects of biochar and/or compost on leaf and root biomass (g of dry tissue weight)*. Data represent the mean (n = 6) ± standard error. Mean values marked with the same letter are not statistically different. One-way ANOVA was applied to weigh the effects of different treatments ($p \leq 0.05$). PS = lettuce plants grown in unfertilized soil; PSB = lettuce plants grown in unfertilized soil plus biochar; PSC = lettuce plants grown in unfertilized soil plus compost; and PSCB = lettuce plants grown in unfertilized soil plus compost and biochar.

compost induced specific enzymatic variations. In detail, in PSB, compared to PS, the activities of alkaline phosphatase, acid phosphatase, chymotrypsin, trypsin, phosphohydrolase, lipase-esterase, and esterase were increased, while lipase was

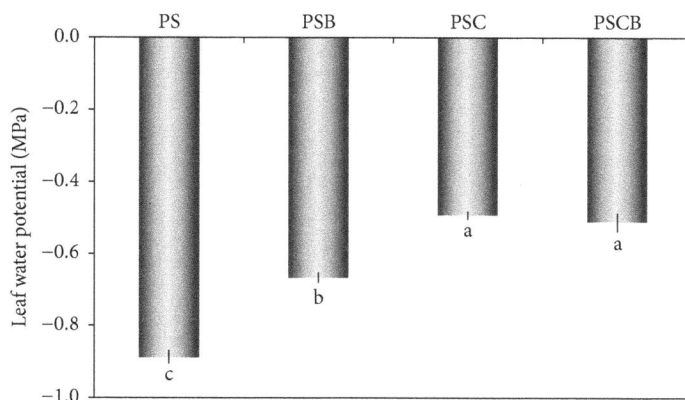

FIGURE 4: *Leaf water potential at midday of plants grown in different substrates.* Values are means (n = 5) ± standard error; significant differences between the means (at least $p \leq 0.05$, according to ANOVA) appear with different letters. PS = lettuce plants grown in unfertilized soil; PSB = lettuce plants grown in unfertilized soil plus biochar; PSC = lettuce plants grown in unfertilized soil plus compost; and PSCB = lettuce plants grown in unfertilized soil plus compost and biochar.

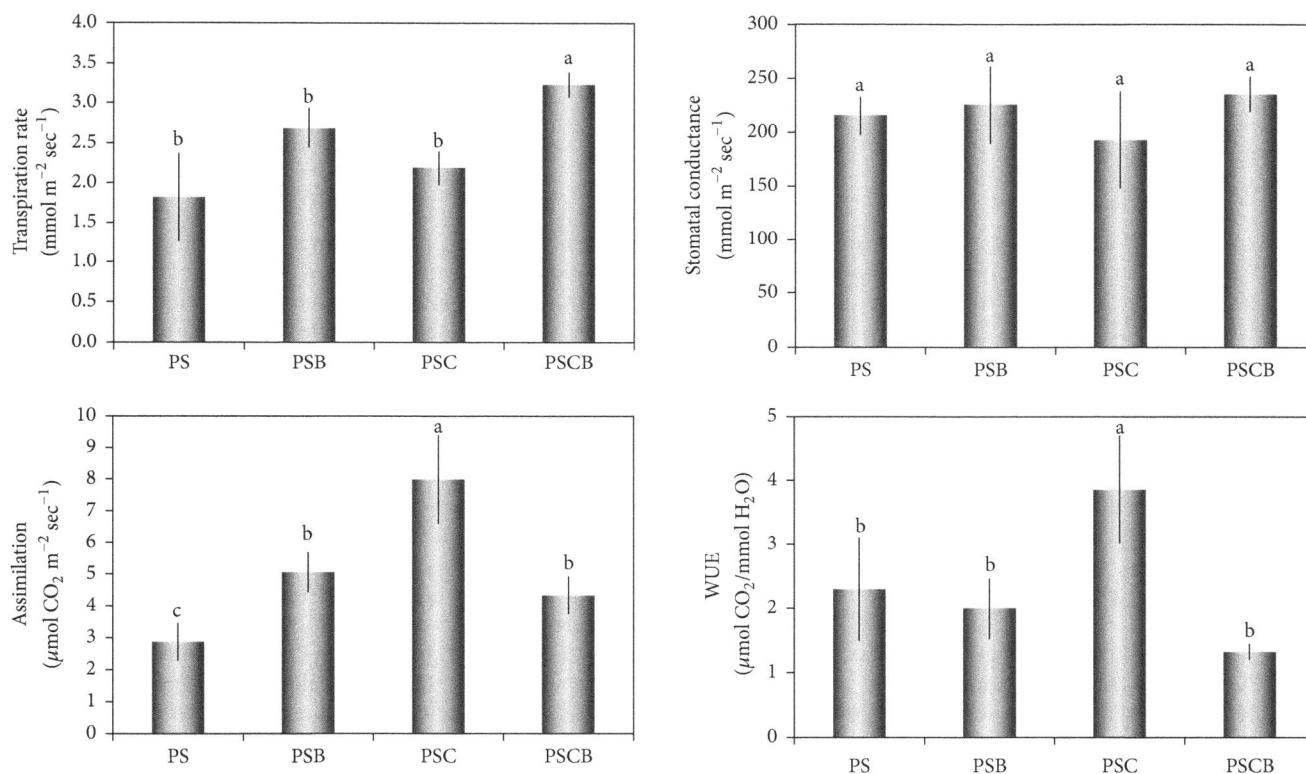

FIGURE 5: *Transpiration rate, stomatal conductance, assimilation, and water use efficiency (WUE) of plants grown in different soils.* Values are means (n = 5) ± standard error; significant differences between the means (at least $p \leq 0.05$, according to ANOVA) appear with different letters. PS = lettuce plants grown in unfertilized soil; PSB = lettuce plants grown in unfertilized soil plus biochar; PSC = lettuce plants grown in unfertilized soil plus compost; and PSCB = lettuce plants grown in unfertilized soil plus compost and biochar.

unchanged (Table 4). In PSBC, compared to PSC, alkaline phosphatase, acid phosphatase, lipase-esterase, and esterase strongly increased ($p \leq 0.01$), whereas chymotrypsin and trypsin activities decreased ($p \leq 0.05$) and phosphohydrolase and lipase activities were lost (Table 4). In PSCB, compared to PSB, we recorded that alkaline phosphatase and acid phosphatase increased and trypsin, phosphohydrolase, and

esterase decreased, while chymotrypsin, lipase-esterase, and lipase were unchanged ($p \leq 0.05$).

4. Discussion

The study showed that both biochar amendment and compost addition to a soil poor in nutrients induced a positive effect on

TABLE 3: *Soil microbiological characteristics.* Values are expressed as log CFU/g dry weight (DW). Data represent the mean (n = 3) ± standard error. Mean values marked with the same letter are not statistically different. One-way ANOVA was applied to weigh the effects of different treatments ($p < 0.05$). PS = unfertilized soil; PSB = unfertilized soil plus biochar; PSC = unfertilized soil plus compost; and PSCB = unfertilized soil plus compost and biochar.

	PS	PSB	PSC	PSCB
Cultivable aerobic bacteria	10.97 ± 0.17^a	7.88 ± 0.18^c	8.84 ± 0.20^b	8.35 ± 0.21^b
Actinomycetes	8.67 ± 0.15^a	5.2 ± 0.12^c	6.4 ± 0.3^b	6.1 ± 0.2^b
Eumycetes	8.85 ± 0.12^a	7.89 ± 0.08^b	8.74 ± 10.40^a	8.52 ± 0.14^a

TABLE 4: *Soil enzymatic activities.* Soil enzymatic activities were determined by the API-ZYM system (bio-Mérieux Italia). Results are expressed as nanomoles/g of dry weight of substrate; the data have been corrected by the dilution factor. PS = unfertilized soil; PSB = unfertilized soil plus biochar; PSC = unfertilized soil plus compost; and PSCB = unfertilized soil plus compost and biochar.

Enzyme	PS	PSB	PSC	PSCB
Alkaline phosphatase	250 ± 25^c	500 ± 25^b	500 ± 50^b	1000 ± 50^a
Acid phosphatase	250 ± 25^c	500 ± 50^b	500 ± 25^b	1000 ± 100^a
Chymotrypsin	10 ± 5^c	250 ± 25^b	500 ± 25^a	250 ± 25^b
Trypsin	10 ± 5^d	250 ± 50^b	500 ± 50^a	100 ± 25^c
Phosphohydrolase	10 ± 5^c	250 ± 25^b	500 ± 50^a	10 ± 5^c
Lipase-esterase	10 ± 5^b	500 ± 50^a	10 ± 5^b	500 ± 25^a
Esterase	10 ± 5^b	500 ± 25^a	10 ± 5^b	10 ± 5^b
Lipase	10 ± 5^b	10 ± 5^b	250 ± 25^a	10 ± 5^b

lettuce plant growth and physiology and on soil chemical and microbiological characteristics; however, no positive synergic or summative effects exerted by compost and biochar in combination were observed.

In detail, the biochar alone induced a positive lettuce yield response, although transpiration, stomatal conductance, and assimilation rate did not show relevant variations. Positive yield responses to biochar addition have been reported for a wide variety of crops. For example, it is reported that maize yield increased by 98–150% as a result of manure biochar addition [53], lettuce and *Arabidopsis* plant biomass increased by 111% after poplar wood chips biochar addition [54], and wheat grain yield increased by 18% with the use of oil mallee biochar [55].

A possible explanation is that the biochar, increasing the pH, CEC, N_{tot}, C_{tot}, P_{tot}, and water content, could enhance available nutrients for plants and, consequently, biomass accumulation [22, 56]. In fact, the increase in CEC value could be driven by the presence of cation exchange sites on the biochar surface [10, 57], and, as also reported in Vaccari et al. [58], this could contribute to retaining NH_4^+, leading to improved N nutrition in biochar-amended soils [59–61]. This would confirm a direct biochar role in the nutrient supply to plants [20, 62]. The increased pH in biochar treated soil could also be indirectly related to growth stimulation in lettuce. Indeed, Beesley and Dickinson [63] hypothesized that soil alkalization caused by biochar addition might positively influence earthworms, as also observed in our study, with a subsequent positive effect on dissolved organic carbon content. The pH value has also been found to influence the soil microbial population and enzymatic activities; indeed, a high pH might enhance bacteria abundance,

whereas it did not change fungi total biomass or dramatically reduce their growth [64, 65]. The activity of alkaline phosphatase, aminopeptidase, and N-acetylglucosaminidase enzymes has also been reported to increase after biochar applications [66, 67]. In accordance with this evidence, our results showed that the biochar alone decreased cultivable microorganisms abundance, while it enhanced the activity of enzymes involved in phosphorus, nitrogen, and carbon cycling (alkaline phosphatase, acid phosphatase, phosphohydrolase, lipase-esterase, esterase, chymotrypsin, and trypsin). These results could indicate that although the bacteria abundance could be reduced by biochar soil alkalization, the microbial communities related to nitrogen, phosphorus, and carbon cycling could be stimulated by biochar-induced increasing of soil P_{tot}, N_{tot}, and C_{tot} availability [68, 69].

Nevertheless, the compost alone amendment showed the best clear and positive effects on plant growth and yield and on soil chemical characteristics. Indeed, according to data reported in the literature [70, 71], in compost amended soil lettuce plants showed the maximum total biomass accumulation, assimilation rate, and water use efficiency, probably due to the high soil nutrients availability (soil C_{tot}, N_{tot}, P_{tot}, and P_{av} content was increased). This high soil nutrient status could also have enhanced the activity of enzymes involved in phosphorus and nitrogen cycling (phosphohydrolase, chymotrypsin, and trypsin), which increased compared to those in the unamended soils; on the other hand, the slight pH increase could be responsible for the decrease of cultivable bacteria.

No synergic or positive effects exerted by compost and biochar combination were observed here compared to the compost alone treatment. Indeed, we showed that lettuce growth changes were negligible combining compost and biochar amendment, although, as shown above, the amount of compost applied and the nutrients supplied were adequate to produce the highest plant benefits in the compost alone treatment. Furthermore, compared to the addition of compost alone, the compost and biochar combination did not improve soil chemical characteristics, except for an increase in C_{tot} and P_{av} content. These increases could be related to biochar capacity to enhance C accumulation and sequestration and to retain and exchange phosphate ions by its positively charged surface sites [59, 60]. Additionally, in compost and biochar-amended soil, microorganism abundance was unchanged while the activity of enzymes involved in N and P mineralization (chymotrypsin, trypsin, and phosphohydrolase) was reduced or completely lost compared to those in the compost alone treatment.

It is reported that biochar can have significant effects on microbially mediated transformation of nutrients by its surface interaction with substrate and soil microbial enzymes [72, 73] or by inducing soil alkalization [74]. Indeed, we hypothesized that soil microbial abundance and activities were shaped by nutrient availability and pH, which in turn could be balanced by biochar-compost combination. However, the biochar benefits could be amplified over time through surface oxidation and bioactivation with soil microbes and fungi [75, 76]. In addition, given that the beneficial effects of biochar were found to increase more in sandy than in loamy substrates [25], we hypothesized that, in PSCB, the high nutrient status, due to the compost, could have masked biochar effects [77].

In summary, our short-term potting experiment clearly demonstrated that compost addition provided the best solution regarding soil quality and fertility, which were also reflected in best plant growth and biomass yield.

Furthermore, taking into account that the soil used in this study had low nutrient status, suboptimal for plant growth without additions of organic and/or inorganic amendments, these results demonstrate that the application of biochar alone could also be effectively used to enhance soil fertility and plant growth and biomass yield. This may have important implications for sustainable low-input agriculture, with economic and environmental benefits for both marginal and productive cropland.

Nevertheless, unexpectedly, combined application of biochar and compost did not outperform compost amendment in terms of biomass yield and soil fertility. However, it may enhance and sustain soil biophysical and chemical characteristics over time, given that most of compost will disappear through mineralization within 5 years after application whereas most of the biochar will stay in the soil for decades [20, 78].

Further long-term and large-scale field experiments are required to analyze differences over time and in particular to quantify the amount of recalcitrant carbon supplied and sequestered in the soil by both biochar alone and the combination of biochar and compost. Their benefits and effects in terms of improving and sustaining soil fertility, crop productivity, and economic returns to users should be also evaluated over time.

Conflicts of Interest

The authors declare that there are no conflicts of interest regarding the publication of this paper.

Authors' Contributions

Gabriella Stefania Scippa coordinated the project. Gabriella Stefania Scippa, Dalila Trupiano, Claudia Cocozza, and Roberto Tognetti conceived and designed the experiments. Dalila Trupiano, Carla Amendola, Claudia Cocozza, Francesca Fantasma, Silvia Baronti, Sara Di Lonardo, and Giuseppe Lustrato performed the experiments. Dalila Trupiano, Carla Amendola, Claudia Cocozza, Silvia Baronti, Sara Di Lonardo, and Giuseppe Lustrato analyzed data. Gabriella Stefania Scippa contributed reagents/materials/analysis tools. Dalila Trupiano wrote the manuscript. Claudia Cocozza, Roberto Tognetti, Silvia Baronti, and Francesco Primo Vaccari revised the manuscript. All authors approved the manuscript.

Acknowledgments

The authors thank Dr. Luisa Andrenelli and Dr. Adriano Baglio (University of Florence), Italian Biochar Association (ICHAR http://www.ichar.org), and Federica Oliva (University of Molise) for their technical support during laboratory measurements. Research was supported by grants from Molise Region (PSR Molise 2007/2013-Misura 124) through the ProSEEAA Project (CUP: D95F14000030007) and it contributes to the EuroCHAR Project (FP7-ENV-2010 ID-265179).

References

[1] J. A. Foley, R. DeFries, G. P. Asner et al., "Global consequences of land use," *Science*, vol. 309, no. 5734, pp. 570–574, 2005.

[2] E. Liu, C. Yan, X. Mei et al., "Long-term effect of chemical fertilizer, straw, and manure on soil chemical and biological properties in northwest China," *Geoderma*, vol. 158, no. 3-4, pp. 173–180, 2010.

[3] B. Vanlauwe, A. Bationo, J. Chianu et al., "Integrated soil fertility management: operational definition and consequences for implementation and dissemination," *Outlook on Agriculture*, vol. 39, no. 1, pp. 17–24, 2010.

[4] J. C. Aciego Pietri and P. C. Brookes, "Relationships between soil pH and microbial properties in a UK arable soil," *Soil Biology and Biochemistry*, vol. 40, no. 7, pp. 1856–1861, 2008.

[5] S. Sohi, E. Lopez-Capel, E. Krull, and R. Bol, "Biochar, climate change and soil: a review to guide future research," CSIRO Land and Water Science Report 05/09, 2009.

[6] C. J. Ennis, A. G. Evans, M. Islam, T. K. Ralebitso-Senior, and E. Senior, "Biochar: carbon sequestration, land remediation, and impacts on soil microbiology," *Critical Reviews in Environmental Science and Technology*, vol. 42, no. 22, pp. 2311–2364, 2012.

[7] S. Malghani, G. Gleixner, and S. E. Trumbore, "Chars produced by slow pyrolysis and hydrothermal carbonization vary in carbon sequestration potential and greenhouse gases emissions," *Soil Biology and Biochemistry*, vol. 62, pp. 137–146, 2013.

[8] C. E. Stewart, J. Zheng, J. Botte, and M. F. Cotrufo, "Co-generated fast pyrolysis biochar mitigates green-house gas emissions and increases carbon sequestration in temperate soils," *GCB Bioenergy*, vol. 5, no. 2, pp. 153–164, 2013.

[9] A. Mukherjee and R. Lal, "Biochar impacts on soil physical properties and greenhouse gas emissions," *Agronomy*, vol. 3, pp. 313–339, 2013.

[10] S. P. Sohi, E. Krull, E. Lopez-Capel, and R. Bol, "A review of biochar and its use and function in soil," *Advances in Agronomy*, vol. 105, no. 1, pp. 47–82, 2010.

[11] L. A. Biederman and W. Stanley Harpole, "Biochar and its effects on plant productivity and nutrient cycling: a meta-analysis," *GCB Bioenergy*, vol. 5, no. 2, pp. 202–214, 2013.

[12] S. Jeffery, F. G. A. Verheijen, M. van der Velde, and A. C. Bastos, "A quantitative review of the effects of biochar application to soils on crop productivity using meta-analysis," *Agriculture, Ecosystems and Environment*, vol. 144, no. 1, pp. 175–187, 2011.

[13] N. M. Jaafar, P. L. Clode, and L. K. Abbott, "Microscopy observations of habitable space in biochar for colonization by fungal hyphae from soil," *Journal of Integrative Agriculture*, vol. 13, no. 3, pp. 483–490, 2014.

[14] R. S. Quilliam, H. C. Glanville, S. C. Wade, and D. L. Jones, "Life in the 'charosphere'—does biochar in agricultural soil provide a significant habitat for microorganisms?" *Soil Biology and Biochemistry*, vol. 65, pp. 287–293, 2013.

[15] J. Lehmann, M. C. Rillig, J. Thies, C. A. Masiello, W. C. Hockaday, and D. Crowley, "Biochar effects on soil biota—a review," *Soil Biology and Biochemistry*, vol. 43, no. 9, pp. 1812–1836, 2011.

[16] K. Crombie, O. Mašek, A. Cross, and S. Sohi, "Biochar—synergies and trade-offs between soil enhancing properties and C sequestration potential," *GCB Bioenergy*, vol. 7, no. 5, pp. 1161–1175, 2015.

[17] J. Thies and M. C. Rillig, "Characteristics of biochar: biological properties," in *Biochar for Environmental Management: Science and Technology*, J. Lehmann and S. Josep, Eds., Earthscan, London, UK, 2009.

[18] K. A. Spokas, K. B. Cantrell, J. M. Novak et al., "Biochar: a synthesis of its agronomic impact beyond carbon sequestration," *Journal of Environmental Quality*, vol. 41, no. 4, pp. 973–989, 2012.

[19] R. D. Lentz and J. A. Ippolito, "Biochar and manure affect calcareous soil and corn silage nutrient concentrations and uptake," *Journal of Environmental Quality*, vol. 41, pp. 1033–1043, 2012.

[20] B. Glaser, J. Lehmann, and W. Zech, "Ameliorating physical and chemical properties of highly weathered soils in the tropics with charcoal—a review," *Biology and Fertility of Soils*, vol. 35, no. 4, pp. 219–230, 2002.

[21] B. Glaser and J. J. Birk, "State of the scientific knowledge on properties and genesis of Anthropogenic Dark Earths in Central Amazonia (*terra preta de índio*)," *Geochimica et Cosmochimica Acta*, vol. 82, pp. 39–51, 2012.

[22] R. Scotti, G. Bonanomi, R. Scelza, A. Zoina, and M. A. Rao, "Organic amendments as sustainable tool to recovery fertility in intensive agricultural systems," *Journal of Soil Science and Plant Nutrition*, vol. 15, no. 2, pp. 333–352, 2015.

[23] K. S. Khan and R. G. Joergensen, "Compost and phosphorus amendments for stimulating microorganisms and growth of ryegrass in a Ferralsol and a Luvisol," *Journal of Plant Nutrition and Soil Science*, vol. 175, no. 1, pp. 108–114, 2012.

[24] D. Fischer and B. Glaser, "Synergisms between compost and biochar for sustainable soil amelioration," in *Management of Organic Waste*, S. Kumar and A. Bharti, Eds., pp. 167–198, InTech, 2012.

[25] H. Schulz and B. Glaser, "Effects of biochar compared to organic and inorganic fertilizers on soil quality and plant growth in a greenhouse experiment," *Journal of Plant Nutrition and Soil Science*, vol. 175, no. 3, pp. 410–422, 2012.

[26] J. Liu, H. Schulz, S. Brandl, H. Miehtke, B. Huwe, and B. Glaser, "Short-term effect of biochar and compost on soil fertility and water status of a Dystric Cambisol in NE Germany under field conditions," *Journal of Plant Nutrition and Soil Science*, vol. 175, no. 5, pp. 698–707, 2012.

[27] G. Agegnehu, M. I. Bird, P. N. Nelson, and A. M. Bass, "The ameliorating effects of biochar and compost on soil quality and plant growth on a Ferralsol," *Soil Research*, vol. 53, no. 1, pp. 1–12, 2015.

[28] H. P. Schmidt, C. Kammann, C. Niggli, M. W. H. Evangelou, K. A. Mackie, and S. Abiven, "Biochar and biochar-compost as soil amendments to a vineyard soil: influences on plant growth, nutrient uptake, plant health and grape quality," *Agriculture, Ecosystems & Environment*, vol. 191, pp. 117–123, 2014.

[29] M. P. Bernal, J. A. Alburquerque, and R. Moral, "Composting of animal manures and chemical criteria for compost maturity assessment. A review," *Bioresource Technology*, vol. 100, no. 22, pp. 5444–5453, 2009.

[30] G. Alfano, G. Lustrato, G. Lima, D. Vitullo, and G. Ranalli, "Characterization of composted olive mill wastes to predict potential plant disease suppressiveness," *Biological Control*, vol. 58, no. 3, pp. 199–207, 2011.

[31] C. Amendola, A. Montagnoli, M. Terzaghi et al., "Short-term effects of biochar on grapevine fine root dynamics and arbuscular mycorrhizae production," *Agriculture, Ecosystems & Environment*, vol. 239, pp. 236–245, 2017.

[32] D. Vitullo, R. Altieri, A. Esposito et al., "Suppressive biomasses and antagonist bacteria for an eco-compatible control of *Verticillium dahliae* on nursery-grown olive plants," *International Journal of Environmental Science and Technology*, vol. 10, no. 2, pp. 209–220, 2013.

[33] ISO, "Soil quality—avoidance test for determining the quality of soils and effects of chemicals on behaviour - part 1: test with earthworms (*Eisenia fetida* and *Eisenia andrei*)," ISO 17512-1, 2008.

[34] USDA, United States Department of Agriculture, Natural Resources Conservation Service, National Soil Survey Handbook, 2005, http://www.nrcs.usda.gov/wps/portal/nrcs/detail/national/home/?cid=nrcs142p2_054241.

[35] S. Carter, S. Shackley, S. Sohi, T. B. Suy, and S. Haefele, "The impact of biochar application on soil properties and plant growth of pot grown lettuce (*Lactuca sativa*) and cabbage (*Brassica chinensis*)," *Agronomy*, vol. 3, no. 2, pp. 404–418, 2013.

[36] H. Sokchea, K. Borin, and T. R. Preston, "Effect of biochar from rice husks (combusted in a downdraft gasifier or a paddy rice dryer) on production of rice fertilized with biodigester effluent or urea," *Livestock Research for Rural Development*, vol. 25, no. 1, 2013.

[37] A. Fabrizio, F. Tambone, and P. Genevini, "Effect of compost application rate on carbon degradation and retention in soils," *Waste Management*, vol. 29, no. 1, pp. 174–179, 2009.

[38] M. Reis, L. Coelho, J. Beltrao, I. Domingos, and M. Moura, "Comparative response of lettuce (*Lactuca sativa*) to inorganic

and organic compost fertilization," in *Recent Advances in Energy, Environment, Economics and Technological Innovation*, pp. 61–68, 2008.

[39] J. Lehmann, J. Gaunt, and M. Rondon, "Bio-char sequestration in terrestrial ecosystems—a review," *Mitigation and Adaptation Strategies for Global Change*, vol. 11, no. 2, pp. 403–427, 2006.

[40] C. A. Black, *Methods of Soil Analysis: Part I Physical and Mineralogical Properties*, American Society of Agronomy, Madison, Wis, USA, 1965.

[41] F. R. Higginson and G. E. Rayment, *Australian Laboratory Handbook of Soil and Water Chemical Methods*, Inkata Press, 1992.

[42] Y. P. Kalra and Soil & Plant Analysis Council, *Handbook on Reference Methods for Soil Analysis*, Soil & Plant Analysis Council, 1992.

[43] ASTM, "Standard test methods for moisture, ash, and organic matter of peat and other organic soils," ASTM D2974-00, ASTM International, West Conshohocken, Pa, USA, 2000.

[44] A. Mehlich, "Use of triethanolamine acetate-barium hydroxide buffer for the determination of some base exchange properties and lime requirement of soil," *Soil Science Society of America Proceedings*, vol. 29, pp. 374–378, 1938.

[45] C. F. H. Liao, "Devarda's alloy method for total nitrogen determination," *Soil Science Society of America Journal*, vol. 45, no. 5, pp. 852–855, 1981.

[46] R. A. Bowman, "A rapid method to determine total phosphorus in soils," *Soil Science Society of America Journal*, vol. 52, no. 5, pp. 1301–1304, 1988.

[47] S. R. Olsen, C. V. Cole, F. S. Watanabe, and L. A. Dean, *Estimation of available phosphorus in soils by extraction with sodium bicarbonate*, Circular 939, USDA, Washington, DC, USA, 1954.

[48] J. B. A. Dumas, "Procedes de l'analyse organique," *Annales de Chimie et de Physique*, vol. 247, pp. 198–213, 1831.

[49] R. Moran and D. Porath, "Chlorophyll determination in intact tissues using N,N-dimethylformamide," *Plant Physiology*, vol. 65, no. 3, pp. 478–479, 1980.

[50] W. P. Inskeep and P. R. Bloom, "Extinction coefficients of chlorophyll a and b in n,n-dimethylformamide and 80% acetone," *Plant Physiology*, vol. 77, no. 2, pp. 483–485, 1985.

[51] C. Viti, A. Mini, G. Ranalli, G. Lustrato, and L. Giovannetti, "Response of microbial communities to different doses of chromate in soil microcosms," *Applied Soil Ecology*, vol. 34, no. 2-3, pp. 125–139, 2006.

[52] S. M. Tiquia, "Evolution of extracellular enzyme activities during manure composting," *Journal of Applied Microbiology*, vol. 92, no. 4, pp. 764–775, 2002.

[53] K. C. Uzoma, M. Inoue, H. Andry, H. Fujimaki, A. Zahoor, and E. Nishihara, "Effect of cow manure biochar on maize productivity under sandy soil condition," *Soil Use and Management*, vol. 27, no. 2, pp. 205–212, 2011.

[54] M. Viger, R. D. Hancock, F. Miglietta, and G. Taylor, "More plant growth but less plant defence? First global gene expression data for plants grown in soil amended with biochar," *GCB Bioenergy*, vol. 7, no. 4, pp. 658–672, 2015.

[55] Z. M. Solaiman, P. Blackwell, L. K. Abbott, and P. Storer, "Direct and residual effect of biochar application on mycorrhizal root colonisation, growth and nutrition of wheat," *Australian Journal of Soil Research*, vol. 48, no. 6-7, pp. 546–554, 2010.

[56] M. Ros, M. T. Hernandez, and C. García, "Soil microbial activity after restoration of a semiarid soil by organic amendments," *Soil Biology and Biochemistry*, vol. 35, no. 3, pp. 463–469, 2003.

[57] D. L. Jones, J. Rousk, G. Edwards-Jones, T. H. DeLuca, and D. V. Murphy, "Biochar-mediated changes in soil quality and plant growth in a three year field trial," *Soil Biology and Biochemistry*, vol. 45, pp. 113–124, 2012.

[58] F. P. Vaccari, A. Maienza, F. Miglietta et al., "Biochar stimulates plant growth but not fruit yield of processing tomato in a fertile soil," *Agriculture, Ecosystems and Environment*, vol. 207, pp. 163–170, 2015.

[59] T. H. De Luca, M. D. MacKenzie, and M. J. Gundale, "Biochar effects on soil nutrient transformations," in *Biochar for Environmental Management: Science and Technology*, J. Lehmann and S. Joseph, Eds., pp. 251–270, Earthscan Publications, London, UK, 2009.

[60] J. Major, C. Steiner, A. Downie, and J. Lehmann, "Biochar effects on nutrient leaching," in *Biochar for Environmental Management: Science and Technology*, J. Lehmann and S. Joseph, Eds., pp. 271–287, Earthscan, London, UK, 2009.

[61] C. C. Hollister, J. J. Bisogni, and J. Lehmann, "Ammonium, nitrate, and phosphate sorption to and solute leaching from biochars prepared from corn stover (*Zea mays* L.) and oak wood (*Quercus* spp.)," *Journal of Environmental Quality*, vol. 42, no. 1, pp. 137–144, 2013.

[62] T. J. Clough and L. M. Condron, "Biochar and the nitrogen cycle: introduction," *Journal of Environmental Quality*, vol. 39, no. 4, pp. 1218–1223, 2010.

[63] L. Beesley and N. Dickinson, "Carbon and trace element fluxes in the pore water of an urban soil following greenwaste compost, woody and biochar amendments, inoculated with the earthworm *Lumbricus terrestris*," *Soil Biology and Biochemistry*, vol. 43, no. 1, pp. 188–196, 2011.

[64] J. Prommer, W. Wanek, F. Hofhansl et al., "Biochar decelerates soil organic nitrogen cycling but stimulates soil nitrification in a temperate arable field trial," *PLoS ONE*, vol. 9, no. 1, Article ID e86388, 2014.

[65] C. R. Anderson, L. M. Condron, T. J. Clough et al., "Biochar induced soil microbial community change: implications for biogeochemical cycling of carbon, nitrogen and phosphorus," *Pedobiologia*, vol. 54, no. 5-6, pp. 309–320, 2011.

[66] J. Rousk, P. C. Brookes, and E. Bååth, "Contrasting soil pH effects on fungal and bacterial growth suggest functional redundancy in carbon mineralization," *Applied and Environmental Microbiology*, vol. 75, no. 6, pp. 1589–1596, 2009.

[67] J. Rousk, P. C. Brookes, and E. Bååth, "Investigating the mechanisms for the opposing pH relationships of fungal and bacterial growth in soil," *Soil Biology and Biochemistry*, vol. 42, no. 6, pp. 926–934, 2010.

[68] V. L. Bailey, S. J. Fansler, J. L. Smith, and H. Bolton Jr., "Reconciling apparent variability in effects of biochar amendment on soil enzyme activities by assay optimization," *Soil Biology and Biochemistry*, vol. 43, no. 2, pp. 296–301, 2011.

[69] H. Jin, *Characterization of microbial life colonizing biochar and biochar amended soils [Ph.D. thesis]*, Cornell University, Ithaca, NY, USA, 2010.

[70] F. Amlinger, S. Peyr, J. Geszti, P. Dreher, W. Karlheinz, and S. Nortcliff, *Beneficial Effects of Compost Application on Fertility and Productivity of Soils*, Federal Ministry for Agricultural and Forestry, Environment and Water Management, Lebensministerium, Vienna, Austria, 2007.

[71] M. Diacono and F. Montemurro, "Long-term effects of organic amendments on soil fertility," *Journal of Sustainable Agriculture*, vol. 2, pp. 761–786, 2011.

[72] V. L. Bailey, S. J. Fansler, J. L. Smith, and H. Bolton, "Reconciling apparent variability in effects of biochar amendment on soil enzyme activities by assay optimization," *Soil Biology and Biochemistry*, vol. 43, no. 2, pp. 296–301, 2011.

[73] C. Lammirato, A. Miltner, and M. Kaestner, "Effects of wood char and activated carbon on the hydrolysis of cellobiose by β-glucosidase from *Aspergillus niger*," *Soil Biology and Biochemistry*, vol. 43, no. 9, pp. 1936–1942, 2011.

[74] Y. Yao, B. Gao, M. Inyang et al., "Biochar derived from anaerobically digested sugar beet tailings: characterization and phosphate removal potential," *Bioresource Technology*, vol. 102, no. 10, pp. 6273–6278, 2011.

[75] H.-J. Xu, X.-H. Wang, H. Li, H.-Y. Yao, J.-Q. Su, and Y.-G. Zhu, "Biochar impacts soil microbial community composition and nitrogen cycling in an acidic soil planted with rape," *Environmental Science and Technology*, vol. 48, no. 16, pp. 9391–9399, 2014.

[76] W.-C. Ding, X.-L. Zeng, Y.-F. Wang, Y. Du, and Q.-X. Zhu, "Characteristics and performances of biofilm carrier prepared from agro-based biochar," *China Environmental Science*, vol. 31, no. 9, pp. 1451–1455, 2011.

[77] B. T. Nguyen, J. Lehmann, W. C. Hockaday, S. Joseph, and C. A. Masiello, "Temperature sensitivity of black carbon decomposition and oxidation," *Environmental Science and Technology*, vol. 44, no. 9, pp. 3324–3331, 2010.

[78] Y. Kuzyakov, I. Subbotina, H. Chen, I. Bogomolova, and X. Xu, "Black carbon decomposition and incorporation into soil microbial biomass estimated by 14C labeling," *Soil Biology and Biochemistry*, vol. 41, no. 2, pp. 210–219, 2009.

17

Response of *Sorghum* (*Sorghum bicolor* L.) to Residual Phosphate in Soybean-*Sorghum* and Maize-*Sorghum* Crop Rotation Schemes on Two Contrasting Nigerian Alfisols

Abdulmajeed Hamza and Ezekiel Akinkunmi Akinrinde

Department of Agronomy, University of Ibadan, Ibadan, Nigeria

Correspondence should be addressed to Abdulmajeed Hamza; hamzaabdulmajeed@yahoo.com and Ezekiel Akinkunmi Akinrinde; akinakinrinde@yahoo.com

Academic Editor: Iskender Tiryaki

The effectiveness of finely ground Sokoto Rock Phosphate and Morocco Rock Phosphate to enhance productivity of maize- (*Zea mays* L.) *Sorghum* (*Sorghum bicolor*) and soybean- (*Glycine max* L.) *Sorghum* crop rotation schemes was evaluated using Single Super Phosphate as reference fertilizer. The experiments were carried out in the screen house of the Department of Agronomy, University of Ibadan, in February and June 2013. The experiments involved 2 × 2 × 4 × 3 factorial in a Completely Randomized Design. In the first and second croppings, the slightly acidic loamy sand still produced higher biomass than the strongly acidic sandy clay loam. On average, MRP was more efficient than SSP for maize dry biomass but, for soybean dry biomass, MRP was less efficient than SSP in the two soils. Sokoto Rock Phosphate was less efficient in the two location soils compared to SSP for the test crops. There was no difference in performance of P-sources in the second cropping. Soybean-*Sorghum* crop rotation scheme produced greater *Sorghum* biomass than maize-*Sorghum* crop rotation scheme. It is evident that pH and clay contents of soils as well as the rotation crop concerned influence the efficiency of finely ground soluble phosphates in crop rotation schemes.

1. Introduction

Soil acidity is a major constraint to crop production throughout the world. It leads to low yields of arable crops [1]. Maize and *Sorghum* cannot do well with soil pH < 5.5. When the soil pH is around 6.7 it is slightly acidic but when it is less than 5.5 it is strongly acidic [2]. Crop performance could be adversely affected by Calcium (Ca), Potassium (K), Phosphorus (P), Magnesium (Mg), Sulphur (S), Zinc (Zn), and Molybdenum (Mo) deficiencies as well as Al, Mn, and Fe toxicities [3]. However, lime and fertilizer application is usually recommended to reduce these adverse effects. Phosphorus is a major plant nutrient (second only to nitrogen). Soils in Nigeria require moderate P-application for optimum crop growth [4]. The direct use of sparingly soluble, ground RP as substitute to the costly, more soluble P fertilizers is gaining widespread acceptance [5]. They provide slow-release P and residual effect for several years [6]. Residual effect is the carryover effect from preceding crops to the succeeding ones. The amount and longevity of the residual effect depends on rate, duration, and frequency of application, solubility, soil properties, crop type, yield level, and extent of P removal [7].

Nitrogen (N) is another important nutrient element needed for crop production. However, in order to ensure sustainability, organic sources of N are preferred. Intercropping or rotation of legumes with cereals is one of the ways of replenishing the soil with organic N, among others. According to Pedersen and Lauer [8], planting of soybean in rotation with cereals consistently gave higher yields than monoculture. Soybean (*Glycine max* L.) belongs to the legume family and is an important source of organic N fertilizer [9] because it fixes atmospheric nitrogen in the soil [10]. Atmospheric N fixation is a cheap, clean, renewable, and environmentally friendly source of nitrogen (N) for the non-N-fixing crop component of the cropping system [11].

The unavailability and high costs of lime and inorganic P fertilizers have led to research into low cost materials like rock phosphates, using SSP as reference. This study was to evaluate the effects of different P-sources, Single Supper Phosphate (SSP), Sokoto Rock Phosphate (SRP), and Morocco Rock Phosphate (MRP), on the performance of maize and soybean in two soil types, with the residual effects of the P-sources and the preceding crop on the performance of *Sorghum* in two crop rotation schemes.

2. Materials and Methods

2.1. Experimental Site. There were two experiments conducted between February and June 2013 in the screen house, Department of Agronomy, Faculty of Agriculture and Forestry, University of Ibadan, Nigeria. The greenhouse conditions were suitable for plants growth, where the wall and the door of the greenhouse were made up of metals with wire nets and roofed with transparent glasses. The floor was well cemented and there were metals tables inside on which the pots were kept. The planting date for Experiment 1 was February 23, 2013, and that of Experiment 2 was May 25, 2013. The university is located at latitude $7°24'$ N and longitude $3°54'$ E.

2.2. Experimental Design and Treatments. The experiments involved 2 (maize and soybean) × 2 (two soil types) × 4 (four levels of Phosphorus) × 3 replications (total of 48 treatments) in a Completely Randomized Design (CRD).

Experiment 1 (effect of different phosphorus fertilizer sources on the performance of maize and soybean grown on two contrasting Nigerian Alfisols). In the first experiment, treatments were as follows.

(i) Phosphorus Fertilizer. Absolute Control (0% P_2O_5), Sokoto Rock Phosphate (34.2% P_2O_5 by weight), Morocco Rock Phosphate (33.3% P_2O_5 by weight), and Single Super Phosphate (18% P_2O_5), the three Phosphorus fertilizers, were collected from Department of Agronomy, Faculty of Agriculture and Forestry, University of Ibadan, Nigeria.

(ii) Soil Locations. The two soils were collected from different locations. The sandy clay loam soil was strong acidic Alfisol collected from Leventis Foundation School of Agriculture, Imoo, Ilesa, Osun State. This area lies within the rainforest vegetation zone of Nigeria. According to Oyedele et al. [12], the parent rocks consist essentially of quartz with small amounts of white micaceous minerals. Also in this area, densely wooded quartz ridges rise abruptly from the surrounding country and are elongated north south following the strike of the rock. Akintoye et al. [13] stated that the soils round Ilesa are classified as Alfisols (mainly Paleustalf). The soil in this area belongs to the Okemesi series [12, 14], while the loamy sand soil was very slightly acidic Alfisol collected from Parry Road, Department of Agronomy Farm, University of Ibadan, Ibadan, Oyo State. According to USDA [15], the soils around Ibadan are classified as Alfisols (Typic Plinthustalf). The soil in this area belongs to the Gambari series.

(iii) Test Crops. Maize (*Zea mays* var. Swan-1) was collected from Department of Agronomy, College of Agriculture and Forestry, University of Ibadan, Nigeria, and soybean (*Glycine*

max var. TGX 1448-2E) was gotten from International Institute of Tropical Agriculture (IITA), Ibadan, Oyo State, for first cropping.

2.3. Methodology. The two soils were crushed with mortar and pestle and passed through a 2 mm sieve separately. The chemical and physical properties of the two soils were determined in the laboratory according to Udo and Ogunwale [16] methods. For each of the soil locations, 4 kg was weighed using weighing balance into 24 bows and made a total of 48 bows of the two soils. 100 mg P_2O_5/kg soil of finely ground Sokoto Rock Phosphate, SRP (34.2% P_2O_5), Morocco Rock Phosphate, MRP (33.3% P_2O_5), Single Super Phosphate, SSP (18% P_2O_5), and Control (0% P_2O_5) was thoroughly mixed with the soil in each bow. The soil in each bow was later filled into forty-eight 4 kg of pots.

2.4. Water Field Capacity. Water field capacities of the two soils were determined in the laboratory. 50 g of each soil was weighed using sensitive scale. Two funnels were placed on two cylindrical flacks. Small tissue papers were put inside the funnels based to prevent soils pour-off. The 50 g of the two soils was separately put into each funnel placed on the cylindrical flasks. 50 mL each of water was poured on the two soils in the funnels. They were allowed to drain into the cylindrical flasks until they stopped dropping. The volumes of drained water in each cylindrical flask were subtracted from the actual volume of water poured. The water field capacity of the strongly acidic sandy clay loam was 2500 mL while the field capacity of slightly acidic loamy sand was 1500 mL. The two soils inside the 4 kg pots containing each phosphate fertilizer and Control were watered to 60% field capacity (FC) and allowed to equilibrate for 3 days.

2.5. Planting of Test Crops. Maize (*Zea mays* var. Swan-1) and soybean (*Glycine max* var. TGX1448-2E) were planted as the first crops. Three seeds of maize (*Zea mays*) and soybean (*Glycine max*) were planted per pot and two seedlings of the test crops were allowed to grow in each pot for 6 weeks.

2.6. Data Collected on Maize (Zea mays L.) and Soybean (Glycine max). Data on the following were taken weekly starting from first week after planting:

(i) Heights (cm): plant heights were measured from the soil level to terminal bud using a measuring tape for six weeks.

(ii) Dry biomass yield (g/kg): after harvest, the fresh biomass yields were thoroughly washed with water and air-dried. They were later oven-dried at 75°C for 24 hours to constant weight and the dry biomass yields were weighed using a sensitive scale.

Experiment 2 (residual effect of phosphorus- (P-) sources on dry biomass yield of *Sorghum* productions on two contrasting Nigerian Alfisols). In order to determine the response of *Sorghum* (*Sorghum bicolor* L. var. Sokoto local) to residual effects of the P-sources (SRP, MRP, SSP, and Control) in the two

TABLE 1: Precropping chemical and particle size analysis of the soils used in the study.

Properties	Soil A (strongly acidic Alfisol)	Soil B (slightly acidic Alfisol)
pH (1 : 1 soil/water ratio)	5.3	6.7
Bray-1-P (mg/kg)	5	9
Total N (g/kg)	2.0	2.0
Organic matter (g/kg)	27.1	26.5
Exchangeable base (cmol/kg)		
K	0.2	0.3
Ca	1.6	1.8
Mg	0.7	0.7
Na	0.1	0.1
Exchangeable acidity (cmol/kg)	0.4	0.6
Exchangeable micronutrients (cmol/kg)		
Fe	0.3	0.2
Mn	0.3	1.3
Cu	0.04	0.01
Zn	0.01	0.01
Particle size distribution (gkg^{-1})		
Sand	628.0	871.0
Silt	150.0	40.0
Clay	222.0	89.0
Texture	Sandy clay loam	Loamy sand

soils, the biomass yield of *Sorghum* (*Sorghum bicolor* L var. Sokoto local) was also evaluated after air-drying and sieving of the first experimental soils and 50 seeds of *Sorghum* were sown inside each pot, that is, maize-*Sorghum* and soybean-*Sorghum* crop rotation schemes. The *Sorghum* plants were cut at 2 WAP and allowed to regenerate before they were cut in another 2 weeks.

2.7. Data Collected on White Sorghum (Sorghum bicolor L.) Fresh biomass yields of *Sorghum* plants were cut 2 cm above the soils surfaces every two weeks of growth. After harvest, the fresh biomass yields were thoroughly washed with water and air-dried. They were later oven-dried at 75°C for 24 hours to constant weight and the dry biomass yields were weighed using a sensitive scale.

2.8. Statistical Analysis. The data collected were subjected to Analysis of Variance (ANOVA) to determine the level of significance of the treatments using SAS (Statistical Analysis System) 2002 computer software (version 9.0). Treatment effects and magnitude of interactions were determined. LSD was used to detect differences between treatment means at 5% significant level.

2.9. Relative Agronomic Efficiency (RAE). The vertical comparison approach was used in this study to measure the relative agronomic efficiency (RAE) index of the Sokoto Rock Phosphate (SRP) and Morocco Rock Phosphate (MRP). This approach defines the RAE index as the ratio of the yield response above Control with the test fertilizer at the same rate [17].

Mathematically,

$$RAE = \frac{\text{Yield of Rock Phosphate} - \text{Yield of Control}}{\text{Yield of Single Super Phosphate} - \text{Yield of Control}} \times 100. \tag{1}$$

3. Results

3.1. Experiment 1

3.1.1. Precropping Soil Characteristics. The particle size analysis and the chemical properties of the two soils used in this study are given in Table 1. The Ilesa location soil (Soil A) was strongly acidic sandy clay loams (pH 5.3) while the Ibadan location soil (Soil B) was slightly acidic loamy sand (pH 6.7). Available P in the strongly acidic sandy clay loam was lower (5 mg/kg) in slightly acidic loamy sand (9 mg/kg). The two soils had adequate amount of total nitrogen (2 g/kg). Organic matter content of the strongly acidic sandy clay loam (27.1 g/kg) was slightly higher than the slightly acidic loamy sand (26.5 g/kg). Exchangeable K was higher in strongly acidic sandy clay loam (0.04 cmol/kg) compared to that of slightly acidic loamy sand (0.03 cmol/kg).

There was no significant difference ($p < 0.05$) between the heights of maize plants grown on strongly acidic sandy clay loam and those grown on slightly acidic loamy sand until 4 weeks after planting (WAP). However, soybean plants grown on the two soils exhibited no significant difference ($p < 0.05$) throughout the growing period. Nevertheless, the

TABLE 2: Influence of soils on heights (cm) of maize and soybean plants (at successive growth periods) in the first cropping.

Treatment	Plants height Weeks after planting					
	1	2	3	4	5	6
Soil acidity level	*Maize plant*					
Strongly acidic	12.07	25.33	25.89	35.82	48.33	57.23
Slightly acidic	12.51	23.09	26.36	41.86	58.79	71.73
LSD (5%)	1.52	1.89	2.51	4.78	5.22	9.58
	Soybean plant					
Strongly acidic	9.52	16.99	19.74	23.95	30.78	35.39
Slightly acidic	10.04	18.31	21.49	27.01	35.05	40.05
LSD (5%)	1.20	1.81	1.72	2.96	3.42	4.27

LSD = least significant difference (5%).

TABLE 3: Influence of P-sources on heights (cm) of maize and soybean plants (at successive growth periods) in the first cropping.

Treatment	Plants height Weeks after planting					
	1	2	3	4	5	6
Phosphate fertilizers	*Maize plant*					
Control	11.78	24.17	24.50	34.47	46.35	59.88
Morocco Rock	12.50	26.00	26.33	41.33	59.05	67.83
Sokoto Rock	12.32	22.22	25.42	37.03	48.10	61.20
Single super	12.55	24.47	28.25	42.52	60.75	69.00
LSD (5%)	2.15	2.67	3.55	6.76	7.38	13.55
	Soybean plant					
Control	9.67	17.67	19.73	24.72	33.55	34.08
Morocco Rock	9.68	17.22	20.50	25.08	30.38	40.25
Sokoto Rock	9.63	16.83	20.23	23.68	30.78	34.23
Single super	10.13	18.88	22.00	28.43	36.93	42.32
LSD (5%)	1.70	2.57	2.43	4.18	4.84	6.04

LSD = least significant difference (5%).

slightly acidic loamy sand tends to produce taller plants compared to the strongly acidic sandy clay loam (Table 2).

The heights of plants treated with different Phosphorus treatments (sources) were significant ($p < 0.05$) different only at 5 WAP for maize, whereas for soybean significant differences were evident at both 5 and 6 WAP. The order of the magnitude of the performance of the P-sources was SSP > MRP > SRP > Control (Table 3).

There was no significant ($p < 0.05$) difference between the dry biomass yields of maize plants treated with SSP and MRP as well as SRP and Control. However, the dry biomass yield of maize plants treated with SSP or MRP was significantly different from those treated with SRP or Control (Table 4). The same trend was observed on dry biomass yield in plant tissue of soybean plants as those of the biomass yield of maize plants. The order of the effectiveness of P-sources on dry biomass yield of maize plants was MRP > SSP > SRP > Control while that of soybean was SSP > MRP > SRP > Control.

TABLE 4: Influence of P-sources on biomass yield (g/pot) and relative agronomic efficiency (RAE) (%) of maize/soybean plants in the first cropping.

Treatment	Dry biomass (g/pot)	RAE (%)
Phosphate fertilizers	*Maize plant*	
Control	11.67	
Morocco Rock	18.43	114.38
Sokoto Rock	12.65	16.58
Single super	17.58	100
LSD (5%)	2.39	
	Soybean plant	
Control	8.24	
Morocco Rock	11.42	85.03
Sokoto Rock	4.56	ND
Single super	11.98	100
LSD (5%)	2.49	

RAE = relative agronomic efficiency of the Phosphorus- (P-) sources = [((yield of GRP − yield of Control)/(yield of SSP − yield of Control)) × 100]%.
NS = nonsignificantly different at $p < 0.05$.
LSD = least significant difference (5%) and ND = not determined.

TABLE 5: Influence of soils and P-sources on dry biomass yield (g/pot) and relative agronomic efficiency (RAE) (%) of maize/soybean plants in the first cropping.

Treatment		Dry biomass (g/pot)	RAE (%)
	Maize plant		
	Control	4.20	
Strongly acidic	Morocco RP	12.40	88.17
	Sokoto RP	8.57	46.99
	Single SP	13.50	100
	Control	19.13	
Slightly acidic	Morocco RP	24.47	210.24
	Sokoto RP	16.73	ND
	Single SP	21.67	100
LSD (5%)		3.38	
	Soybean plant		
	Control	4.65	
Strongly acidic	Morocco RP	9.57	139.77
	Sokoto RP	4.02	ND
	Single SP	8.17	100
	Control	11.83	
Slightly acidic	Morocco RP	13.26	36.2
	Sokoto RP	5.11	ND
	Single SP	15.78	100
LSD (5%)		3.53	

RAE = relative agronomic efficiency of the Phosphorus- (P-) sources = [((yield of GRP − yield of Control)/(yield of SSP − yield of Control)) × 100]%.
NS = nonsignificantly different at $p < 0.05$.
ND = not determined.

There was significant ($p < 0.05$) difference between dry biomass yields of maize plants cut on slightly acidic soils treated with either SSP or MRP and those treated with SRP or strongly acidic soils treated with P-sources (Table 5). The

TABLE 6: Influence of the soils, P-source, or crop effects on biomass yield (g/pot) of *Sorghum* plants in the second cropping.

Treatment	Dry biomass (g/pot)	RY/RAE	Dry biomass (g/pot)	RY/RAE
	First cutting		Second cutting	
Crop effects				
Maize-*Sorghum*	1.68	5.17	0.05	0.15
Soybean-*Sorghum*	2.36	100	0.18	100
LSD (5%)	0.33		0.1	
Soil acidity level				
Strongly acidic	1.7	4.93	0.09	0.26
Slightly acidic	2.34	100	0.13	100
LSD (5%)	0.33		NS	
Phosphate fertilizers				
Control	2.1		0.12	
Morocco Rock	2	ND	0.11	50
Sokoto Rock	1.92	ND	0.13	ND
Single super	2.07	100	0.1	100
LSD (5%)	NS		NS	

RAE = relative agronomic efficiency of the Phosphorus- (P-) sources = [((yield of GRP − yield of Control)/(yield of SSP − yield of Control)) × 100]%.
RY = relative yield of test crop (maize or soybean) = [(yield of the crop on a particular soil type/maximum yield) × 100]%.
NS = nonsignificantly different at $p < 0.05$.
LSD = least significant difference (5%) and ND = not determined.

order of the effectiveness of the soils and P-sources on dry biomass yield of maize plants cut on strongly acidic soil was SSP > MRP > SSP > Control, while on slightly acidic soil it was MRP > SSP > Control > SRP. However, there was no significant ($p < 0.05$) difference between dry biomass yields of soybean plants cut on slightly acidic soil treated with SSP and those treated with MRP, whereas there were significant differences from those treated with SRP as well as those cut on strongly acidic soil. The order of the effectiveness of the soils and P-sources on dry biomass yield of soybean plants was SSP > MRP > Control > SRP on slightly acidic soil, while on strongly acidic soil it was MRP > SSP > SRP > Control (Table 5).

Sokoto Rock Phosphate was less than 50% efficient in the two location soils compared to SSP for the test crops. The MRP (relative agronomic efficiency, RAE, of 210.24%) was more efficient than SSP (100% RAE) in slightly acidic loamy sand but less efficient in strongly acidic sandy clay loam (88.17% RAE) compared to SSP (100% RAE) for maize (Table 5). However, for soybean (*Glycine max* L.), MRP (36.2% RAE) was less efficient than SSP (100% RAE) in slightly acidic loamy sand but more efficient (139.77% RAE) in strongly acidic sandy clay loam compared to SSP (100% RAE) (Table 5).

3.2. Experiment 2. It was also observed that slightly acidic loamy sand produced crops with higher dry biomass yield than strongly acidic sandy clay loam (Table 6). The soybean-*Sorghum* crop rotational scheme constantly produced dry biomass yield of *Sorghum* plants compared to that of maize-*Sorghum* crop rotational scheme (Table 6). Based on the residual effects of the various P fertilizer treatments, the dry biomass yields of *Sorghum* plants at first and second cuttings were not different (Table 6).

For the residual influence of the soils and P-sources, there were no significant differences ($p < 0.05$) among the *Sorghum* dry biomass yield produced on the two soils treated with the various P-sources and untreated ones at the first and second cuttings. However, similar trend was observed on the influence of the crop effects and soils on *Sorghum* dry biomass yield as those produced on soils and P-sources (Table 7).

The relative agronomy efficiency (RAE) of MRP in strongly acidic soil was 137.04% while SRP was 85.19% as efficient as SSP (100%) with regard to dry biomass production at first cutting of *Sorghum* plants while in the second cutting, RAE of P-sources were undefined (Table 7). However, in slightly acidic loamy sand only MRP was 85.71% as efficient as SSP at first cutting of *Sorghum* plants.

The relative yields of dry biomass yield of *Sorghum* in soybean-*Sorghum* crop effect are greater than those of maize-*Sorghum* crop effect at both first and second cuttings.

The various residual effects of crop effects and P-sources on dry biomass yield of *Sorghum* plants were significant ($p < 0.05$) at first cutting. The highest dry biomass yield of *Sorghum* plants was gotten from the influence of soybean-*Sorghum* treated with SSP while the least result was gotten from the influence of maize-*Sorghum* treated with SSP at the first cutting. At the second cutting, the influence of crop effects and P-sources had no significant difference on dry biomass yield of *Sorghum* plants. However, in maize-*Sorghum* only MRP was 78.13% as efficient as SSP at first cutting of *Sorghum* plants (Table 8).

The results of the influence of the various interactions among the experimental factors (soils, crops, and P-sources) show that there were no significant differences. The implication is that plant vigour and biomass yields did not differ at the various levels of each of the experimental factors.

Table 7: Influence of the experimental soils and P-source or crop effects and soils on dry biomass yield (g/pot) of *Sorghum* plants in the second cropping.

Treatment		Mean	Dry biomass (g/pot) Standard deviation	RAE (%)	Mean	Dry biomass (g/pot) Standard deviation	RAE (%)
			First cutting			Second cutting	
Soils acidity level	*Phosphate fertilizer*						
	Control	1.92	±0.52		0.11	±0.16	
Strongly acidic	Morocco Rock Phosphate	1.55	±0.69	137.04	0.07	±0.08	ND
	Sokoto Rock Phosphate	1.69	±0.25	85.19	0.06	±0.09	ND
	Single Super Phosphate	1.65	±0.55	100	0.12	±0.20	100
	Control	2.28	±0.37		0.12	±0.18	
Slightly acidic	Morocco Rock Phosphate	2.46	±0.81	85.71	0.14	±0.14	ND
	Sokoto Rock Phosphate	2.15	±0.60	ND	0.2	±0.28	ND
	Single Super Phosphate	2.49	±1.14	100	0.07	±0.09	100
LSD (5%)		NS			NS		
Crop effects	*Soils acidity level*						
Maize-*Sorghum*	Strongly acidic	1.39	±0.30	4.06	0.02	±0.03	0.06
	Slightly acidic	1.97	±0.29	100	0.07	±0.10	100
Soybean-*Sorghum*	Strongly acidic	2.01	±0.49	5.78	0.16	±0.17	0.46
	Slightly acidic	2.71	±0.88	100	0.2	±0.22	100
LSD (5%)		NS			NS		

RAE = relative agronomic efficiency of the Phosphorus- (P-) sources = [((yield of GRP − yield of Control)/(yield of SSP − yield of Control)) × 100]%.
NS = nonsignificantly different at $p < 0.05$ and ND = not determined.

Table 8: Influence of crop effects and P-sources on biomass yield (g/pot) of *Sorghum* in the second cropping.

Treatment		Dry biomass (g/pot)	RAE (%)	Dry biomass (g/pot)	RAE (%)
		First cutting		Second cutting	
Crop effects	*Phosphate fertilizer*				
Maize-*Sorghum*	Control	1.81		0.03	
	Morocco Rock	1.56	78.13	0.06	ND
	Sokoto Rock	1.86	ND	0.06	ND
	Single Super	1.49	100	0.03	100
Soybean-*Sorghum*	Control	2.39		0.2	
	Morocco Rock	2.44	ND	0.15	ND
	Sokoto Rock	1.98	ND	0.2	ND
	Single Super	2.64	100	0.16	100
LSD (5%)		0.66		NS	

RAE = relative agronomic efficiency of the Phosphorus- (P-) sources = [((yield of GRP − yield of Control)/(yield of SSP − yield of Control)) × 100]%.
NS = nonsignificantly different at $p < 0.05$.
LSD = least significant difference (5%) and ND = not determined.

4. Discussion

4.1. First Cropping

4.1.1. Visual Growth Observation.
Crops (maize and soybean) grown in strongly acidic sandy clay loam were not as abundant in growth and yield as those grown in slightly acidic loamy sand where crops in untreated pots had the least plant height and biomass yield of 59 cm/11.67 g/pot for maize plant and 35.39 cm/8.24 g/pot for soybean plant, respectively. The maize plants grown in untreated pots also developed purple colouration which is a symptom of Phosphorus deficiency [18]. It was observed that SSP treatment supported most growth (plant height 74.83 cm) and biomass 21.67 g/pot in slightly acidic soil, whereas MRP treatment supported most growth (plant height 63.17 cm) and biomass yield 13.50 g/pot in strongly acidic soil for maize plants. However, for soybean plants, SSP treatment supported the most growth and yield with 43.63 cm/15.78 g/pot in slightly acidic soil as well as 41 cm/8.17 g/pot in strongly acidic soil, respectively.

4.1.2. Experimental Data.
From the results above, it was observed that values of P in Table 1 imply that both soils were deficient in P contents since the critical levels range between 10 and 15 mg/kg P [19, 20]. The two soil types were

adequately furnished with the same contents of total nitrogen 0.2% where the critical level of nitrogen is 0.15% [19]. The organic matter in strongly acidic sandy clay loam (27.1 g/kg) was slightly higher compared to that of slightly acidic loamy sand (26.5 g/kg). The exchangeable K was higher in strongly acidic sandy clay loam (0.04 cmol/kg) when compared to that of slightly acidic loamy sand (0.03 cmol/kg); both fall within the critical range 0.01–0.15 cmol/kg K [19]. The strongly acidic sandy clay loam had lower proportion of sand (62.8%) compared to slightly acidic loamy sand which had 87.1% sand, while Soil A had higher proportions of silt (15%) and clay (22.2%) when compared to Soil B with 4% silt and 8.9% clay.

Slightly acidic loamy sand constantly produced crops with higher plant height (Table 2) compared to crops grown on strongly acidic sandy clay loam. For example, crops grown on slightly acidic loamy sand were 58.89 cm, per plant height on the average, compared to crops grown on strongly acidic sandy clay loam with 46.31 cm after 6 weeks of growth. Slightly acidic loamy sand was able to support the growth of the crops as much as strongly acidic sandy clay loam because slightly acidic loamy sand was more fertile than strongly acidic sandy clay loam. The soil pH and clay content values for slightly acidic loamy sand were more suitable for crops growth compared to strongly acidic sandy clay loam. This is in agreement with the statement made by Akinrinde and Adigun [2] that crops performed better in slightly acidic soil when compared to medium acid Alfisol. Also, there is possibility of higher P-fixation of applied phosphate ions in strongly acidic sandy clay loam than slightly acidic loamy sand. This is similar to the experiment carried out by Akinrinde and Adigun, [2] quoting Borggard [21] that close linear relationship exists between clay content and phosphate fixation.

Furthermore, the Control and SRP had lesser values for all the growth component parameters (Tables 3 and 4). The conventional soluble P fertilizer (SSP) and one of the rock phosphates, that is, MRP, almost gave the same result compared to SRP. For instance, in Table 4, applied MRP significantly gave higher values of dry biomass 18.43 (g/pot) for maize plants than applied SSP with 17.58 (g/pot). SRP performed less in this experiment; this could be due to soils type because not all soils and cropping situations are suitable for direct use of the RPs from different sources [22]. For instance, this experiment showed that strongly acidic sandy clay loam treated with SRP gave higher soybean plants than slightly acidic loamy sand treated with the same SRP, though they were not significantly different at 6 WAP (Table 5). SRP's poor performance in this experiment could also be attributed to the higher amount of $CaCO_3$ it contains (79%) compared to SSP (35% $CaCO_3$) and MRP (14% $CaCO_3$) [23, 24]. This could increase the soil pH of the slightly acidic loamy sand from 6.7 to alkaline soil pH which could affect proper functioning of the roots of crops and lead to poor growth and yield. This is similar to the result gotten by Ojo [25] which stated that RPs have more Ca than SP; thus, when applied they tend to make the soil alkaline. While in strongly acidic sandy clay loam, the $CaCO_3$ in SRP helps to increase the soil pH from 5.3 to slightly acidic soil which is favourable to growth of plants. However, slightly acidic loamy sand treated with any of the P fertilizers gave

better results in terms of growth and biomass yield of maize plants than strongly acidic sandy clay loam treated with the same fertilizer. This shows that maize plants could survive in wide range of soil pH compared to soybean plants. It also supports the fact that differences among P-sources enhancing growth and yield components or not are attributed to environmental, plant, and soil characteristic factors [2, 25, 26]. The order of the effectiveness of P-sources for the growth and yield of the crops (maize and soybean) is MRP ≥ SSP > SRP in the first cropping. This shows the P-sources superiority of P released and availability for plants metabolism.

For efficient utilization of RP, marked differences have been found in the ability of plant species to extract P from PRs [25, 27, 28]. Similar results were observed in this study where, on average, MRP (114.38% RAE) was more efficient than SSP (100% RAE) for dry biomass yield of maize (Zea mays L.) but, for soybean (Glycine max L.) dry biomass, MRP (85.03% RAE) was less efficient than SSP (100% RAE) in the two soils. Sokoto Rock Phosphate was less than 50% efficient in the two location soils compared to SSP for the test crops. The MRP (relative agronomic efficiency, RAE of 210.24%) was more efficient than SSP (100% RAE) in slightly acidic loamy sand but less efficient in strongly acidic sandy clay loam (88.17% RAE) compared to SSP (100% RAE) for maize. However, for soybean (Glycine max L.), MRP (36.2% RAE) was less efficient than SSP (100% RAE) in slightly acidic loamy sand but more efficient (139.77% RAE) in strongly acidic sandy clay loam compared to SSP (100% RAE).

4.2. Second Cropping. It was also observed that slightly acidic loamy sand produced crops with higher dry biomass yield than strongly acidic sandy clay loam. It could be due to similar reasons given in the first cropping of this experiment.

The soybean-*Sorghum* crop rotational scheme constantly produced biomass yield of *Sorghum* plants compared to maize-*Sorghum* crop rotational scheme. This might be as a result of nitrogen fixed by the leguminous plants which was used by the following *Sorghum* plants while cereal-cereal crop rotational scheme is nitrogen demanding. Legumes are used commonly in agricultural systems as a source of atmospheric N through symbiotic N_2 fixation for subsequent crops, maintaining soil nitrogen levels, and through subsoil retrieved [29]. Rotation of cereals and legumes is usually preferred to sole cropping of either crop because of higher yield [30]. Therefore, it is beneficial to alternate soybean with cereals and other plants that require nitrogen.

Based on the residual effects of P fertilizer treatment, the residual effects of the various P fertilizer treatments on the dry biomass yield of *Sorghum* plants at first and second cuttings did not differ. According to the experiment carried out by Akinrinde and Adigun, [2] stated that the P-sources produced significant differences in the height and fresh biomass yield but not in the dry matter production.

The relative agronomy efficiency (RAE) of MRP in strongly acidic soil was 137.04% while SRP was 85.19% as efficient as SSP (100%) with regard to dry biomass production at first cutting of *Sorghum* plants while in the second cutting, RAE of P-sources were undefined (Table 7). However, in slightly acidic loamy sand only MRP was 85.71% as efficient as SSP at first cutting of *Sorghum* plants. Rock phosphate of P dissociation improved with time which in turn improves P availability as well as increased yield [25].

The results of the influence of the various interactions among the experimental factors (soils, crops, and P-sources) showed that there were no significant differences. The implication is that plant vigour and biomass yields were not different at the various levels of each of the experimental factors.

5. Conclusions

The strongly acidic sandy clay loam produced crops with lower plant height than crops grown in slightly acidic loamy sand.

The relative agronomic efficiency (RAE) of MRP was more efficient than that of SSP in slightly acidic loamy sand but less efficient in strongly acidic sandy clay loam compared to SSP as reference fertilizer for maize plants. However, for soybean plants, MRP was less efficient than SSP in slightly acidic loamy sand but more efficient in strongly acidic sandy clay loam than SSP.

The residual effects of the various P fertilizer treatments on the dry biomass yield of *Sorghum* plants at first and second cuttings were not different.

The soybean-*Sorghum* crop rotational scheme constantly produced biomass yield of *Sorghum* plants compared to maize-*Sorghum* crop rotational scheme.

However, based on points made above, it is evident that pH and clay contents of soils as well as the crop concerned determine the efficiency of finely ground soluble phosphates in crop production as well as positive effects of the crop rotation schemes. It can serve as means of production of forage or hay for ruminant animal.

Competing Interests

The authors declare that there are no competing interests regarding the publication of this paper.

References

[1] M. E. Sumner and A. D. Noble, "Soil acidification: the world story," in *Handbook of Soil Acidity*, Z. Rengel, Ed., pp. 1–28, Marcel Dekker, New York, NY, USA, 2003.

[2] E. A. Akinrinde and I. O. Adigun, "Phosphorus-use efficiency by pepper (*Capsicum frutescens*) and okra (*Abelmoschus esculentum*) at different phosphorus fertilizer application levels on two tropical soils," *Journal of Applied Sciences*, vol. 5, no. 10, pp. 1785–1791, 2005.

[3] K. N. Fageria, "Soil acidity affects availability of Nitrogen, Phosphorus and Potassium," *Better Crops International*, vol. 10, pp. 8–9, 1994.

[4] A. Rashid, "Phosphorus use efficiency in soils of Pakistan," in *Proceedings of the 4th National Congress of Soil Science*, Soil Science Society of Pakistan, Islamabad, Pakistan, May 1992.

[5] E. A. Akinrinde and K. A. Okeleye, "Short- and long-term effects of sparingly soluble phosphates on crop production in two contrasting Nigerian Alfisols," *West African Journal of Applied Ecology*, vol. 8, no. 1, pp. 141–149, 2005.

[6] P. A. Sanchez, A. U. Mokwunye, F. R. Kwesiga, C. G. Ndiritu, and P. L. Woomer, "Soil fertility replenishment in africa. An investment in natural resource capital," in *Replenishment Soil Fertility in Africa*, R. J. Buresh, Ed., vol. 51, pp. 1–46, SSSA Special, 1997.

[7] H. L. S. Tandon, *Phosphorus Research and Agricultural Production in India*, 1987.

[8] P. Pedersen and J. G. Lauer, "Influence of rotation sequence and tillage system on the optimum plant population for corn and soybean," *Agronomy Journal*, vol. 94, pp. 968–974, 2002.

[9] A. R. J. Eaglesham, F. R. Minchin, R. J. Summerfield, P. J. Dart, P. A. Huxley, and J. M. Day, "Nitrogen nutrition of cowpea (*Vigna unguiculata*). 3: distribution of nitrogen within effectively nodulated plants," *Experimental Agriculture*, vol. 13, no. 4, pp. 369–380, 1977.

[10] A. Bationo and A. U. Mokwunye, "Alleviating soil fertility constraints to increased crop production in West Africa: the experience in the Sahel," *Fertilizer Research*, vol. 29, no. 1, pp. 95–115, 1991.

[11] C. P. Vance, P. H. Graham, and D. L. Allan, "Biological nitrogen fixation: phosphorus Ba critical future need?" in *Nitrogen Fixation from Molecules to Crop Productivity*, F. O. Pederosa, M. Hungria, M. G. Yates, and W. E. Newton, Eds., pp. 509–518, Kluwer Academic, Dordrecht, The Netherlands, 2000.

[12] D. J. Oyedele, O. O. Awotoye, and S. E. Popoola, "Soil physical and chemical properties under continuous maize cultivation as influenced by hedgerow trees species on an alfisol in South Western Nigeria," *African Journal of Agricultural Research*, vol. 4, no. 8, pp. 736–739, 2009.

[13] H. A. Akintoye, A. A. Adekunle, and A. A. Kintomo, "The role of training in urban and peri-urban vegetable production: the case study of Leventis Foundation Agricultural Schools in Nigeria," *Learning Publics Journal of Agriculture and Environmental Studies*, vol. 2, no. 2, pp. 21–40, 2011.

[14] A. J. Smyth and R. F. Montgomery, *Soils and Land Use in Central Western Nigeria*, Government Printer, Ibadan, Nigeria, 1962.

[15] United States Grain Council November, Sorghum, 2010, http://www.grains.org/sorghum.

[16] E. J. Udo and J. A. Ogunwale, *Laboratory Manual for the Analysis of Soil, Plants and Water Samples*, Department of Agronomy, University of Ibadan, Ibadan, Nigeria, 1981.

[17] O. P. Engelstad, A. Jugsujinda, and S. K. De Datta, "Response by flooded rice to phosphate rocks varying in citrate solubility," *Soil Science Society of America Journal*, vol. 38, no. 3, pp. 524–529, 1974.

[18] K. Mengel and E. A. Kirkby, *Principle of Plant Nutrition*, International Potash Institute Publisher, 1987.

[19] G. O. Adeoye and A. A. Agboola, "Critical levels for soil pH, available P, K, Zn and Mn and maize ear-leaf content of P, Cu and Mn in sedimentary soils of South-Western Nigeria," *Fertilizer Research*, vol. 6, no. 1, pp. 65–71, 1985.

[20] R. A. Solubo and A. O. Osiname, *Soils and Fertilizer Use in Western Nigeria*, Research Bullrtin no. 11, Institute of Agriculture Research and Training, University of Ife, Ife, Nigeria, 1981.

[21] O. K. Borggard, "Iron oxides in relation to phosphate adsorption by soils," *Acta Agriculturae Scandinavica*, vol. 36, no. 1, pp. 107–118, 1986.

[22] G. O. Obigbesan and E. A. Akinrinde, "Evaluation of the performance of Nigerian rock phosphates applied to millet in

selected benchmark soils," *Nigerian Journal of Soil Science*, vol. 12, pp. 88–99, 2000.

[23] E. A. Akinrinde and G. O. Obigbesan, "Benefits of phosphate rocks in crop production: experience on benchmark tropical soil areas in Nigeria," *Journal of Biological Sciences*, vol. 6, no. 6, pp. 999–1004, 2006.

[24] P. Van Straaten, *Rocks for Crops: Agro-Minerals of Sub-Saharan Africa*, ICRAF, Nairobi, Kenya, 2002.

[25] O. D. Ojo, *Growth development and yield of amaranth (*Amaranthus cruentus *L) varieties in response to different sources of phosphorus [Ph.D. thesis]*, University of Ibadan, Ibadan, Nigeria, 2001.

[26] D. P. Schachtman, R. J. Reid, and S. M. Ayling, "Phosphorus uptake by plants: from soil to cell," *Plant Physiology*, vol. 116, no. 2, pp. 447–453, 1998.

[27] E. N. Flach, "A comparison of the rock phosphate mobilizing capacities of various crop species," *Tropical Agriculture*, vol. 64, pp. 347–352, 1987.

[28] S. S. S. Rajan, J. H. Watkinson, and A. G. Sinclair, "Phosphate rocks for direct application to soils," *Advances in Agronomy*, vol. 57, pp. 77–159, 1996.

[29] S. M. Gathumbi, G. Cadisch, and K. E. Giller, "15N natural abundance as a tool for assessing N_2-fixation of herbaceous, shrub and tree legumes in improved fallows," *Soil Biology and Biochemistry*, vol. 34, no. 8, pp. 1059–1071, 2002.

[30] B. O. Baldock, R. L. Higgs, W. H. Paulson, J. A. Jackobs, and W. D. Shader, "Legume and mineral fertilizer effects on crop yields in several crop sequences in the upper Mississipi Valley," *Agronomy Journal*, vol. 73, no. 5, pp. 885–890, 1981.

18

Effects of Short-Day and Gibberellic Acid Treatments on Summer Vegetative Propagation of Napier Grass (*Pennisetum purpureum* Schumach)

Yasuyuki Ishii,[1] **Asuka Yamano,**[2] **and Sachiko Idota**[1]

[1]*Faculty of Agriculture, University of Miyazaki, Miyazaki 889-2192, Japan*
[2]*Graduate School of Agriculture, University of Miyazaki, Miyazaki 889-2192, Japan*

Correspondence should be addressed to Yasuyuki Ishii; yishii@cc.miyazaki-u.ac.jp

Academic Editor: Kent Burkey

The effects of short-day (SD) and gibberellic acid (GA$_3$) treatments on promoting vegetative propagation during the summer were examined in Napier grass (*Pennisetum purpureum* Schumach). A dwarf variety of late heading type (DL) Napier grass was exposed to three SD treatments (5, 10, and 20 short days plus a spray of 400 ppm GA$_3$ solution following each SD treatment, GASD) or no treatment (control). Additionally, then, a dwarf variety of early heading (DE) and the normal variety of Merkeron (ME) were exposed to 10 days of GA-SD treatment together with nontreated controls. For DL and DE, GA-SD treatments showed the following effects: 10-day GA-SD treatment increased significantly ($P < 0.05$) the length of lateral tiller buds, maintained a high rooting percentage, and increased the diameter of the tiller buds. This resulted in a taller plant, one with enhanced tiller numbers, and thus a greater number of established nursery plants for the two dwarf varieties. In contrast, there was only a limited positive effect of the GA-SD treatments on the normal variety, ME. Thus, 10 days of GA-SD treatment was judged to be the most effective treatment for promoting lateral tiller bud elongation and early maturation in tiller buds for the two dwarf varieties of Napier grass.

1. Introduction

Napier grass (*Pennisetum purpureum* Schumach) belongs to a tropical C$_4$ grass with high dry matter productivity [1, 2] and it can elongate its stem internode without phase transition from the vegetative to the reproductive state. Napier grass is now used for multiple purposes, including cut-and-carry fodder for herbivores [3], rotational grazing use [4], and feedstock for bioenergy production [5] and for phytoremediation [6] in the Kyushu area of Japan. Napier grass rarely produces viable seed and thus is commonly propagated vegetatively using stem cuttings and rooted tillers [1, 7, 8]. However, because protocols for efficient nursery production of Napier grass have not yet been established and a stable supply of nursery plants cannot be assured during all of the growing seasons, cultivation of this grass species remains limited mainly due to the shortage of nursery-grown rooted propagules [9].

In our previous study, although vegetative propagation of Napier grass was accomplished with cell-tray nursery plants [10, 11], the propagation season was strictly limited to the late autumn. It is difficult for these late-autumn propagated nursery plants to be transplanted in the field soon after propagation due to the frost damage. Thus, nursery plants needed to be maintained in the glasshouse over the wintering season.

Hare et al. [12] conducted short-day (SD) treatment on another tropical grass, *Paspalum atratum*, to induce the transition of growth stage to the reproductive phase, resulting in the heading and flowering in summer season, which is unusual since *P. atratum* normally never heads until late September, after being subjected to the naturally occurring

SD photoperiod. It has been reported that tropical grasses have a suppressed vegetative growth in response to application of growth retardants [13]. Dwarf varieties of Napier grass are blocked in the pathway of gibberellin biosynthesis and thus exhibit severely suppressed stem elongation before the phase transition to the reproductive state [7]. Thus, it is considered as feasible that SD and/or GA_3 treatments could elongate the stem internode [14–17], which may facilitate the successful vegetative propagation and production of nursery plants for dwarf Napier grass.

The aim of the present study, therefore, was to determine the optimal length of SD and GA_3 treatments in dwarf Napier grass (Experiment 1) and also difference of Napier grass in SD and GA treatments (Experiment 2) on the effect of the nursery propagation in the summer.

2. Materials and Methods

2.1. Optimal Growth Stages of SD and GA_3 Treatments (Experiment 1). The experiment was conducted at the University of Miyazaki in southern Kyushu, Japan (31.83°N and 131.41°E, 22 m asl) from 2 July 2010 to 16 November 2010. The aboveground stems in the dwarf variety of late heading type (DL) Napier grass, which was overwintered in a glasshouse from December 2009 to April 2010, were cut into single-node stem cuttings and transplanted into a cell tray (36 cells and $4 \times 4 \times 4$ cm depth/cell), filled with "raising seedling" medium (Takii Co. Ltd., Kyoto) on 4 May 2010. After emergence of tillers, the cell tray was fertilized with a chemical fertilizer (14(%) : 14(%) : 14(%) in N, P_2O_5, and K_2O, resp.) at 2 g per cell tray on 9 June, 22 June, and 17 July, respectively. Cell-tray DL Napier grass plants were irrigated as required and thinned to one main shoot only for each cell at the start of treatment.

Treatments were applied in a split plot design with 3 replications under 3 levels of SD, namely, 5, 10, and 20 days, combined with and without foliar spray of 400 ppm GA_3 solution and surfactant (HitenPower, Hokko Chemical Industry, Co., Ltd., Tokyo, 0.1% v/v), which was determined to be the most effective concentration in preliminary experiments with the DL variety of Napier grass, at the start of treatments (designated as GA-SD and SD). Plants designated as controls (C) were grown under natural LD and received no GA-SD or SD treatment. The SD treatment was conducted by setting cell trays of DL Napier grasses into a plastic tunnel at 1.75 m height and 0.5 m base diameter with ventilation by an air pump at 2 L/min to reduce overheating of the inside temperature. The tunnel was shaded with black cloth for about 14 h, thus giving a 10 h SD. Initially, three lengths of SD were utilized, 5, 10, and 20 days, starting on 12 July 2010. The natural day length when the SD treatments were imposed ranged from 13 h 43 mins to 14 h 6 mins.

Height of the node where the uppermost expanded leaf was attached was measured 5 days before treatment and 0, 2, and 5 weeks after treatment for 4 plants (shoots) per replication within each of the cell trays. At 5 weeks after treatment, internode length and length, diameter, and rooting of lateral tiller buds were determined from the bottom to the uppermost expanded leaf nodal position. However, immature stem nodes with less than 10 mm of internode length were omitted from the observation due to the likelihood of poor vegetative propagation. After these measurements each shoot was cut into the single-node cuttings, which were transplanted into the cell tray in the same size as mentioned before, incubated for 4 weeks outside under natural (long) day lengths, and fertilized with 0.33 L per cell tray of a 1/500 Hyponex solution, containing 6(%), 10(%), and 5(%) of N, P_2O_5, and K_2O, respectively, at 3 weeks after transplanting. At the end of 4-week cell culture, rooted stem cuttings were transplanted to the Andosols (soils of active volcanic areas) at 10 cm $\times$ 20 cm of intrarow and interrow spacing, respectively, and fertilized at 5 g/m^2, each of N, P_2O_5, and K_2O, respectively, using the chemical fertilizer. On 16 November 2010, that is, 8, 7, and 6 weeks after the termination of 5, 10, and 20 days of SD treatments, respectively, the percentage of fully established plants was determined, and plant height and lateral tiller number for the DL Napier grass were measured for all vegetative (not flowering) nursery plants.

2.2. Varietal Differences in SD and GA Treatments (Experiment 2). The experiment was also conducted at the University of Miyazaki from 26 March 2011 to 22 October 2011. Varieties utilized were DL, another dwarf variety which was an early heading type (DE) and a normal (tall) variety of Merkeron (ME). Single-node cuttings were obtained from the overwintered plants in the glasshouse on 26 March 2011 for ME and on 21 April 2011 for DL and DE. These cuttings were transplanted into cell tray of the same size as in Experiment 1 and incubated outside. Fertilizer was supplied by 0.33 L of a 1/500 diluted Hyponex solution per cell tray on 26 April and additionally with 5 g of chemical fertilizer per cell tray on both 7 June and 5 July 2011 for ME and on 18 May, 7 June, and 5 July, each for DL and DE. Cell-tray Napier grasses were irrigated as required and thinned to one main shoot only for each cell at the start of treatment.

Treatments were applied in a split plot design with 3 replications under 10 days of SD treatments for 14 hours of dark condition, combined with foliar spray of 400 ppm GA_3 solution at the start of treatments (designated as GA-SD), and started on 5 July 2011. The natural day length when the SD treatments were imposed was ranging from 14 hr 4 mins to 14 hr 9 mins. No treatment of GA-SD was applied to Napier grass plants which were designated as control (C). Heights were measured as detailed for Experiment 1 (above). At 5 weeks after treatment on 23 August 2011, internode length and the length, diameter, rooting of tiller buds, and rooting of stem nodes were determined. Heights were measured from the soil surface to the uppermost expanded leaf node position. After making these measurements, each shoot was cut into the single-node cuttings, which were transplanted into the cell tray, incubated for 2 weeks outside, and fertilized with 0.33 L per cell tray of a 1/500 diluted Hyponex solution at 1 week after transplanting on 30 August 2011. Nursery plants with leaves were counted and transplanted to the Andosols at 15 cm and 25 cm of intrarow and interrow spacing, respectively, on 13 September 2011.

(a) 5-day treatment

(b) 10-day treatment

(c) 20-day treatment

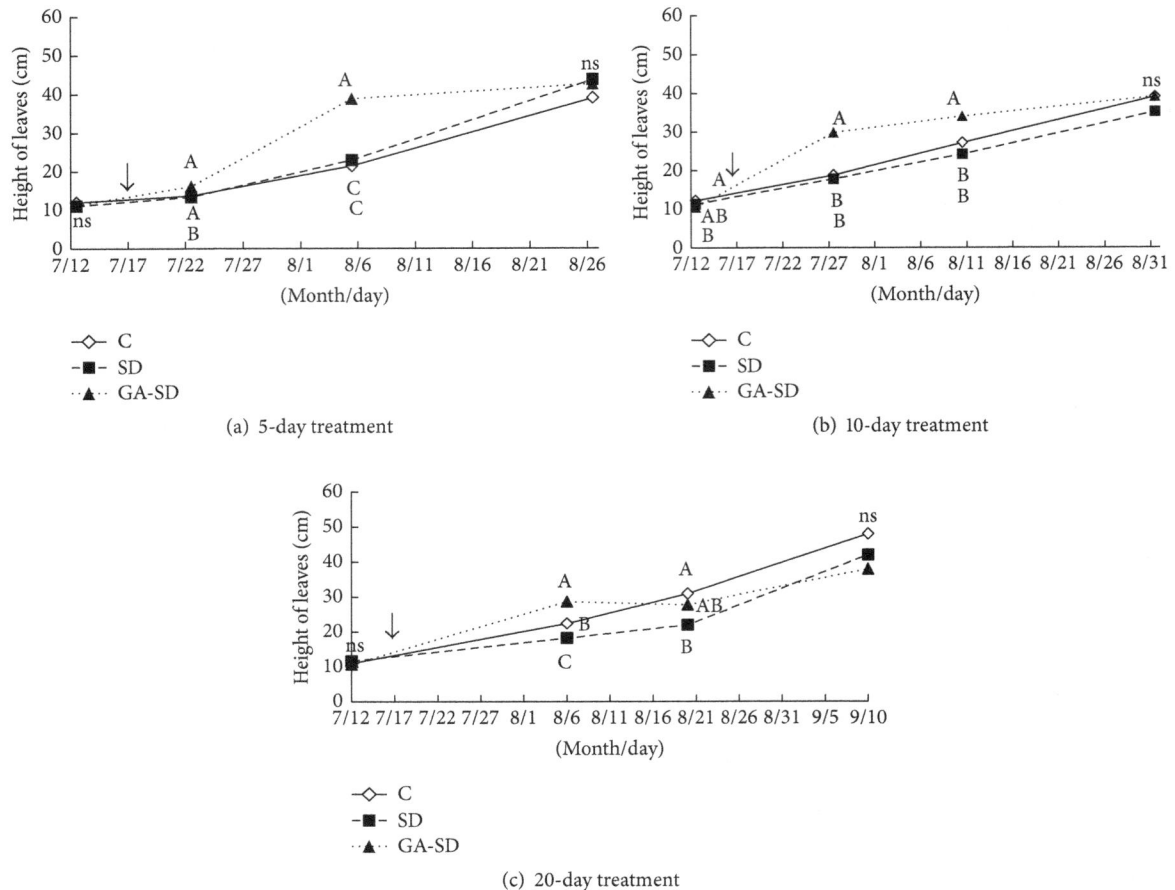

FIGURE 1: Changes in height of the uppermost expanded leaves in dwarf (DL) Napier grass for 5 weeks after treatment (Exp. 1). Treatment: control (C), short-day (SD), and foliar spray of gibberellin acid (GA$_3$), followed by SD (GA-SD). An arrow indicates the date of the start of treatment. Figures with different letters indicate the significant difference at the same date by the Bonferroni test at the 5% level.

2.3. Statistical Analysis. For both experiments, Student's *t*-test and one-way analysis of variance (ANOVA) for a split plot design were performed using software (Excel statistics 2010). Mean separations were tested using the Bonferroni method at the 5% level of probability.

3. Results

3.1. Optimal Growth Stages of SD and GA Treatments (Experiment 1). Height of the uppermost expanded leaf node was significantly higher in GA-SD than in SD and C plots irrespective of SD treatment periods at 1 and 2 weeks after treatment. The height increased gradually with time for plants given SD treatment (Figure 1). However, 20-day SD treatment tended to retard the stem elongation at 5 weeks after treatment, relative to control plants.

Internode length and length, diameter, and rooting of tiller buds showed a significant effect of GA-SD and/or SD treatments for all of 5, 10, and 20 days of treatment periods at 5% level (Table 1). Length, diameter, and rooting of tiller buds were significantly higher in 5 and 10 days of GA-SD

treatments than in control and the degree of enhancement was larger for the 10-day treatment than the 5-day treatment. Effect of treatments was determined in each nodal position under the 10-day long SD treatment, which was the most effective SD treatment (Figure 2). The GA-SD treatments significantly increased the internode length at the 3rd node, tiller bud length at the 1st to the 5th nodes, tiller bud diameter at the 3rd node, and tiller bud rooting at the 1st node, all relative to the control plants and also relative to the SD plants. The SD treatments plants had increased internode lengths at the 1st to the 3rd nodal positions and tiller bud lengths at the 8th node, relative to the control plants.

Established percentage of stem cuttings with rooted tiller buds achieved more than 80% among all the treatments. Five days of GA-SD treatment increased the plant height of nursery plants and 10 days of GA-SD treatment increased both plant height and tiller number of nursery plants, compared with control (Figure 3).

3.2. Varietal Differences in Responses to SD and GA Treatments (Experiment 2). Height of the uppermost expanded leaf

TABLE 1: Effect of treatment and nodal position on nursery plant attributes in dwarf (DL) Napier grass (Exp. 1).

(a) 5-day treatment

	Plant attribute			
	Internode length (mm)	Tiller bud length (mm)	Tiller bud diameter (mm)	Rooting (%)
Treatment	ns	**	**	**
Nodal position	**	**	**	**
Interaction of treatment × nodal position	**	**	*	**
C	28.64	8.28[b]	3.24[b]	0.00[b]
SD	32.73	8.54[b]	3.31[ab]	0.00[b]
GA-SD	35.48	15.27[a]	4.02[a]	6.25[a]

(b) 10-day treatment

	Plant attribute			
	Internode length (mm)	Tiller bud length (mm)	Tiller bud diameter (mm)	Rooting (%)
Treatment	ns	**	**	**
Nodal position	**	**	**	**
Interaction of treatment × nodal position	**	**	ns	**
C	28.12	7.40[c]	3.27[b]	0.00[b]
SD	31.10	8.67[b]	3.88[ab]	1.85[b]
GA-SD	28.09	12.89[a]	4.13[a]	9.26[a]

(c) 20-day treatment

	Plant attribute			
	Internode length (mm)	Tiller bud length (mm)	Tiller bud diameter (mm)	Rooting (%)
Treatment	**	**	**	ns
Nodal position	**	*	**	*
Interaction of treatment × nodal position	**	**	ns	**
C	31.97[a]	9.69[a]	3.57[a]	2.27
SD	27.35[ab]	8.48[ab]	3.06[b]	0.00
GA-SD	21.27[b]	8.44[b]	2.93[b]	0.00

$^*P < 0.05$, $^{**}P < 0.01$, and $^{ns}P > 0.05$.
Values with different letters indicate the significant difference by the Bonferroni test at the 5% level.

node was significantly higher in GA-SD treatment than in C plots at 1 week after treatment and thereafter across all varieties. Heights were 26, 40, and 19 cm higher than the control for DL, DE, and ME, respectively (Figure 4).

Internode length and length, diameter, and rooting from stems showed significant effect of GA-SD treatments for almost all the examined varieties at 5% level (Table 2), except for several nonsignificant effects (Table 2). Internode length and tiller bud length in GA-SD treatment were significantly higher than those in the control for DL and DE, and the enhancement was more effective in DE than in DL. In addition, tiller bud diameter in DL, stem rooting in DE, and internode length in ME were significantly higher in the GA-SD treatment, compared with the control.

Effect of treatments was measured in each nodal position in the most responsive variety of DE (Figure 5). The internode length at the 4th to 7th nodes, the tiller bud length at the 3rd to 7th nodes, the tiller bud diameter at the 3rd to 4th nodes, and stem rooting at the 4th to 6th nodes were significantly higher in the GA-SD treatment than in the control. It is shown that the positively promoted effects on tiller bud maturation appeared at the central to the upper nodal positions (Figure 5).

Established percentage of stem cuttings with leaved tiller buds showed more than 95% across all the varieties and treatments, except for the GA-SD treatment in ME where it was at 85.3%. No significant differences in plant height and tiller number were detected in any variety (Figure 6).

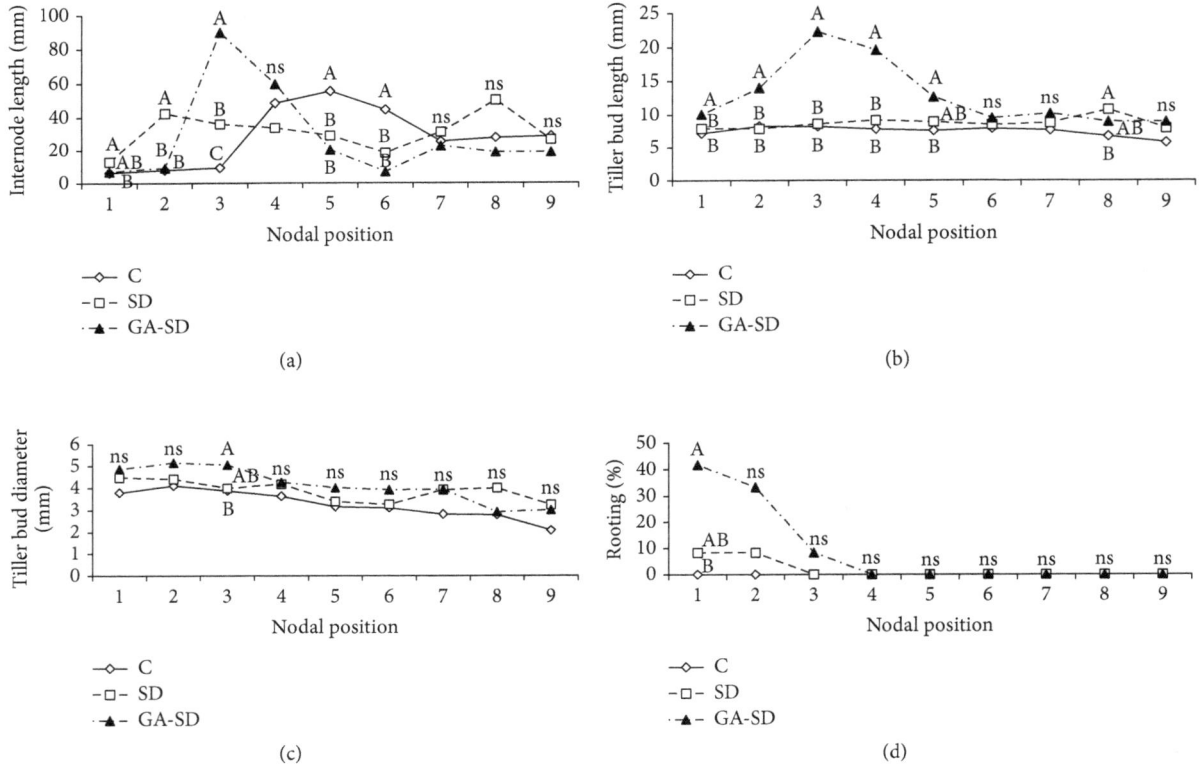

(a)

(b)

(c)

(d)

FIGURE 2: Changes in plant attributes of internode length (a), tiller bud length (b), tiller bud diameter (c), and tiller bud rooting (d) among nodal positions in dwarf (DL) Napier grass, subjected to 10-day treatment (Exp. 1). Nodal position was counted from the bottom of shoots. For treatments, refer to Figure 1. Figures with different letters indicate the significant difference at the same nodal position by the Bonferroni test at the 5% level.

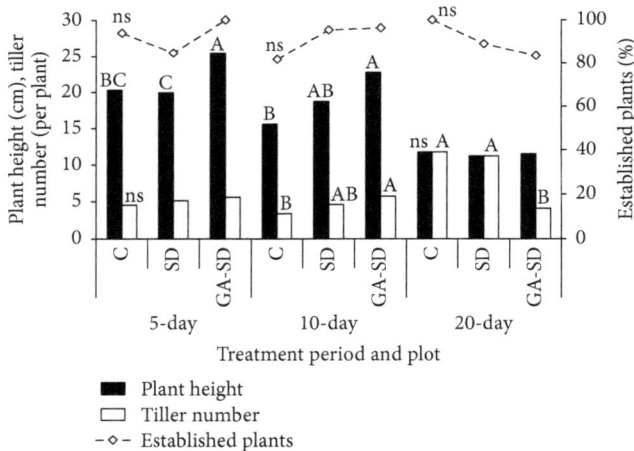

FIGURE 3: Plant growth attributes in transplanted regrowth plants after treatments in dwarf (DL) Napier grass (Exp. 1). For treatments, refer to Figure 1. Figures with different letters indicate the significant difference at the same date by the Bonferroni test at the 5% level.

4. Discussion

Height of the uppermost expanded leaf node in DL reached 30 cm in late July for the 10-day GA-SD treatment and in early August for the 5-day GA-SD treatment, compared with late August in the control plot. This may be brought about by two factors; the first was due to the rapid enhancement of stem elongation in the basal several stem nodes by the foliar GA_3 spray [18] as shown in Figures 1 and 2. The second was due to the gradual enhancement of stem elongation by SD treatment [14, 17]. Therefore, GA_3 treatment showed the rapid effect and SD treatment showed gradual effect on stem elongation in DL Napier grass. Vegetative nursery propagation of Napier grass requires the use of the single nodal stem cuttings and thus the stem elongation in the dwarf varieties is essential for saving labor time and costs in vegetative propagation. In the present study, both the 5-day and 10-day GA-SD treatments were demonstrated as being effective treatments in vegetative propagation by enhancing stem elongation in the early summer season.

In Experiment 2, the 10-day GA-SD treatment induced a similar enhancement of stem elongation by about one month across the three varieties of Napier grass. Therefore, 10-day GA-SD treatment allows for significant promotion of stem elongation early in the summer season.

The SD and GA-SD treatments increased almost all of the growth attributes effective for the vegetative propagation and the additive positive effect of GA-SD treatment makes it a preferred treatment, relative to SD only treatments for 5 days

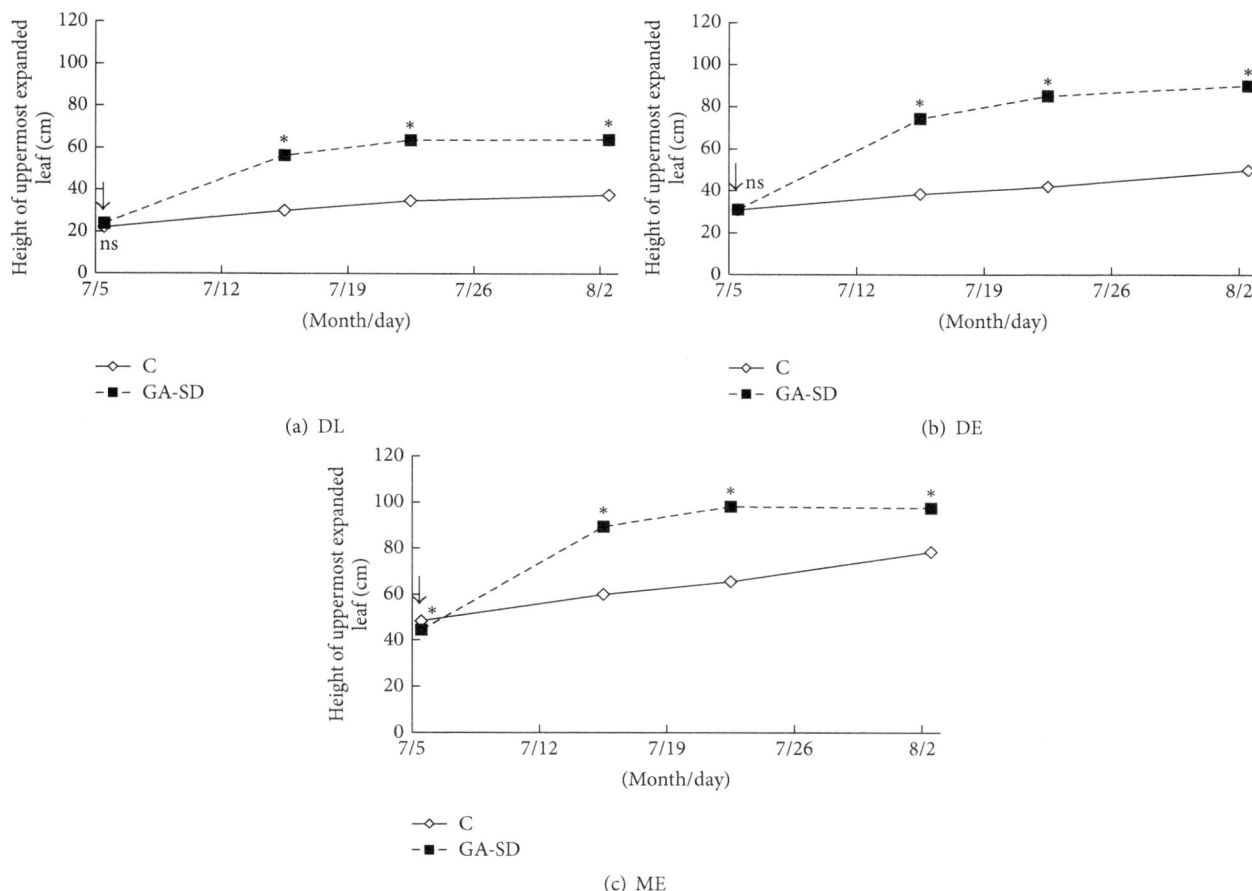

(a) DL

(b) DE

(c) ME

FIGURE 4: Changes in height of the uppermost expanded leaves in 3 varieties of Napier grass for 5 weeks after treatment (Exp. 2). Variety: dwarf-late heading type (DL), dwarf-early heading type (DE), and Merkeron (ME). Treatment: control (C) and foliar spray of gibberellin acid (GA_3), followed by SD (GA-SD). An arrow indicates the date of the start of treatments. $^{*}P < 0.05$ and $^{ns}P > 0.05$.

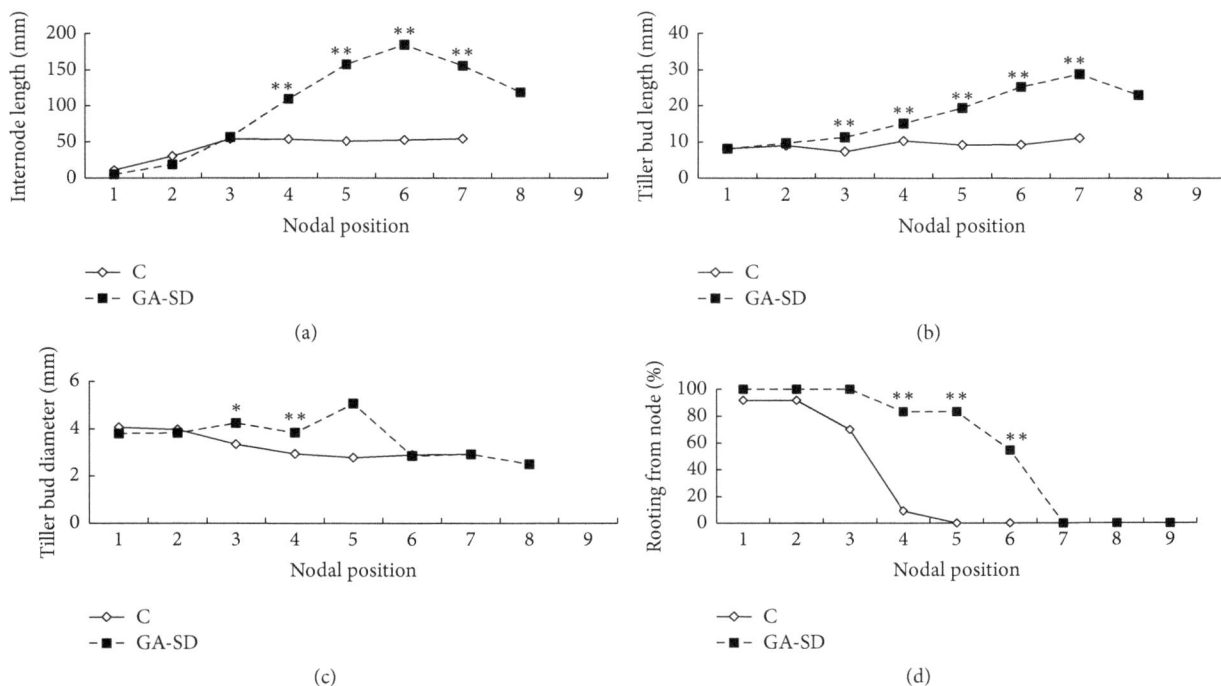

(a)

(b)

(c)

(d)

FIGURE 5: Changes in plant attributes of internode length (a), tiller bud length (b), tiller bud diameter (c), and rooting from nodes (d) among nodal positions in dwarf (DE) Napier grass, subjected to 10-day treatment (Exp. 2). Nodal position was counted from the bottom of shoots. For treatments, refer to Figure 4. $^{*}P < 0.05$ and $^{**}P < 0.01$.

TABLE 2: Effect of treatment and nodal position on nursery plant attributes in three varieties of Napier grass (Exp. 2).

(a) Variety DL

	Plant attribute			
	Internode length (mm)	Tiller bud length (mm)	Tiller bud diameter (mm)	Rooting (%)
Treatment	**	**	*	ns
Nodal position	**	**	ns	**
Interaction of treatment × nodal position	**	**	ns	ns
C	33.56	8.39	3.26	13.55
GA-SD	65.65**	14.45**	3.55*	24.07

(b) Variety DE

	Plant attribute			
	Internode length (mm)	Tiller bud length (mm)	Tiller bud diameter (mm)	Rooting (%)
Treatment	**	**	ns	**
Nodal position	**	**	ns	**
Interaction of treatment × nodal position	**	**	ns	**
C	44.03	9.22	3.26	29.16
GA-SD	101.11**	17.62**	3.63	57.91**

(c) Variety ME

	Plant attribute			
	Internode length (mm)	Tiller bud length (mm)	Tiller bud diameter (mm)	Rooting (%)
Treatment	**	ns	**	ns
Nodal position	**	**	**	**
Interaction of treatment × nodal position	**	**	ns	ns
C	61.88	12.12	4.45**	53.79
GA-SD	95.34**	15.05	3.89	50.68

$^*P < 0.05$, $^{**}P < 0.01$, and $^{ns}P > 0.05$.

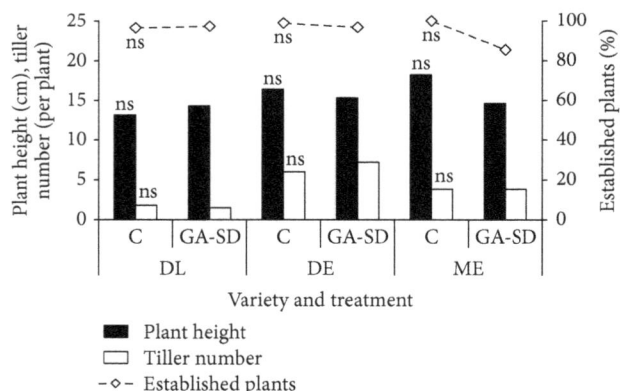

FIGURE 6: Plant growth attributes in transplanted regrowth plants after treatments of 3 Napier grass varieties (Exp. 2). For variety and treatment, refer to Figure 4. $^{ns}P > 0.05$.

and 10 days (Tables 1(a) and 1(b)). The 20-day long treatment suppressed positive effects of shorter length (5-day and 10-day long SD and GA-SD treatments) (Table 1(c)).

In Experiment 2, the 10-day GASD treatment is the most effective in increasing the internode length, length and diameter of tiller buds, and stem rooting to enhance maturation of tiller buds and facilitate the emergence of tiller buds after transplanting to the field, which enables maintaining the enough biomass for overwintering before below-freezing temperatures occur [19].

Ito et al. [13] demonstrated that the foliar spray of growth retardant regulator of paclobutrazol to ME decreased the height of the uppermost expanded leaf node significantly and the suppression of plant growth continued for 2 months after the treatment. The suppressing effect on the growth attributes appeared at the lowest stem nodal position for internode length, followed by leaf sheath length and the leaf length and width at the higher nodal positions [13]. In Experiment 1, the 10-day GA-SD treatment promoted the internode length at the 3rd to 4th nodal positions, while length, diameter, and rooting of tiller buds tended to be promoted at the 1st to the 3rd or 4th nodal positions. In Experiment 2, DE showed the promoting effect on the internode length and stem rooting at the 4th to 7th nodal positions, while length and diameter of tiller buds increased from the 3rd nodal positions. Therefore, in both Experiments 1 and 2, growth attributes of tiller buds, which are considered as the next generation once they form roots, were significantly affected by the treatments in their lower nodal positions, a finding which is consistent with the growth retardant effect on the normal variety of ME [13].

However, GA-SD treatment in the normal variety of ME promoted stem elongation much more compared to dwarf varieties of DL and DE. In doing so, the GA-SD treatment led to excessively succulent growth in ME, which suffered badly from lodging, breakage of stems, and insect damage, events which were quite rare in the control ME plants. It will thus be necessary to determine whether there is an optimal dose of foliar GA_3 spray to the normal variety (ME) of Napier grass.

In conclusion, the most effective treatment for promoting maturity of tiller buds and establishment of nursery plants in the field is the 10-day GA-SD treatment, which was suitable for dwarf varieties DL and DE; that is, it increased internode length and tiller bud length, thereby reducing labor time and cost for vegetative propagation.

Competing Interests

The authors declare that there are no competing interests regarding the publication of this paper.

References

[1] W. R. Ocumpaugh and L. E. Sollenberger, "Elephantgrass," in *Forages Volume 1: An Introduction to Grassland Agriculture*, R. F. Barns, D. A. Miller, and C. J. Nelson, Eds., pp. 443–445, Iowa State University Press, Ames, Iowa, USA, 1995.

[2] M. J. Williams and W. W. Hanna, "Performance and nutritive quality of dwarf and semi-dwarf elephantgrass genotypes in the south-eastern USA," *Tropical Grassands*, vol. 29, no. 2, pp. 122–123, 1995.

[3] C. J. Chaparro and L. E. Sollenberger, "Nutritive value of clipped 'Mott' elephantgrass herbage," *Agronomy Journal*, vol. 89, no. 5, pp. 789–793, 1997.

[4] L. E. Sollenberger, G. M. Prine, W. R. Ocumpaugh, S. C. Schank, R. S. Kalmbacher, and C. S. Jones, "Dwarf napiergrass: a high quality forage with potential in Florida and the tropics," *Soil and Crop Science Society of Florida Proceedings*, vol. 46, no. 1, pp. 42–46, 1987.

[5] C.-I. Na, L. E. Sollenberger, J. E. Erickson, K. R. Woodard, J. M. B. Vendramini, and M. L. Silveira, "Management of perennial warm-season bioenergy grasses. i. biomass harvested, nutrient removal, and persistence responses of elephantgrass and energycane to harvest frequency and timing," *Bioenergy Research*, vol. 8, no. 2, pp. 581–589, 2015.

[6] Y. Ishii, K. Hamano, D.-J. Kang, S. Idota, and A. Nishiwaki, "Cadmium phytoremediation potential of napiergrass cultivated in Kyushu, Japan," *Applied and Environmental Soil Science*, vol. 2015, Article ID 756270, 6 pages, 2015.

[7] L. E. Sollenberger, G. M. Prine, W. R. Ocumpaugh et al., *Mott' Dwarf Elephantgrass: A High Quality Forage for the Subtropics and Tropics*, Circular S-356, Agricultural Experiment Station, Institute of Food and Agricultural Sciences, University of Florida, Gainsville, Fla, USA, 1988.

[8] W. W. Hanna and L. E. Sollenberger, "Tropical and subtropical grasses," in *Forages Volume II, The Science of Grassland Agriculture*, R. F. Barner, C. J. Nelson, K. J. Moore, and M. Collons, Eds., pp. 245–255, Blackwell, Ames, Iowa, USA, 2007.

[9] T. Kipnis and S. Bnei-Moshe, "Improved vegetative propagation of napiergrass and pearl millet × napiergrass interspecific hybrids," *Tropical Agriculture (Trinidad)*, vol. 65, no. 3, pp. 158–160, 1988.

[10] A. Yamano, Y. Ishii, K. Mori, R. F. Utamy, and S. Idota, "Effect of density of mother plants on efficiency of nursery production in dwarf napiergrass (*Pennisetum purpureum* Schumach)," in *Proceedings of the 7th Asian Crop Science Association Conference*, pp. 345–348, Bogor, Indonesia, September 2011.

[11] R. F. Utamy, Y. Ishii, L. Khairani, S. Idota, and K. Fukuyama, "Development of mechanical methods for cell-tray propagation and field transplanting of dwarf napiergrass (*Pennisetum purpureum* Schumach)," *Journal of Agriculture and Rural Development in the Tropics and Subtropics*, vol. 117, no. 1, pp. 11–19, 2015.

[12] M. D. Hare, K. Wongpichet, M. Saengkham, K. Thummasaeng, and W. Suriyajantratong, "Juvenility and long-short day requirement in relation to flowering of *Paspalum atratum* in Thailand," *Tropical Grasslands*, vol. 35, no. 2, pp. 139–143, 2001.

[13] K. Ito, Y. Ishii, M. Misumi, and H. Iwakiri, "Studies on the dry matter production of napiergrass. VI. Effect of a plant growth retardant, paclobutrazol on growth and dry matter of shoots," *Japanese Journal of Crop Science*, vol. 59, no. 3, pp. 469–474, 1990.

[14] P. Yañez, S. Chinone, R. Hirohata, H. Ohno, and K. Ohkawa, "Effects of time and duration of short-day treatments under long-day conditions on flowering of a quantitative short-day sunflower (*Helianthus annuus* L.) 'Sunrich Orange'," *Scientia Horticulturae*, vol. 140, pp. 8–11, 2012.

[15] V. Jokela, P. Virkajärvi, J. Tanskanen, and M. M. Seppänen, "Vernalization, gibberellic acid and photo period are important signals of yield formation in timothy (*Phleum pratense*)," *Physiologia Plantarum*, vol. 152, no. 1, pp. 152–163, 2014.

[16] R. W. King, L. N. Mander, T. Asp, C. P. MacMillan, C. A. Blundell, and L. T. Evans, "Selective deactivation of gibberellins below the shoot apex is critical to flowering but not to stem elongation of *Lolium*," *Molecular Plant*, vol. 1, no. 2, pp. 295–307, 2008.

[17] M. Talon and J. A. D. Zeevaart, "Stem elongation and changes in the levels of gibberellins in shoot tips induced by differential photoperiodic treatments in the long-day plant *Silene armeria*," *Planta*, vol. 188, no. 4, pp. 457–461, 1992.

[18] C. Matthew, W. A. Hofmann, and M. A. Osborne, "Pasture response to gibberellins: a review and recommendations," *New Zealand Journal of Agricultural Research*, vol. 52, no. 2, pp. 213–225, 2009.

[19] Y. Ishii, K. Ito, and H. Numaguchi, "Effects of cutting date and cutting height before overwintering on the spring regrowth of summer-planted napiergrass (*Pennisetum purpureum* Schumach)," *Journal of Japanese Grassland Science*, vol. 40, no. 4, pp. 396–409, 1995.

19

Varietal Evaluation of Potato Microtuber and Plantlet in Seed Tuber Production

Md. Sadek Hossain,[1] M. Mofazzal Hossain,[2] M. Moynul Haque,[3] Md. Mahabubul Haque,[4] and Md. Dulal Sarkar[5]

[1]*Seed Distribution Division, Bangladesh Agricultural Development Corporation, Dhaka, Bangladesh*
[2]*Department of Horticulture, Bangabandhu Sheikh Mujibur Rahman Agricultural University, Gazipur 1703, Bangladesh*
[3]*Department of Agronomy, Bangabandhu Sheikh Mujibur Rahman Agricultural University, Gazipur 1703, Bangladesh*
[4]*Farm Division, Bangladesh Agricultural Research Institute, Gazipur, Bangladesh*
[5]*Department of Horticulture, Sher-e-Bangla Agricultural University, Dhaka 1207, Bangladesh*

Correspondence should be addressed to Md. Dulal Sarkar; dulalsau_121@yahoo.com

Academic Editor: Maria Serrano

Diamant, Asterix, and Granola varieties differed significantly in foliage coverage, plant height, and yield. They produced lower graded minituber (67.62%, 78.16% ha^{-1}, and 66.27% of Asterix, Granola, and Diamant varieties, resp.) as per seed rule of the National Seed Board of Bangladesh, while foliage coverage (74.38%) was the maximum in Diamant. Microtuber in field condition showed the maximum survivability, plant height, foliage coverage, number of stems plant^{-1}, and SPAD value as well as yield of minituber compared to plantlet. On the contrary, microtuber derived plants of the three varieties gave the maximum yield (20.49 t ha^{-1}, 19.12 t/ha^{-1}, and 19.98 t ha^{-1} of Asterix, Granola, and Diamant varieties, resp.) and it was the minimum in plants of plantlets derived from all varieties (9.50 t ha^{-1}, 7.88 t ha^{-1}, and 9.70 t ha^{-1} of Asterix, Granola, and Diamant varieties, resp.). Microtuber derived plants produced a minimum percentage of <28 mm size of minituber compared to plantlet derived plants in case of all varieties.

1. Introduction

Potato (*Solanum tuberosum* L.) is a winter vegetable crop mainly in Bangladesh [1]. Ideal production protocols are wanted regarding microtuber induction and development, especially for Bangladeshi potato cultivars. Also, their comparative growth and yield efficiency study is the consensus at open field condition in relation to other propagules for minituber or seed tuber production. Utilization of microtubers for the speedy proliferation of seed tuber is becoming an important technique [2]. There is significant uncertainty regarding the utility of microtubers for evaluation of agronomic characters to the commercial growers. However, the application of microtubers in germplasm conservation is extensively recognized. Microtubers in different grades have different dormancy requirements and differ widely in relative growth potential and productivity. Again, the key problems of traditional seed tuber production systems are the stumpy

growth rate of field-grown potatoes and the susceptibility to viral, bacterial, or fungal diseases increases with field multiplication, which are conveyed to progeny through the seed tubers. This is especially significant because the countries are unable to produce high-quality potato seed tubers [3–5]. In recent years, alternative seed production program has been developed in which the first multiplication steps are speeded up by using in vitro plantlets [6–8], microtubers [9–11], and minitubers [12]. Microtubers are particularly convenient for handling, storage, and transportation of germplasm and for the development of disease-free materials [9, 13]. Microtuber needs to grow at least once and preferably two or three times under protected or field conditions to produce desired lot of quality seed [14]. Although there are some protocols for in vitro microtuberization, there is also no comparative information regarding their field performance especially for Bangladeshi potato cultivars. So the usefulness of microtubers will be contingent on their yield potentiality under open field conditions. Therefore, the present study was undertaken to evaluate the field performance of in vitro microtuber in comparison with in vitro plantlets.

2. Materials and Methods

2.1. Experimental Site. The experiment was conducted at Bangabandhu Sheikh Mujibur Rahman Agricultural University during the period from November 2012 to March 2013. The location of the experimental site was 24.09°N latitude and 90.26°E longitude with an elevation of 8.2 m above sea level.

2.2. Seed Source. Diseases-free in vitro plantlets of three potato varieties, namely, Asterix, Granola, and Diamant, were collected from Bangladesh Agricultural Research Institute and Bangladesh Agricultural Development Corporation Tissue Culture Laboratory which were prepared through meristem culture earlier.

2.3. In Vitro Multiplication of Plantlets. In vitro plantlets of three potato varieties were multiplied as per routine by subculturing of single-stem nodes at every three weeks' interval for growing the explants 68-node stage for experimentation. The multiplication medium contained minerals, salts, and vitamins [15] which were supplemented with $0.1\,mg\,l^{-1}$ Gibberellic acid (GA_3), $0.01\,mg\,l^{-1}$ naphthaleneacetic acid (NAA), $4\,mg\,l^{-1}$ D-calcium pantothenate, and $30\,g\,l^{-1}$ sucrose. The medium was solidified with $8\,g\,l^{-1}$ agar and pH was adjusted to 5.7 prior to autoclaving. The temperature in the growth chamber was $20 \pm 1°C$ with 16-hour photoperiod and the light was supplied by fluorescent tubes at an intensity of 3000 Lux.

2.4. In Vitro Production of Microtuber

2.4.1. Step I. Eight stem segments (each with 3 nodes) of in vitro subcultured plantlets were again cultured in liquid medium in 250 mL Erlenmeyer flasks containing mineral salts and vitamins [15] supplemented with $0.1\,mg\,l^{-1}$ GA_3, $0.01\,mg\,l^{-1}$ NAA, $4\,mg\,l^{-1}$ D-calcium pantothenate, and $30\,g\,l^{-1}$ sucrose for 28 days.

2.4.2. Step II. After 28 days, the liquid media were decanted off and $40\,mL$ microtuber induction medium based on MS medium [15] was supplemented with $10\,mg\,l^{-1}$ benzyl adenine (BA) and different concentrations of sucrose (0, 3, 4, 6, 8, 10, 12, and 14%). Then the microtuber induction cultures were incubated in the dark at 20°C [16]. All cultures in Erlenmeyer flask were closed with a cotton cap.

2.5. Harvest of Microtuber. Cultures with microtuber were kept in full light after 63 days of incubation for greening. After 70 days of incubation, microtubers were harvested aseptically and washed properly and then treated with Bavistin. Those microtubers were stored in a refrigerator at 4°C temperature.

2.6. Propagule Preparation. 21-day-old in vitro rooted plantlets were removed from culture and washed to remove

agar and treated with fungicides. The plantlets were kept for 48 hours in poly bags for hardening and after that plantlets were planted in the field. At the same time, fungicide treated sprouted microtubers (>250 mg) were planted in the field. Watering was done to keep the soil moist accordingly. Pseudostem of banana was used as mulch every day from 9:00 am to 3:00 pm for 10 days to protect the plantlets and sprouted microtubers from direct sunshine after planting.

2.7. Treatments and Design of the Experiment. The factorial experiment was designed as a randomized complete block design with four replications. Treatments were consisting of microtuber and in vitro plantlets of three varieties, namely, Asterix, Granola, and Diamant. Those were planted at 60 cm× 25 cm spacing and plot size was 1.8 m × 2.0 m.

2.8. Land Preparation and Planting of Propagules. The propagules were planted on 1 December in the field. The soil of the plot was light clay loam supplemented with well rotten cow dung. The entire doses of $Ca(H_2PO_4)_2$ (220 kg/ha), KCl (270 kg/ha), $CaSO_4 \cdot 2H_2O$ (120 kg/ha), $ZnSO_4$ (5 kg/ha), and 50% $CO(NH_2)_2$ (180 kg/ha) were made at the time of land preparation. The remaining portion of $CO(NH_2)_2$ (180 kg/ha) was side-dressed at the time of last earthing-up [17]. Intercultural operations like earthing-up and weeding were done as and when required. The experimental plots were irrigated frequently to maintain adequate soil moisture and also to keep the soil cool. Secure and Dimecron were sprayed to protect the crop from late blight and aphid infestation, respectively.

2.9. Collection of Data. Data were collected on plant stand at 45 days after planting and foliage coverage (%) at 45 days after planting and plant height (cm), stem plant^{-1}, number of minitubers plant^{-1}, yield of minitubers plant^{-1} (g), yield of minitubers (t ha^{-1}), and the grade of minitubers (%) were collected during harvest at 90 days after planting.

Analysis of Data. All the collected data were analyzed by analysis of variance and the means were compared according to Duncan's Multiple Range Test at 5% level of probability.

3. Results and Discussion

3.1. Varietal Performance of Potato in Field Condition. The varieties did not vary significantly in plant stand at 45 DAP, stems plant^{-1}, SPAD value, and number of minitubers plant^{-1} (Table 1). They differed significantly in foliage coverage, plant height, yield of minitubers plant^{-1}, and yield of minitubers (t ha^{-1}). Zakaria [18] found no significant difference in main stems plant^{-1} and plant stand at 45 DAP, while there were significant differences in the number of tubers plant^{-1}. Foliage coverage (74.38%) was the maximum in Diamant, while plant height, yield of minitubers plant^{-1}, and yield of minitubers (t ha^{-1}) were statistically similar to the variety Asterix. All of the three varieties produced lower graded minituber as per seed rule of NSB, while Diamant and Asterix

TABLE 1: Varietal performance of potato in field condition.

Variety	Plant stand (%) at 45 DAP	Foliage coverage (%) at 45 DAP	Plant height (cm) at 90 DAP	Number of main stems plant^{-1} at 90 DAP	SPAD value of leaves at 90 DAP	Number of minitubers plant^{-1} at 90 DAP	Yield of minitubers (g plant^{-1}) at 90 DAP	Yield of minitubers (t ha^{-1}) at 90 DAP
Asterix	92.7	66.25[b]	32.25[a]	2.00	51.55	13.11	224.90[a]	14.99[a]
Granola	88.5	52.13[c]	30.34[b]	1.88	52.42	13.47	202.50[b]	13.50[b]
Diamant	90.6	74.38[a]	31.63[ab]	2.00	50.25	14.26	222.10[a]	14.84[a]

Means followed by same letter(s) in a column are not significantly different by DMRT at 5% level of probability.

TABLE 2: Planting materials influencing potato production in field condition.

Treatments	Plant stand (%) at 45 DAP	Foliage coverage (%) at 45 DAP	Plant height (cm) at 90 DAP	Number of main stems plant^{-1} at 90 DAP	SPAD value of leaves at 90 DAP	Number of minitubers plant^{-1} at 90 DAP	Yield of minitubers (g plant^{-1}) at 90 DAP	Yield of minitubers (t ha^{-1}) at 90 DAP
Microtuber	97.23	72.50	38.43	2.50	52.08	12.59	297.94	19.86
Plantlet	84.02	56.00	24.39	1.42	50.748	14.64	135.08	9.03
SE	1.49	1.11	0.56	0.14	0.61	0.3866	4.94	0.34

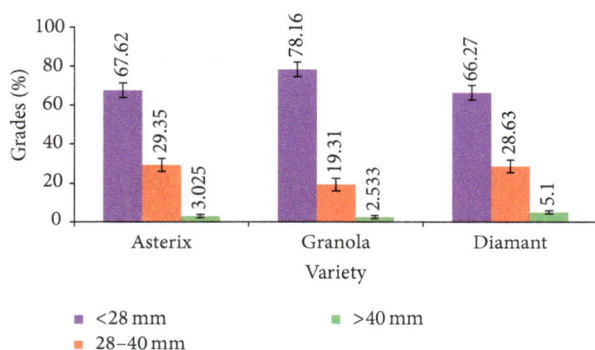

FIGURE 1: Varietal performance on minituber grades according to size.

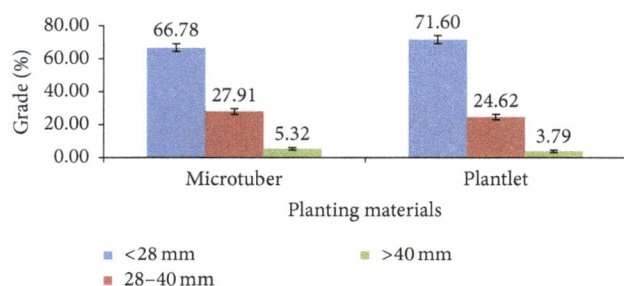

FIGURE 2: Effect of planting materials on production of minituber grades according to size.

produced a higher percentage of 28–40 mm grade minituber than Granola (Figure 1).

3.2. Performance of Potato Microtuber and Plantlet in Field Condition. Planting materials widely varied from one another regarding plant stand at 45 DAP (Table 2). Plants derived from microtuber sharply showed the maximum survivability, plant height, foliage coverage, number of stems plant^{-1}, and SPAD value as well as yield of minituber compared to plant derived from plantlet in field condition, while the number of minitubers plant^{-1} was more in plantlets derived plants than in microtuber derived plants. Kawakami et al. [19] reported that, unlike micropropagated plantlets, microtubers do not require time-consuming hardening periods in greenhouses and can be handled much like normal seed tubers. These have the potential to be used for field planting as the fresh tuber yields from microtuber plants were 82% that of conventional tuber plants. Pérez-Alonso et al. [20] reported also that, fifteen days after field planting,

89% of the microtubers sprouted and subsequently produced vigorous plants, whereas plants derived from microtuber of the cardinal variety showed about 85% survivability [21], with minimum foliage coverage with in vitro plantlets [22]. Microtubers produced a relatively higher number of stems plant^{-1} and yield plant^{-1} compared to plantlets [23]. Microtubers produced a higher number of middle-grade minitubers than plantlet derived plants and also produced a similar number of minitubers of <28 mm grade, but the minituber was heavier compared to plantlet derived plant (Figure 2). Microtubers and minitubers produced statistically similar yields [18]. Dodds et al. [24] concluded that in vitro microtubers behave similarly to in vitro plantlets with the advantage of being easy to handle and show more resistance to stressful environmental conditions. So current findings agreed with them.

3.3. Interaction Effect of Variety and Planting Materials of Potato in Field Condition. Microtubers of all the three varieties had the highest plant stand at 45 DAP compared to the plant stands of plantlet derived plants and the lowest plant

TABLE 3: Interaction effect of variety and planting materials in field condition.

Treatments		Plant stand (%) at 45 DAP	Foliage coverage (%) at 45 DAP	Plant height (cm) at 90 DAP	Number of main stems plant^{-1} at 90 DAP	SPAD value of leaves at 90 DAP	Number of minitubers plant^{-1} at 90 DAP	Yield of minitubers (g plant^{-1}) at 90 DAP	Yield of minitubers (t ha^{-1}) at 90 DAP
Asterix	Microtuber	97.93^a	70.00^b	40.13^a	2.50^a	52.07^{ab}	11.93^c	307.30^a	20.49^a
	Plantlet	87.50^b	62.50^c	24.38^c	1.50^b	51.03^{ab}	14.30^{ab}	142.5^b	9.50^b
Granola	Microtuber	95.85^a	65.00^{bc}	35.38^b	2.50^a	52.85^a	12.46^{bc}	299.7^a	19.12^a
	Plantlet	81.23^b	39.25^d	25.31^c	1.25^b	52.00^{ab}	14.49^{ab}	118.2^b	7.88^b
Diamant	Microtuber	97.94^a	82.50^a	39.78^a	2.50^a	51.30^{ab}	13.39^{abc}	286.8^a	19.98^a
	Plantlet	83.32^b	66.25^{bc}	23.48^c	1.50^b	49.20^b	15.14^a	144.5^b	9.70^b

Means followed by same letter(s) in a column are not significantly different by DMRT at 5% level of probability.

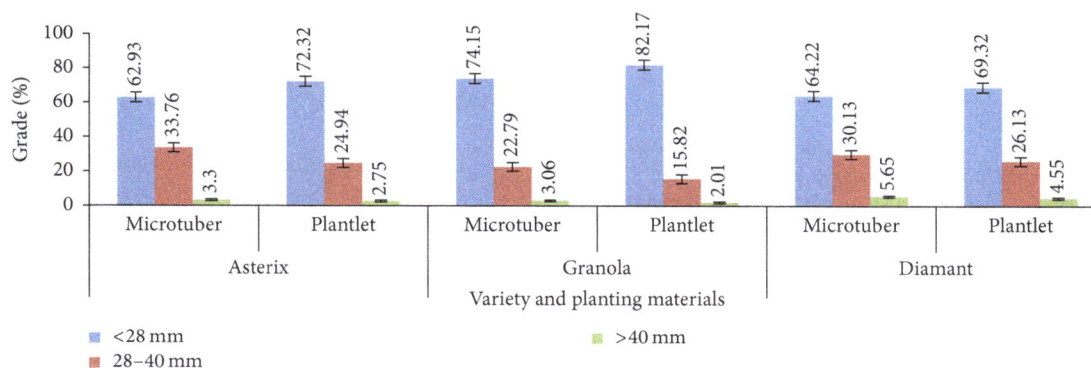

FIGURE 3: Interaction effects of variety and planting materials of potato on minituber grades according to size.

stand was observed in plantlets derived plants of Granola which was statistically similar with Diamant for the trait (Table 3). The plant height was biggest (40.13 cm) in the plants propagated from microtubers of Asterix which was statistically similar to Diamant followed by microtuber propagated plants of Granola, while all the three varieties produced short plants in plantlets propagated plants (Table 3). Main stems plant^{-1} also were higher in microtuber derived plants than those derived from in vitro plantlet (Table 3). SPAD value was statistically the same in both planting materials derived plants except plants of plantlet derived from Diamant (Table 3). The in vitro plantlets of the variety Diamant produced the maximum number of minitubers plant^{-1} and it was statistically similar with minitubers plant^{-1} of Asterix and Granola of the same planting materials. Again, statistically, the number of minitubers plant^{-1} was lowest in those plants which are derived from microtuber in case for all varieties (Table 3). On the contrary, yield of minitubers plant^{-1} (g) was the highest in plants which are derived from microtuber in case for all varieties. Microtuber derived plants of the three varieties gave the maximum yield of minitubers (20.49, 19.12, and 19.98 t ha^{-1} of Asterix, Granola, and Diamant varieties, resp.) and it was the minimum in plantlets derived plants of all varieties (9.50, 7.88, and 9.70 t ha^{-1} of Asterix, Granola, and Diamant varieties, resp.) (Table 3). Zakaria [18] reported that in vitro plantlets showed very poor yield

performance regarding minituber or normal seed tuber production compared to microtubers derived plants. Ranalli et al. [22] reported that microtubers and minitubers of cv. Monalisa were produced in the laboratory and compared with normal seed tubers in a field experiment. These tubers were planted at similar plant densities (13.6 sprouts m^{-2}) with two distances between rows (60 and 90 cm). Normal seed and minitubers and microtubers yielded, respectively, 50.8, 31.7, and 17.0 t ha^{-1} which corroborate with the present findings. In all three varieties, microtuber derived plants produced a minimum percentage of <28 mm size of minituber and higher maximum percentage of 28–40 mm and >40 mm size of minituber compared to plantlet derived plants (Figure 3).

4. Conclusion

Microtuber using about 250–500 mg size is the best for maximizing minituber yield compared to plantlet. Also, the use of microtubers instead of in vitro plantlets is advantageous as they are easier to store and handle and do not require any acclimatization treatment.

Competing Interests

The authors have declared that no competing interests exist.

Authors' Contributions

This work was carried out in collaboration between all authors. Author Md. Sadek Hossain designed the study, wrote the protocol, and wrote the first draft of the manuscript. Authors M. Mofazzal Hossain and M. Moynul Haque reviewed the study design and all drafts of the manuscript. Authors Md. Mahabubul Haque and Md. Dulal Sarkar performed the statistical analysis and also managed the literature searches. All authors read and approved the final manuscript.

Acknowledgments

The authors express profound gratitude to the Ministry of Science and Technology, Bangladesh, for financial support during conducting this research.

References

[1] M. H. Rashid, S. Akhter, M. Elias, M. G. Rasul, and M. H. Kabir, "Seedling tubers for ware potato production: Influence of size and plant spacing," *Chinese Potato Journal*, vol. 3, pp. 14–17, 1993.

[2] M. Zakaria, M. Hossain, M. K. Mian, T. Hossain, and M. Uddin, "*In vitro* tuberization of potato influenced by benzyl adenine and chloro choline chloride," *Bangladesh Journal of Agricultural Research*, vol. 33, no. 3, pp. 419–425, 2008.

[3] E. Baldacci, "Comarzione dei risultati di coltivazione nei primi due annie conclusion realeivel," *Genetica Agaria*, vol. 14, pp. 213–219, 1956.

[4] G. Faccioli and C. Rubies-Autonell, "PVX and PVY distribution in potato meristem tips and their eradication by the use of thermotherapy and meristem-tip culture," *Journal of Phytopathology*, vol. 103, no. 1, pp. 66–76, 1982.

[5] P. Ranalli, E. Forti, G. Mandolino, and B. Casarini, "Improving production and health of seed potato stocks in Italy," *Potato Research*, vol. 33, no. 3, pp. 377–387, 1990.

[6] W. M. Roca, N. O. Espinoza, M. R. Roca, and J. E. Bryan, "A tissue culture method for the rapid propagation of potatoes," *American Potato Journal*, vol. 55, no. 12, pp. 691–701, 1978.

[7] G. Hussey and N. J. Stacey, "In Vitro propagation of potato (*Solanum tuberosum* L.)," *Annals of Botany*, vol. 48, no. 6, pp. 787–796, 1981.

[8] G. Wattimena, B. McCown, and G. Weis, "Comparative field performance of potatoes from microculture," *American Potato Journal*, vol. 60, no. 1, pp. 27–33, 1983.

[9] G. Hussey and N. J. Stacey, "Factors affecting the formation of *in vitro* tubers of potato (*Solanum tuberosum* L.)," *Annals of Botany*, vol. 53, no. 4, pp. 565–578, 1984.

[10] G. Rosell, F. G. De Bertoldi, and R. Tizio, "In vitro mass tuberisation as a contribution to potato micropropagation," *Potato Research*, vol. 30, no. 1, pp. 111–116, 1987.

[11] E. Forti, G. Mandolino, and P. Ranalli, "*In vitro* tuber induction: influence of the variety and of the media," *Acta Horticulturae*, no. 300, pp. 127–132, 1992.

[12] P. C. Struik and W. J. M. Lommen, "Production, storage and use of micro-and minitubers," in *Proceedings of the 11th Triennial Coference of the European Association for Potato Research*, pp. 122–141, Edinburgh, UK, 1990.

[13] P.-J. Wang and C.-Y. Hu, "*In vitro* mass tuberization and virus-free seed-potato production in Taiwan," *American Potato Journal*, vol. 59, no. 1, pp. 33–37, 1982.

[14] D. E. Van der Zaag, "The implication of tissue culture micropropagation for the future of seed potato production system in Europe," in *Proceedings of the 11th Triennial Conference of the European Association for Potato Research*, pp. 28–45, Edinburgh, UK, 1991.

[15] T. Murashige and F. Skoog, "A revised medium for rapid growth and bio assays with tobacco tissue cultures," *Physiologia Plantarum*, vol. 15, no. 3, pp. 473–497, 1962.

[16] P. S. Naik and D. Sarkar, "Influence of light-induced greening on storage of potato microtubers," *Biologia Plantarum*, vol. 39, no. 1, pp. 31–34, 1997.

[17] Anonymous, *Fertilizer Recommendation Guide-2012*, Bangladesh Agricultural Research Council, Dhaka, Bangladesh, 2012.

[18] M. Zakaria, *Induction and performance of potato microtuber [Ph.D. thesis]*, Bangabandhu Sheikh Mujibur Rahman Agriculture University, Gazipur City, Bangladesh, 2003.

[19] J. Kawakami, K. Iwama, Y. Jitsuyama, and X. Zheng, "Effect of cultivar maturity period on the growth and yield of potato plants grown from microtubers and conventional seed tubers," *American Journal of Potato Research*, vol. 81, no. 5, pp. 327–333, 2004.

[20] N. Pérez-Alonso, E. Jiménez, M. de Feria et al., "Effect of inoculum density and immersion time on the production of potato microtubers in temporary immersion systems and field studies," *Biotecnología Vegetal*, vol. 3, pp. 149–154, 2007.

[21] M. S. Islam, *Indigenous potato varieties of Bangladesh: characterization by RAPD markers and production of virus free stock [Ph.D. thesis]*, Bangabandhu Sheikh Mujibur Rahman Agricultural University, Gazipur City, Bangladesh, 1995.

[22] P. Ranalli, F. Bassi, G. Ruaro, P. Del Re, M. Di Candilo, and G. Mandolino, "Microtuber and minituber production and field performance compared with normal tubers," *Potato Research*, vol. 37, no. 4, pp. 383–391, 1994.

[23] A. J. Haverkort, M. Van De Waart, and J. Marinus, "Field performance of potato microtubers as propagation material," *Potato Research*, vol. 34, no. 3, pp. 353–364, 1991.

[24] J. H. Dodds, P. Tover, R. Chandra, E. Estrella, and R. Cabello, "Improved methods for *in vitro* tuber induction and use of *in vitro* tubers in seed programs," in *Proceedings of the Symposium on Improved Potato Planting Material*, pp. 157–158, Asia and Pacific Seed Association (APSA), Kunming, China, June 1988.

20

Efficacy of a Phosphate-Charged Soil Material in Supplying Phosphate for Plant Growth in Soilless Root Media

Young-Mi Oh,[1] **Paul V. Nelson,**[1] **Dean L. Hesterberg,**[2] **and Carl E. Niedziela Jr.**[3]

[1]*Department of Horticultural Science, North Carolina State University, Raleigh, NC 27695-7609, USA*
[2]*Department of Soil Science, North Carolina State University, Raleigh, NC 27695-7619, USA*
[3]*Department of Biology, Elon University, Elon, NC 27244, USA*

Correspondence should be addressed to Carl E. Niedziela Jr.; cniedziela@elon.edu

Academic Editor: Manuel Tejada

A soil material high in crystalline Fe hydrous oxides and noncrystalline Al hydrous oxides collected from the Bw horizon of a Hemcross soil containing allophane from the state of Oregon was charged with phosphate-P at rates of 0, 2.2, and 6.5 mg·g^{-1}, added to a soilless root medium at 5% and 10% by volume, and evaluated for its potential to supply phosphate at a low, stable concentration during 14 weeks of tomato (*Solanum esculentum* L.) seedling growth. Incorporation of the soil material improved pH stability, whether it was charged with phosphate or not. Bulk solution phosphate-P concentrations in the range of 0.13 to 0.34 mg·dm^{-3} were associated with P deficiency. The only treatment that sustained an adequate bulk solution concentration of phosphate-P above 0.34 mg·dm^{-3} for the 14 weeks of testing contained 10% soil material charged with 6.5 mg·g^{-1} P, but initial dissolved P concentrations were too high (>5 mg·g^{-1} phosphate-P) from the standpoint of phosphate leaching. The treatment amended with 10% soil material charged with 2.2 mg·g^{-1} P maintained phosphate-P within an acceptable range of 0.4 to 2.3 mg·dm^{-3} for 48 d in a medium receiving no postplant phosphate fertilization.

1. Introduction

Containerized soilless media have low phosphate sorption capacities [1] and are subject to extensive leaching of applied nutrients due to high hydraulic conductivity. To assure adequate phosphate for plant growth, phosphate is typically applied multiple times at high concentrations, which leads to environmental problems [2, 3]. Fertilizer solution phosphate-P concentrations of 14–44 mg·dm^{-3} are common, which leads to bulk solution concentrations ≥15 mg·dm^{-3} [4]. However, equilibrium bulk solution concentrations as low as 0.18 mg·dm^{-3} phosphate-P for field grown chrysanthemums [5] and 0.2 mg·dm^{-3} phosphate-P for most plant species have been reported to be adequate when provided constantly to plant roots at the rate of plant uptake [6].

Efforts to alleviate phosphate leaching in soilless root media have been directed toward establishment of a phosphate reserve that provides a low, sustained equilibrium (dissolved) concentration. Studies have included incubation with superphosphate [7] and amendment with anion exchange resins [8]. These treatments, however, were time-consuming and expensive or failed to reduce phosphate leaching. As an alternative, phosphate-charged alumina (P-Al$_2$O$_3$) was tested as an amendment in an experimental sand culture system [9]. Combining P-Al$_2$O$_3$ with sand successfully maintained constant and steady phosphate concentrations for plant growth [9–12]. The phosphate-charged alumina amendment in a 70 peat moss : 30 perlite medium (by volume) effectively retained and slowly released phosphate during *Chrysanthemum* growth [13]. But alumina is an expensive media amendment in terms of crop production cost for the greenhouse industry. Phosphate-charged calcined arcillite and attapulgite clays added to a 70 peat moss : 30 perlite medium provided a less expensive source of phosphate that fully met the needs of a *Chrysanthemum* crop [13]. However, the rate of phosphate charging caused the initial phosphate equilibrium concentrations to be undesirably high in order to insure a sufficient concentration at the end of the crop. This system promoted phosphate leaching.

Metal oxides and noncrystalline aluminosilicates like allophane have a high specific surface area and surface metal hydroxyl groups (>Fe-OH or >Al-OH) that adsorb oxyanions and metal cations as inner-sphere surface complexes, with a degree of covalent bonding that makes the ions less exchangeable [14, 15]. The high phosphate adsorption capacity of these minerals could potentially be advantageous for reducing phosphate leaching and increasing efficiency of phosphate uptake in soilless root media.

In our previous studies to evaluate synthesized minerals for their phosphate adsorption capacities [16] and desorption characteristics [17], allophane (noncrystalline $Si_3Al_4O_{12} \cdot nH_2O$) showed the most favorable potential among the four minerals tested (allophane, alumina (Al_2O_3), goethite (α-FeOOH), and hematite (α-Fe_2O_3)). Allophane had the highest phosphate adsorption capacity and released the largest quantity of phosphate. Alumina, goethite, and hematite had lower phosphate adsorption capacities and required a greater quantity of these minerals to be mixed with soilless media to supply sufficient phosphate for plant growth. Since synthesis of minerals requires high energy, using a natural soil material high in allophane or metal oxides would be more economically feasible in terms of production cost for the industry.

The objective of this study was to determine if phosphate could be supplied at an adequate, low, relatively constant concentration during 14 weeks of tomato seedling growth when a phosphate-charged natural soil material high in metal oxides and allophanic aluminosilicates was incorporated into a soilless root medium.

2. Materials and Methods

2.1. Characterization of Soil Properties. A soil material was collected from the Bw horizon of a Hemcross soil (medial, ferrihydritic, mesic Alic Hapludands) in the state of Oregon. This soil material was selected because it was of young volcanic origin, typically high in allophane and Fe and Al oxides. The soil material was stored at 4°C prior to testing to maintain field moisture and minimize any alteration of surface chemical properties by air-drying [18]. The sample was passed through a 2 mm sieve prior to measurement of soil properties. Soil moisture content was determined by oven-drying a subsample at 105°C to constant weight. Soil pH was measured on a 1 soil : 1 water (w/v) sample [19]. Organic carbon content was measured by Walkley-Black method [20]. The content of Al, Si, and Fe in the soil material was measured by ammonium-oxalate extraction, pyrophosphate extraction, and dithionite-citrate extraction methods [21, 22]. Plant-available P was measured by extraction with buffered alkaline solution [23].

2.2. Characterization of Phosphate Adsorption. A stock suspension of soil material was prepared in a 10 mM KCl background solution for use in developing phosphate adsorption isotherms. Potassium chloride saturation was used to avoid precipitation of phosphate with divalent cations such as Ca^{2+} that would interfere with adsorption measurements.

The stock suspension was prepared as follows: Six 250 mL centrifuge bottles, each containing 5.88 g of sieved, field-moist soil material (equal to 4.5 g dry solids) and 219.12 mL of deionized water (1 solid : 50 suspension), were shaken for 24 h. Then, KCl salt was weighed into the suspensions to give a 1 M concentration, and the suspensions were shaken for 2 additional hours. The suspensions were centrifuged at 21,500 RCF (Relative Centrifugal Force) for 20 min after which the supernatant solutions were discarded. The soil samples were washed two more times by shaking with 1 M KCl solution for 2 h. They were then washed four times with 10 mM KCl solution to bring the background electrolyte to this concentration. After the final centrifugation, the six soil samples were rinsed into one bottle using a minimal amount of 10 mM KCl solution. The solids concentration of the final suspension was 0.052 g·g^{-1} as determined by oven-drying subsamples at 105°C to constant weight.

To characterize the adsorption envelope as shown in Figure 1(a) (plot of adsorbed phosphate-P versus pH for a given level of added phosphate [24]), six concentrations of KH_2PO_4 solution were formulated in a 10 mM KCl background solution, each at several pH levels varying from 4.0 to 9.0. While vigorously mixing, 1.54 g of the stock suspension (equal to 0.08 g dry solids) was weighed into each of a series of 50 mL centrifuge tubes, and 38.46 g of KH_2PO_4 solution was added to obtain a 40.00 g suspension sample containing 2 g dry solids per kg of suspension. The centrifuge tubes containing the sample suspensions were tightly capped and equilibrated by shaking for 48 h at 1.9 cycles per second in a water bath at 24°C. The equilibrated samples were centrifuged at 20,210 RCF for 10 min and the supernatant solutions were collected. The pH level was measured on unfiltered subsamples of each of the supernatant solutions, and the remaining solution was vacuum-filtered through a 0.2 μm polycarbonate membrane filter. Using the Murphy-Riley method [25], phosphate-P was measured in the filtered supernatant solutions and in the original series of KH_2PO_4 solutions added to each centrifuge tube. The amount of adsorbed phosphate-P was calculated as the difference between added phosphate-P and equilibrium dissolved phosphate-P (i.e., loss from solution). An adsorption isotherm at pH 6.0 was constructed using best-fit statistical curves to interpolate between data points on the phosphate adsorption envelopes and to determine the maximum adsorption capacity of the soil material.

2.3. Greenhouse Experiments. To evaluate the soil material for supplying phosphate for plant growth as a preplant source, the soil material was charged with phosphate at three concentrations based on the adsorption isotherm (Figure 1(b)). The first concentration was selected from the upper end of the curve and contained an adsorbed phosphate-P concentration of 6.5 mg·g^{-1} soil material with dissolved phosphate-P concentration of 400 mg·dm^{-3}. The second concentration contained an adsorbed phosphate-P concentration of 2.2 mg·g^{-1} soil material with dissolved phosphate-P concentration of 50 mg·dm^{-3}. The third concentration was 0 mg of adsorbed phosphate-P and was prepared in a similar manner to

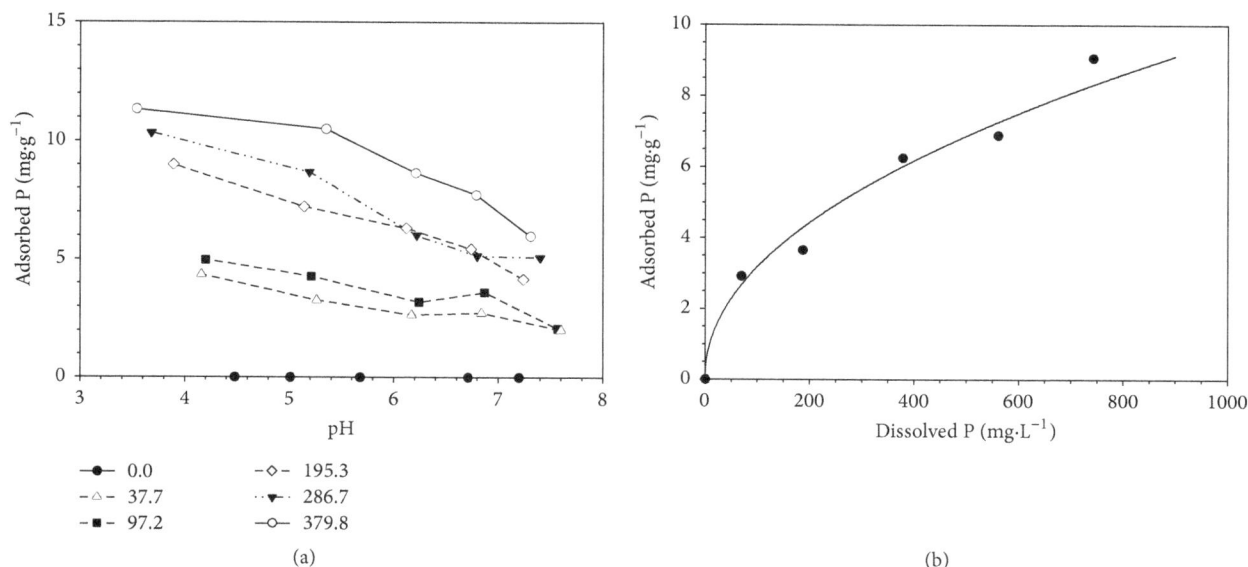

FIGURE 1: (a) Phosphate adsorption envelope at five pH levels of the soil material for various concentrations of added phosphate-P (10 mM KCl background solution, 24°C). Legend indicates the milligrams of phosphate-P added per gram of soil material. Lines connecting data points were included to aid in visualizing trends in the data. (b) Phosphate adsorption isotherm of soil at pH 6.0 with a van Bemmelen-Freundlich model fit.

the two previous rates. Phosphate charging was accomplished as follows. The field-moist soil material was passed through a 2 mm sieve and 2,614 g of soil material (equal to 2,000 g of solids) was weighed into a 12 L plastic bucket. To yield a 1 solid : 5 suspension ratio, 7,386 mL of deionized water was added to the soil material to obtain a 10 L suspension. The soil suspension was mixed with a stirrer for 2 h; then KCl salt was added to the suspension to give 1 M concentration and the suspension was stirred for an additional 2 h. The suspension was then allowed to flocculate and settle before siphoning off the supernatant solution. The suspension was washed one more time with 1 M KCl solution and then four times with 10 mM KCl solution. After the last wash, KH_2PO_4 solution prepared to yield the desired adsorption was added to the soil material and KOH or HCl was added during 48 h of equilibration to obtain the final solution pH of 6.0 ± 0.1. After equilibration, the supernatant solution was siphoned off and collected for analysis to determine the amount of phosphate-P adsorbed on the soil material. The soil material was washed again with 10 mM KCl solution and then with deionized water by stirring for 1 h to remove excess phosphate in the solution before centrifugation at 21,500 RCF for 20 min. The phosphate-charged soil material was then freeze-dried and ground through a 250 μm sieve using a ball mill.

There were seven treatments in the greenhouse experiment including a control. The base soilless root medium consisted of 3 sphagnum peat moss (St-Raphael Peat Moss Ltd., Haut-Lameque, NB, Canada) : 1 perlite (Carolina Perlite Company Inc., Gold Hill, NC) by volume amended with 6 and 0.45 g·dm^{-3} of dolomitic limestone and Micromax (The Scotts Co., Marysville, OH) micronutrient mix, respectively. No soil material was added to the root medium for

the control treatment. In two treatments, uncharged soil material (0 mg·g^{-1} P) was incorporated into the base medium at 5% and 10% of the final volume (designated as 0 P-5% and 0 P-10%, resp.). These treatments served to show the influence of the soil material without the effect of presorbed phosphate. The remaining four treatments were a factorial of two phosphate-P charging concentrations (2.2 and 6.5 mg·g^{-1} P) and two incorporation rates of the charged soil material into the root medium (5% and 10%). These four treatments of 5% soil material charged with 2.2 mg·g^{-1} P, 10% charged with 2.2 mg·g^{-1} P, 5% charged with 6.5 mg·g^{-1} P, and 10% charged with 6.5 mg·g^{-1} P were designated 2.2 P-5%, 2.2 P-10%, 6.5 P-5%, and 6.5 P-10%, respectively. The incorporation percentages equated to 40 and 80 g of the charged soil in 13.6-cm (1.0 L) standard plastic pots at 5 and 10% of the final volume, respectively. The former three treatments (control, 0 P-5%, and 0 P-10%) were fertilized with an N-P-K solution at each irrigation whereas the latter four treatments received an N-K fertilizer. Three successive crops were grown over a combined period of 14 weeks in the same pots containing the same media to determine the longevity of phosphate supply from the phosphate-charged soil material. In crops 1 and 2, a fertilizer containing 100 mg·dm^{-3} N (2 NH$_4^+$-N : 3 NO$_3^-$-N) was applied as an N-P-K fertilizer consisting of 0.7 mM NH$_4$H$_2$PO$_4$, 2.1 mM KNO$_3$, and 2.1 mM NH$_4$NO$_3$ and with an N-K fertilizer consisting of 0.35 mM (NH$_4$)$_2$SO$_4$, 2.1 mM KNO$_3$, and 2.1 mM NH$_4$NO$_3$. In crop 3, these fertilizer formulations were increased by 50%. In addition to the previously described lime and micronutrient amendments contained in the base soilless root medium, those treatments that received the N-P-K fertilizer (control, 0 P-5%, and 0 P-10%) were amended with 0.9 g·dm^{-3} of gypsum to compensate for lack of SO_4^{2-} that was contained in the N-K fertilizer formulation.

Tomato seeds ("Colonial," Petoseed Co., Saticoy, CA) were germinated in the base soilless root medium. Twenty-day-old tomato seedlings were then transplanted into 13.6 cm pots filled with 1.0 L of base soilless root medium amended as previously described for each treatment. Day and night temperatures were set at 25 and 20°C, respectively. Fertilizer was applied at each watering (fertigation) with a 20% leaching fraction. Irrigation occurred when the surface of the medium in the pots started to dry.

Each week, bulk solutions were collected for pH, phosphate-P, and K measurements by fertilizing the pots, allowing them to equilibrate for 1 h, and then pouring 30 mL of water into each pot to displace a sample of bulk solution. The pH of bulk solutions was measured using a pH meter (Extech Instruments Corp., Waltham, MA). Phosphate-P was measured by the Murphy-Riley procedure [25] using a Perkin-Elmer Lambda 3 UV/VIS spectrophotometer (Perkin-Elmer, Norwalk, CT), and K was determined using the Perkin-Elmer 373 atomic absorption spectrophotometer (Perkin-Elmer, Norwalk, CT).

Plant height was measured weekly. Total above-ground shoots of plants were harvested to measure fresh weight and dry weight when the first cluster started to bloom. Harvested shoots were washed in 0.5 N HCl solution for 1 min and rinsed with deionized water before drying at 70°C for tissue analysis. Dried tissue was ground in a stainless steel Wiley mill (Thomas Scientific, Philadelphia) to pass through a 1 mm screen (20 mesh). Nitrogen content in the tissue was analyzed using the Perkin-Elmer PE 2400 CHN elemental analyzer (Perkin-Elmer, Norwalk, CT). Tissue used for P and K analyses was dry-ashed at 500°C, dehydrated in 3 N HCl, and dissolved in 0.5 N HCl. Phosphorus and K were measured using an inductively couple plasma emission spectrometer, the Perkin-Elmer Plasma 2000 system (Perkin-Elmer, Norwalk, CT).

After plant shoots were removed from the first crop and later from the second crop, 20-day-old seedlings for the next crop were transplanted into the same pots and root medium used for the previous crop without removal of roots from the previous crop and with minimal disturbance of the root medium. The same cultural practices and data collection were employed during each crop.

2.4. Experimental Design and Statistical Analysis. Treatments were arranged in a randomized complete block design with three blocks and three pots per experimental unit. The bulk solution data were analyzed separately by individual crops as a two-way factorial with seven treatments in each crop and six sampling dates in crop 1, five sampling dates in crop 2, and five sampling dates in crop 3. With three blocks in each crop, there were 126, 105, and 105 experimental units in crops 1, 2, and 3, respectively. The growth and tissue analysis data were analyzed as a two-way factorial with three crops and seven treatments with three blocks (63 experimental units). Data were subjected to analysis of variance using PROC ANOVA (SAS 9.4, SAS Inst., Cary, NC). Means were separated using a protected LSD test at the 0.05 level of significance.

3. Results and Discussion

3.1. Characterization of Soil Properties. The soil material used in this study was acidic with a pH of 5.1, contained only 0.23% concentration of organic carbon as expected since it was sampled from a B horizon, and had a water content of 0.235 g·g⁻¹ soil. The amount of Al extracted by ammonium-oxalate extraction (Al_O), which dissolves Al in allophane, imogolite, organic-matter complexes, and noncrystalline hydrous Al oxides, was 0.79% (w/w). In contrast, 1.48% Al was extracted by a dithionite-citrate extraction (Al_D), which represents Al in organic complexes, noncrystalline Al hydrous oxides, and Al-substituted Fe oxides. These results suggested that this soil material contained a low level of allophane. A low concentration of allophane in this soil material was further indicated by the low oxalate-extractable Si (Si_O) concentration of 0.07%. The Fe extracted by ammonium-oxalate extraction (Fe_O) represents Fe in organic complexes and noncrystalline Fe hydrous oxides, and Fe extracted by dithionite-citrate extraction (Fe_D) represents Fe in organic complexes and both are found in noncrystalline and crystalline Fe hydrous oxides. Since Fe_D (7.73%) was much higher than Fe_O (0.46%), this soil material contained mainly crystalline Fe hydrous oxides even though the soil series was described as ferrihydritic. Therefore, we concluded that the main phosphate-adsorbing solids in this soil material were crystalline Fe hydrous oxides and noncrystalline Al hydrous oxides, suggesting that the soil sample was collected above the more allophonic-rich zone of the B horizon. However, no direct measurements of allophane were made. The soil material was found to contain almost no plant-available P, 0.001 mg·g⁻¹ P, by the buffered alkaline solution extraction method [23].

3.2. Phosphate Adsorption Characteristics. The phosphate adsorption envelope of the soil material for different concentrations of added P had typical patterns of decreased adsorption with increasing pH (Figure 1(a)). The adsorption isotherm indicated that the maximum phosphate-P adsorption capacity of the soil material approached 10 mg·g⁻¹ P for a dissolved phosphate-P concentration of 900 mg·dm⁻³ at pH 6.0 (Figure 1(b)), consistent with the adsorption envelopes. The phosphate adsorption capacity of this soil material was much lower than those of minerals synthesized and tested at this same pH by Oh et al. [16]. Adsorbed phosphate capacities for allophane, goethite, and alumina reported in the previous study were >39, 15.7, and 14.8 mg·g⁻¹ P. However, oxide and allophanic minerals represent only a fraction of the total mass of the soil material, which also constitutes sand and silt particles of lower specific surface area. For example, the dithionite-citrate extractable Fe and Al concentrations of 7.73 and 1.48% are equivalent of 12.3% FeOOH (goethite) and 3.23% AlOOH (alumina), which would produce a soil phosphate adsorption capacity of only 2.4 mg·g⁻¹ soil. This calculation suggests that the soil contained considerable high-P adsorbing mineral such as allophane that was not necessarily extracted by oxalate and dithionite extractions.

3.3. Greenhouse Experiments: Uncharged Soil Material. Both the 0 P-5% and 0 P-10% treatments developed P deficiency

TABLE 1: Tomato shoot height, fresh weight, and dry weight at the end of three successive seedling crops grown in root media amended with no soil material (control) and six soil material treatments combinations containing 5 or 10% soil material (v/v) charged with 0, 2.2, or 6.5 $mg \cdot g^{-1}$ phosphate-P[y].

Treatment	Ht. (cm)			Fresh wt. (g)			Dry wt. (g)		
	Crop 1	Crop 2	Crop 3	Crop 1	Crop 2	Crop 3	Crop 1	Crop 2	Crop 3
Control	41.7	42.9	47.3	72.6	89.7	101.9	9.9	10.1	9.8
0 P-5% soil	43.4	43.2	47.3	78.1	88.7	98.2	10.2	10.2	9.7
0 P-10% soil	29.3	42.3	47.8	24.1	90.1	97.8	2.4	9.9	9.9
2.2 P-5% soil	47.4	38.7	17.9	95.3	55.2	11.3	12.7	6.2	1.4
2.2 P-10% soil	48.4	44.8	33.6	103.1	95.7	48.1	13.4	10.9	6.1
6.5 P-5% soil	48.7	43.7	42.1	98.0	89.7	73.5	12.9	10.4	8.8
6.5 P-10% soil	48.6	45.1	47.4	101.9	98.0	106.3	13.3	11.1	11.3
LSD$_{0.05}$[z]		2.20			5.52			0.86	

[y]Crops 1, 2, and 3 were harvested at first cluster bloom (37, 27, and 36 days after transplanting 20-day-old seedlings, resp.).
[z]LSD values are for comparing soil material treatments within a given crop or crops within a given soil material treatment.

symptoms by 10 d into crop 1, even though they received phosphate in the fertigation program equivalent to that in the 0 P-0% control treatment which was free of P deficiency. These results indicate that phosphate in the added nutrient solutions was adsorbed by the soil material added in the 0 P-5% and 0 P-10% treatments. Plants in the 0 P-5% treatment recovered from the symptoms after two weeks of growth in crop 1 while 0 P-10% plants did not recover during crop 1. Plants grown in crop 2 developed deficiency symptoms from a few days up to 10 d. Symptoms were consistent with P deficiency, which included stunting, exceptionally deep green foliage, and purple pigmentation on the lower surface of leaves. The plants in the 0 P-5% treatment completely regained size relative to control plants by the end of crop 1, as seen in the growth measurements of height and fresh and dry weight (Table 1). The 0 P-10% plants were much smaller than the control plants at the end of crop 1. At the end of crops 2 and 3, plants in both 0 P soil material treatments did not exhibit P deficiency symptoms or difference in size from control plants.

Bulk solution phosphate-P concentrations in the 0 P-5% treatment ranged from 0.2 to 0.6 $mg \cdot dm^{-3}$ in crop 1, from 1.2 to 3.3 $mg \cdot dm^{-3}$ in crop 2, and from approximately 2.7 to 17.3 $mg \cdot dm^{-3}$ in crop 3 (Table 2). All of these concentrations were lower than in the control treatment medium. Root medium phosphate-P concentrations were consistently lower in the 0 P-10% compared to the 0 P-5% treatments. These results were consistent with phosphate being sorbed by the soil material and that sorption was greater with the higher amount of soil material incorporated into the root medium, as would be expected from the adsorption isotherm in Figure 1(b). While depletion of applied phosphate was less at the end of the experiment, it still occurred. These effects can be seen likewise in the shoot P concentrations (Table 3). Tissue P was lower, compared to the control plants, at the end of the first two crops in plants in the 0 P-5% treatment and during all three crops in the 0 P-10% plants. These data point to the problem that could ensue if high phosphate fixation capacity soil material without supplemental phosphate was added to root medium. The soil material did not appear to

have adverse effects on the crops other than P deficiency since after the P deficiency dissipated, growth did not differ in the control versus the 0 P-5% and 0 P-10% treated plants.

3.4. Greenhouse Experiments: Charged Soil Material. Within each crop, tissue P concentrations related well to the quantity of phosphate provided in the form of phosphate-charged soil material (Table 3). Throughout the three crops, tissue P concentrations declined over time as phosphate was released from the soil material and dissolved concentrations were lower. Plant growth in crop 1, measured as height, fresh weight, and dry weight, in the 2.2 P and 6.5 P treatments exceeded growth of control plants (Table 1). Compared to the control plants, the final crop 1 shoot concentration of P was higher in all of these treatments except the 2.2 P-5% treated plants (Table 3). However, that lower P concentration was concluded to be adequate because growth parameters exceeded those of the control. A positive relationship has been reported between P tissue concentration and growth of young plants, including tomato, up to P levels three or more times higher than the minimum adequate levels established for mature crops [26]. However, the greater levels of growth associated with higher-than-adequate phosphate concentrations were not deemed desirable for container grown plants due to etiolation within the constrained space required for commercial container production. Although the normal range of tomato tissue P has been reported to be 0.3 to 0.8% [27], it was not possible to determine adequacy of phosphate from our shoot P concentrations alone. Foliar tissue standards were developed for specific, relatively young leaves, whereas we analyzed the entire shoots that also contained stems and older leaves. Stems and older leaves are typically lower in P than younger leaves. Tissue P concentrations at the end of the crop in plants that exhibited P deficiency symptoms were 0.13%, 0.11%, 0.12%, and 0.14% for plants in treatments 2.2 P-5% in crop 2, 2.2 P-5% in crop 3, 2.2 P-10% in crop 3, and 6.5 P-5% in crop 3, respectively (Table 3).

Two weeks after the start of crop 2, plants in 2.2 P-5% treatment started to exhibit P deficiency symptoms. These

TABLE 2: Phosphorus concentration in bulk solutions of root media amended with no soil material (control) and six soil material treatment combinations containing 5 or 10% soil material (v : v) charged with 0, 2.2, or 6.5 mg·g^{-1} phosphate-P during three successive tomato seedling crops[x].

Trt	Crop 1 (DAT)[y]						Crop 2 (DAT)					Crop 3 (DAT)				
	4	11	16	21	24	37	0	11	16	25	27	0	15	21	27	36
	Bulk solution P (mg·dm^{-3})															
Control	8.85	6.19	13.72	4.39	5.87	18.66	16.79	13.53	13.39	8.83	17.48	15.97	22.05	21.85	29.71	33.42
0 P-5%	0.39	*0.57*	0.46	0.34	0.21	0.53	1.18	2.25	3.29	1.43	1.86	2.71	4.36	5.48	7.68	17.34
0 P-10%	0.33	0.34	*0.33*	*0.30*	*0.25*	*0.18*	0.14	0.47	0.41	0.49	0.57	0.67	1.01	1.19	2.28	7.58
2.2 P-5%	2.40	1.90	0.97	0.51	0.33	0.20	0.23	0.34	*0.23*	*0.27*	*0.19*	0.30	*0.16*	*0.15*	*0.16*	*0.13*
2.2 P-10%	2.29	1.77	1.28	0.75	0.51	0.28	0.38	0.49	0.32	0.32	0.23	0.37	*0.19*	*0.18*	*0.20*	*0.21*
6.5 P-5%	43.30	29.18	20.06	5.98	4.08	0.78	1.02	2.26	1.34	0.46	0.35	0.69	0.43	*0.21*	*0.17*	*0.23*
6.5 P-10%	58.00	41.24	30.65	15.69	10.47	2.89	5.02	8.73	4.81	1.66	1.08	3.64	1.43	0.81	0.54	0.45
LSD$_{0.05}$[z]	1.27						1.32					1.15				

[x]Bold italic values within treatments were associated with symptoms of P deficiency when the sample was collected.
[y]Days after transplanting of 20-day-old tomato seedlings.
[z]LSD values are for comparing soil material treatments within a given crop or crops within a given soil material treatment.

TABLE 3: Nutrient concentrations in tomato shoots at the end of three successive seedling crops grown in root media amended with no soil material (control) and six soil material treatments combinations containing 5 or 10% soil material (v/v) charged with 0, 2.2, or 6.5 mg·g^{-1} phosphate-P[y].

Treatment	Nutrient (%)								
	N			P			K		
	Crop 1	Crop 2	Crop 3	Crop 1	Crop 2	Crop 3	Crop 1	Crop 2	Crop 3
Control	2.29	3.30	4.77	0.31	0.52	0.90	2.38	3.23	4.09
0 P-5% soil	2.39	3.27	4.73	0.22	0.47	0.88	3.34	3.29	4.17
0 P-10% soil	3.90	4.88	4.45	0.26	0.38	0.68	4.80	4.86	4.12
2.2 P-5% soil	2.25	4.52	3.97	0.24	0.13	0.11	3.34	4.50	3.42
2.2 P-10% soil	2.19	4.99	3.62	0.38	0.19	0.12	3.50	4.95	3.78
6.5 P-5% soil	2.09	4.22	3.64	0.70	0.30	0.14	3.32	4.15	4.02
6.5 P-10% soil	1.89	5.00	4.16	0.82	0.63	0.34	3.44	4.94	4.37
LSD$_{0.05}$[z]	0.37			0.05			0.35		

[y]Crops 1, 2, and 3 were harvested at first cluster bloom (37, 27, and 36 days after transplanting 20-day-old seedlings, resp.).
[z]LSD values are for comparing soil material treatments within a given crop or crops within a given soil material treatment.

plants were smaller at the end of this crop than the control plants (Table 1). Plants grown in the 2.2 P-10%, 6.5 P-5%, and 6.5 P-10% treatments did not differ in growth from the control plants except in the 2.2 P-10% treatment where fresh weight was greater than the control plants and in the 6.5 P-10% treatment where fresh weight and dry weight were greater than the control plants. In crop 3, P deficiency symptoms developed after two days in the 2.2 P-10% treatment and at 20 d in the crop in the 6.5 P-5% treatment. These plants finished smaller in height, fresh weight, and dry weight than the control plants. Treatment 6.5 P-10% plants were the only plants that received adequate phosphate throughout the three crops. These plants ended similar in height and fresh weight but higher in dry weight than the control plants.

The minimum critical bulk solution phosphate-P concentration for tomato in peat moss-perlite medium can be estimated from data in Table 2. Average bulk solution phosphate-P concentrations during the periods of P deficiency for the 0 P-10%, 2.2 P-5%, 2.2 P-10%, and 6.5 P-5% treatments

were 0.22, 0.20, 0.20, and 0.20 mg·dm^{-3}, respectively. Bulk solution phosphate-P concentrations throughout these periods ranged from 0.13 to 0.34 mg·dm^{-3}, an indication that this range is in the deficient zone. Nishimoto et al. [5] placed the minimum critical concentration of P in bulk solution to attain 95% of maximum growth of field grown *Chrysanthemum* at 0.18 mg·dm^{-3}. Their soil type was a Typic Eutrandept Kula which has a high phosphate fixation capacity. The soil was charged with superphosphate in accordance with its phosphate adsorption isotherm to achieve various initial bulk solution equilibrium concentrations. The higher required phosphate level in the soilless medium of our study may be due to the coarser texture with its lower unsaturated hydraulic conductivity and thus more tortuous water pathway for phosphate diffusion to roots.

Because phosphate-adsorbing minerals in soils serve as a reservoir for phosphate-P in solution, dissolved phosphate-P concentrations in root media were mainly controlled by the phosphate concentration at which the soil material

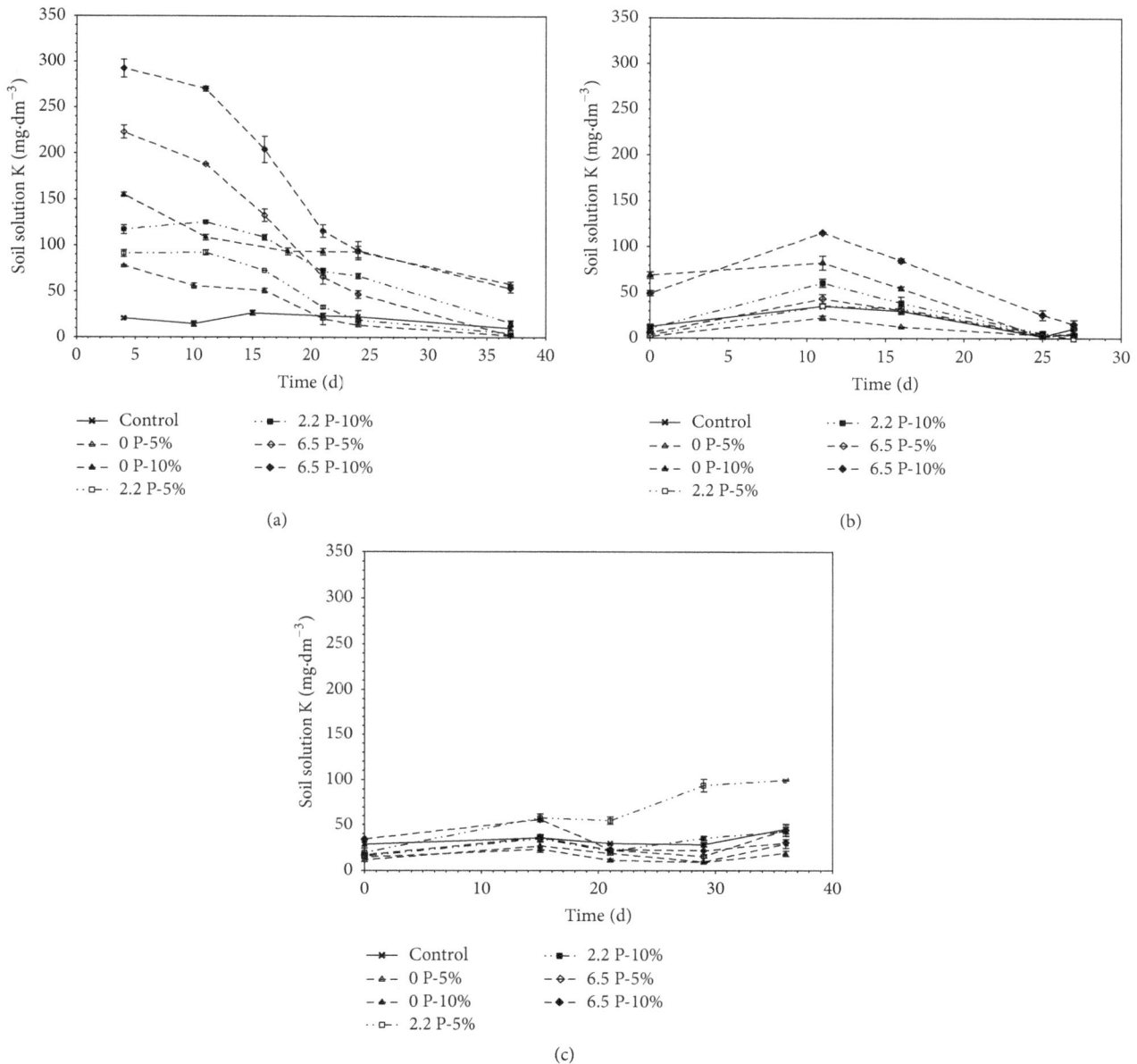

FIGURE 2: Potassium concentration in bulk solutions of root media amended with no soil material (control) and six soil material treatment combinations containing 5 or 10% soil material (v : v) charged with 0, 2.2, or 6.5 mg·g^{-1} phosphate-P during the (a) first, (b) second, and (c) third of three successive tomato seedling crops. Bars indicate ± SE (n = 3). Lines connecting data points were included to aid in visualizing trends in the data.

was charged. This was seen in the large difference in bulk solution phosphate-P concentrations between the pair of 2.2 P treatments and the pair of 6.5 P treatments and the relatively smaller differences within each pair of treatments (Table 2). The increased longevity of phosphate release was strongly under the control of both the quantity of phosphate-charged soil material added to the root medium and the phosphate concentration at which the soil material was charged.

Increased growth in some treatments with phosphate-charged soil material compared to the control plants may have been due in part to the increased supply of K from the phosphate-charged soil. The shoot concentrations of K were higher in all treatments compared to the control in crop 1 and higher in all treatments except 0 P-5% in crop 2 (Table 3). There were no shoot K concentrations higher than in the control in crop 3. This same pattern is seen in root medium dissolved K concentration (Figure 2). All soil material treatments in the first crop resulted in higher root medium K concentrations than the concentration in the control. The curves for the second crop illustrate the contribution of K from soil material becoming depleted. Soil material contributed K whether it was charged with KH_2PO_4

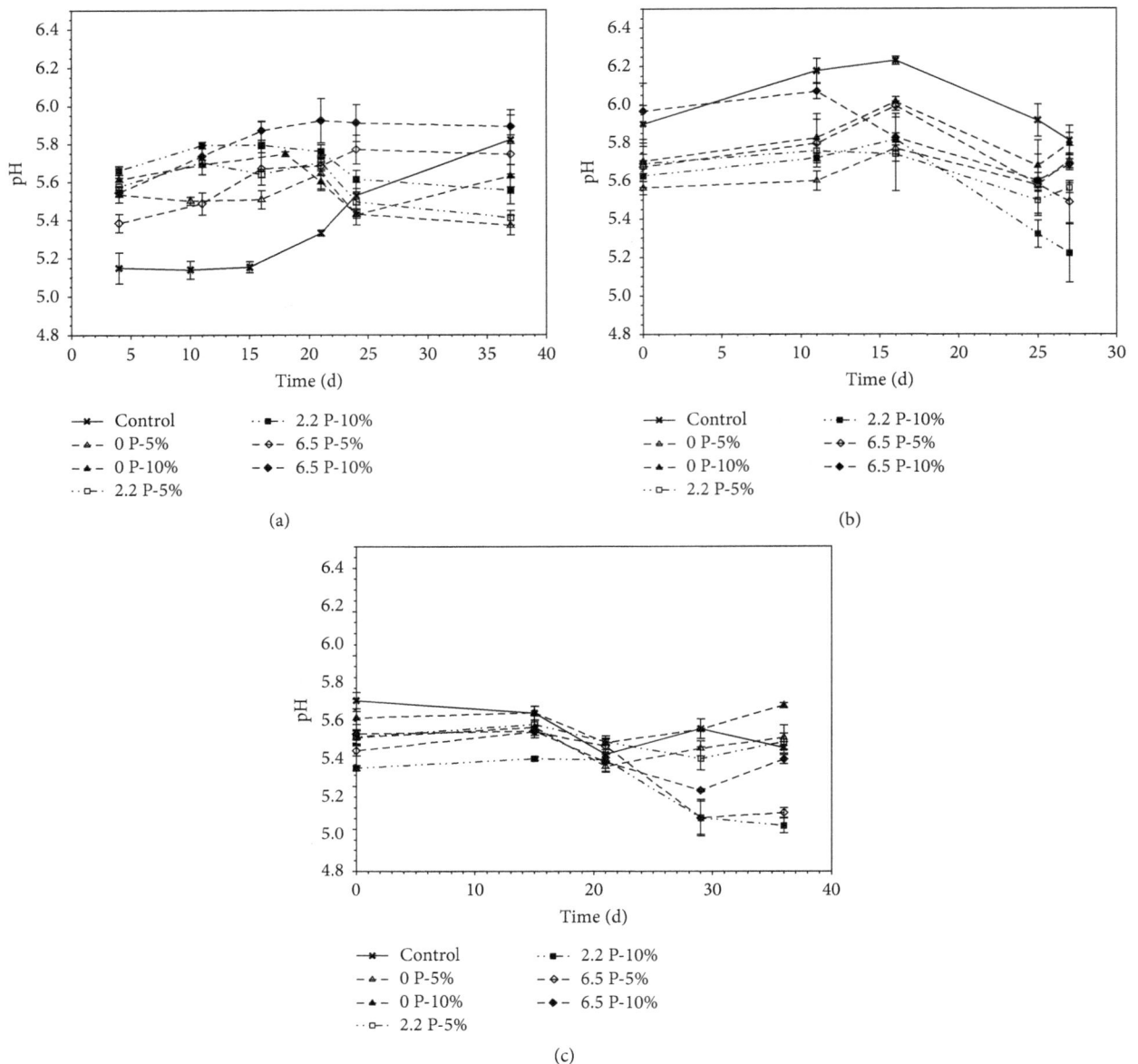

FIGURE 3: The pH of root media amended with no soil (control) and six soil treatment combinations containing 5 or 10% soil material (v:v) charged with 0, 2.2, or 6.5 mg·g^{-1} phosphate-P during the (a) first, (b) second, and (c) third of three successive tomato seedling crops. Bars indicate $\pm$ SE (n = 3). Lines connecting data points were included to aid in visualizing trends in the data.

or not. This can be attributed to K sorbed on the soil material during washing with KCl and in the case of soils charged with phosphate, from the additional equilibration with KH_2PO_4. However, the possible growth stimulatory effect of K would have been weak because no stimulation was recorded in half of the treatments where tissue levels of K were elevated.

Treatments with soil material (including 0 P treatments) generally showed less fluctuation in root medium pH than the control during growth of the three crops, especially during crops 1 and 2 (Figure 3). Stabilization of root medium pH would be a valuable property for any phosphate-charged soil material intended for incorporation in soilless media for commercial crop production.

4. Conclusions

Adding soil material that has not been charged with phosphate to a soilless root medium receiving phosphate from a continuous postplant liquid fertilization program can result in P deficiency due to adsorption of added P by soil solids. Phosphate-charged soil material resulted in low, relatively stable bulk solution phosphate-P concentrations. Phosphate-P concentration was most heavily affected by the level of phosphate charging on the soil material while bulk solution phosphate-P longevity depended extensively on both the level of phosphate charging and the amount of charged soil material. These parameters could be varied to optimize the amount of charging and mass to mix with soilless media, given that the soil may deteriorate physical properties of the media such as hydraulic conductivity and aeration. In our study, the only treatment that sustained an adequate bulk solution concentration of phosphate-P for the 14 weeks of testing was 6.5 P-10%, but concentrations during crop 1 were too high from the standpoint of phosphate leaching. The objective of the study, to provide a low, adequate phosphate-P concentration (above $0.34 \, \text{mg·dm}^{-3}$), was best met in the 2.2 P-10% treatment where bulk solution phosphate-P ranged from an initial concentration of $2.29 \, \text{mg·dm}^{-3}$ to a concentration of $0.49 \, \text{mg·dm}^{-3}$ 48 d later. Initially high phosphate-P concentrations in bulk solution might be reduced by using a natural soil material with the higher phosphate sorption capacity of allophane and equilibrating it with a lower solution phosphate concentration. This would extend the length of time within a desirable bulk solution phosphate concentration and lower phosphate leaching.

Disclosure

The use of trade names in this publication does not imply endorsement by the North Carolina Agricultural Research Service of the products named nor criticism of similar ones not mentioned.

Competing Interests

The authors declare that they have no competing interests.

Acknowledgments

Appreciation is expressed to Sun Gro Horticulture, Bellevue, WA, and the North Carolina Agricultural Research Service (NCARS), Raleigh, NC, for support.

References

[1] D. J. Marconi and P. V. Nelson, "Leaching of applied phosphorus in container media," *Scientia Horticulturae*, vol. 22, no. 3, pp. 275–285, 1984.

[2] C. S. M. Ku and D. R. Hershey, "Growth response, nutrient leaching, and mass balance for potted poinsettia. II. Phosphorus," *Journal of the American Society for Horticultural Science*, vol. 122, no. 3, pp. 459–464, 1997.

[3] K. A. Williams and P. V. Nelson, "Low, controlled nutrient availability provided by organic waste materials for chrysanthemum," *Journal of the American Society for Horticultural Science*, vol. 117, no. 3, pp. 422–429, 1992.

[4] Southern Nursery Association, *Best Management Practices: Guide for Producing Nursery Crops*, Southern Nursery Association, Acworth, Ga, USA, 3rd edition, 2013.

[5] R. K. Nishimoto, R. L. Fox, and P. E. Parvin, "External and internal phosphate requirements of field grown chrysanthemums," *HortScience*, vol. 10, no. 3, pp. 279–280, 1975.

[6] R. S. Beckwith, "Sorbed phosphate at standard supernatant concentration as an estimate of the phosphate needs of soils," *Australian Journal of Experimental Agriculture and Animal Husbandry*, vol. 5, no. 16, pp. 52–58, 1965.

[7] T. H. Yeager and J. E. Barrett, "Phosphorus and sulfur leaching from an incubated superphosphate-amended soilless container medium," *HortScience*, vol. 20, no. 4, pp. 671–672, 1985.

[8] T. H. Yeager and J. E. Barrett, "Influence of an anion exchange resin on phosphorus and sulfur leaching from a soilless container medium," *HortScience*, vol. 21, no. 1, article 152, 1986.

[9] R. R. Coltman, G. C. Gerloff, and W. H. Gabelman, "A sand culture system for simulating plant responses to phosphorus in soil," *Journal of the American Society for Horticultural Science*, vol. 107, pp. 938–942, 1982.

[10] G. C. Elliott, "Evaluation of sand-alumina-P media for studies of P nutrition," *Journal of Plant Nutrition*, vol. 12, no. 3, pp. 265–278, 1989.

[11] J. Lynch, E. Epstein, A. Lauchli, and G. I. Weigt, "An automated greenhouse sand culture system suitable for studies of P nutrition," *Plant, Cell & Environment*, vol. 13, no. 6, pp. 547–554, 1990.

[12] Y.-L. P. Lin, E. J. Holcomb, and J. P. Lynch, "Marigold growth and phosphorus leaching in a soilless medium amended with phosphorus-charged alumina," *HortScience*, vol. 31, no. 1, pp. 94–98, 1996.

[13] K. A. Williams, *Increasing phosphate and potassium retention of soilless container media through the use of clay, aluminum, and zeolite amendments [Ph.D. thesis]*, North Carolina State University, Raleigh, NC, USA, 1995.

[14] M. B. McBride, "Surface chemistry of soil minerals," in *Minerals in Soil Environments*, J. B. Dixon and S. B. Weed, Eds., pp. 35–88, Soil Science Society of America, Madison, Wis, USA, 2nd edition, 1989.

[15] G. E. Brown Jr. and N. C. Strurchio, "An overview of synchrotron radiation applications to low temperature geochemistry and environmental science," *Reviews in Mineralogy and Geochemistry*, vol. 49, no. 1, pp. 1–115, 2002.

[16] Y.-M. Oh, D. L. Hesterberg, and P. V. Nelson, "Comparison of phosphate adsorption on clay minerals for soilless root media," *Communications in Soil Science and Plant Analysis*, vol. 30, no. 5-6, pp. 747–756, 1999.

[17] Y.-M. Oh, *Development of a pre-plant source of phosphorus for soilless root media [Ph.D. thesis]*, North Carolina State University, Raleigh, NC, USA, 2000.

[18] R. J. Haynes and R. S. Swift, "Effects of air-drying on the adsorption and desorption of phosphate and levels of extractable phosphate in a group of acid soils, New Zealand," *Geoderma*, vol. 35, no. 2, pp. 145–157, 1985.

[19] G. W. Thomas, "Soil pH and soil acidity," in *Methods of Soil Analysis Part 3*, D. L. Sparks, Ed., pp. 475–490, American Society of Agronomy and Soil Science Society of America, Madison, Wis, USA, 1996.

[20] D. W. Nelson and L. E. Sommers, "Total carbon, organic carbon, and organic matter," in *Methods of Soil Analysis Part 3*, D. L. Sparks, Ed., pp. 961–1010, American Society of Agronomy and Soil Science Society of America, Madison, Wis, USA, 1996.

[21] M. L. Jackson, C. H. Lim, and L. W. Zelazny, "Oxides, hydroxides, and aluminosilicates," in *Methods of Soil Analysis Part 1*, A. Klute, Ed., pp. 101–150, American Society of Agronomy and Soil Science Society of America, Madison, Wis, USA, 2nd edition, 1986.

[22] R. A. Dahlgren, "Quantification of allophane and imogolite," in *Quantitative Methods in Soil Mineralogy*, J. E. Amonette and J. W. Stucici, Eds., pp. 430–451, Soil Science Society of America, Madison, Wis, USA, 1994.

[23] S. Kuo, "Phosphorus," in *Methods of Soil Analysis Part 3*, D. L. Sparks, Ed., pp. 869–919, American Society of Agronomy and Soil Science Society of America, Madison, Wis, USA, 1996.

[24] G. Sposito, *The Chemistry of Soils*, Oxford University Press, New York, NY, USA, 1989.

[25] J. Murphy and J. P. Riley, "A modified single solution method for the determination of phosphate in natural waters," *Analytica Chimica Acta*, vol. 27, pp. 31–36, 1962.

[26] P. V. Nelson, C.-Y. Song, J. Huang, C. E. Niedziela Jr., and W. H. Swallow, "Relative effects of fertilizer nitrogen form and phosphate level on control of bedding plant seedling growth," *HortScience*, vol. 47, no. 2, pp. 249–253, 2012.

[27] J. B. Jones Jr., *Tomato Plant Culture*, CRC Press, Boca Raton, Fla, USA, 1999.

21

Environmentally Smart Nitrogen Performance in Northern Great Plains' Spring Wheat Production Systems

Olga S. Walsh and Kefyalew Girma

Southwest Research & Extension Center, University of Idaho, 29603 U of I Lane, Parma, ID 83660-6699, USA

Correspondence should be addressed to Olga S. Walsh; owalsh@uidaho.edu

Academic Editor: Glaciela Kaschuk

Experiments were conducted in Montana to evaluate Environmentally Smart Nitrogen (ESN) as a nitrogen (N) source in wheat. Plots were arranged in a split-plot design with ESN, urea, and a 50%-50% urea-ESN blend at low, medium, and high at-seeding N rates in the subplot, with four replications. Measurements included grain yield (GY), protein (GP), and N uptake (GNU). A partial budget economic analysis was performed to assess the net benefits of the three sources. Average GY varied from 1816 to 5583 kg ha^{-1} and grain protein (GP) content ranged from 9.1 to 17.3% among site-years. Urea, ESN, and the blend resulted in higher GYs at 3, 2, and 2 site-years out of 8 evaluated site-years, respectively. Topdressing N improved GY for all sources. No trend in GP associated with N source was observed. With GP-adjusted revenue, farmer would not recover investment costs from ESN or blend compared with urea. With ESN costing consistently more than urea per unit of N, we recommend urea as N source for spring wheat in Northern Great Plains.

1. Introduction

To address the nutritional needs of the fast-growing human population, the annual cereal production will have to increase to about 3 billion tones by 2050 from currently produced 2.1 billion tones [1]. It was shown that 80 percent of the required crop production increases will have to come from improved GY and crop intensification versus only 20 percent from expansion of agricultural land [1, 2]. Furthermore, the concern is that the rate of growth in cereal GY has been steadily declining worldwide.

The reports indicate that the rate of growth of major cereals has dropped from 3.2 percent per year in the 1960s to only 1.5 percent in 2000 [1, 2]. Although the genetics and biotech industries estimated their ability to increase cereal GYs due to genetic GY potential of 3 to 4% per year [3], history indicates that the genetic advances alone may not be sufficient to solve the world's food shortage issues [4]. At least 50% of food currently produced in the world is possible due to use of commercial N, phosphorus (P), and potassium (K) fertilizer [5]. Roberts [4] pointed out the critical role the fertilizers play in the world's food security and underlined the close correlation between global cereal crop production and fertilizer consumption. Consequently, the fertilizer use is estimated to increase dramatically worldwide at an expected rate of 2.5 million metric tons per year [6].

Nitrogen use efficiency (NUE), the ratio of the difference of N uptake in the N treated plot and N uptake in the check and the total applied N rate, is only at 40 to 50% for most crop production systems [7]. A considerable improvement from the previously estimated 33% NUE in the late 1990s [8] is mainly due to advances in precision agriculture and novel fertilizer technologies [9]. Developing an effective and efficient N management strategy and improving N guidelines are the key challenges that must be addressed to maintain and enhance the sustainability of cereal crop production. Identifying practical solutions for reducing agricultural input costs such as fertilizer is critical for maintaining soil fertility and improving productivity for crop growers [10].

The Northern Great Plains is the key region for production of top quality spring wheat grain. Spring wheat continues to be one of the major cereal crops for Montana and North Dakota, which account approximately for 20 and 44%, respectively, of all spring wheat produced in the United States

[11]. Not considering water, inadequate N nutrition is considered to be the most GY-limiting factor in wheat production [12].

The awareness and interest to enhanced-efficiency fertilizer technologies among the Northern Great Plains producers are apparent. This trend is mainly due to steady efforts to improve the efficiency of fertilizer use and to minimize the negative impact of intensive agricultural production on the environment. The key to both increasing NUE and reducing N losses is the synchronization of N availability with the crop requirements [10], which has been the basis of many enhanced-efficiency fertilizers [13]. Shaviv [14] has proposed that enhanced-efficiency fertilizers could be effectively used to improve synchrony between soil N availability and crop uptake requirements for N. The enhanced-efficiency fertilizer products can reduce nutrient losses to the environment and deliver improved availability of nutrients to the crop [15].

Environmentally Smart Nitrogen (ESN), a common enhanced-efficiency fertilizer, is produced by coating urea granules with a polymer shell that allows for a slow-release of N to the soil. Literature review points to numerous studies showing that higher GYs and better crop quality could be achieved with ESN compared to conventional urea providing crop producers with a higher return on their fertilizer investment [16, 17]. Furthermore, due to the slow-release technology, the increase in NUE is achieved by reduction of N losses through denitrification, leaching, and volatilization. In the recent years, the interest in use of the enhanced-efficiency fertilizers has increased due to their potential to reduce negative environmental impact [18, 19] observed improved GY and significantly lower N losses with sidedress ESN application to corn.

Recently, enhanced-efficiency fertilizers have been defined as "fertilizer products with characteristics that allow increased plant uptake and reduce the potential of nutrient losses to the environment (e.g., gaseous losses, leaching, or runoff) when compared to an appropriate reference product" by the Association of American Plant Food Control Officials [20]. Numerous reports showed that enhanced-efficiency fertilizer products can decrease the nitrous oxide (N_2O) emission rate from soils, especially immediately after fertilizer application [18].

The Environmental Protection Agency (EPA) estimated that, in 2012, N_2O accounted for about 6% of all US greenhouse gas emissions associated with human activities; agricultural activities account for approximately 75% of all N_2O emissions [21]. Halvorson et al. [22] observed a 42% reduction in N_2O emissions compared to traditional urea (46-0-0). Hatfield and Parkin [23] suggested that the increase in GY could be due to prolonged period of greenness in crops, especially during the grain-filling period. Furthermore, the use of enhanced-efficiency fertilizers such as ESN resulted in improved NUE, demonstrating positive environmental and agronomic advantages. Because the ESN was formulated to minimize N loss, ESN application is considered an environmentally sound farming practice.

In the USA the growers using ESN are eligible for payments through the Environmental Quality Incentives Program (EQIP), which helps to offset the higher cost of ESN compared to traditional uncoated urea. The ESN also may be beneficial for producers in complying with tighter environmental regulations, shorter crop seeding windows, and adverse weather conditions [24].

Some consider the ESN as a controlled release N fertilizer [25, 26] that supplies the crop with N all throughout the growing season [13]. Moreover, the manufacturer affirms that ESN offers a predictable release of N due to the coating characteristics. The polymer coating represents a semipermeable membrane that allows water to diffuse into the granule and dissolve N; encapsulated N remains inside and is released over time at a controlled rate [25].

The N release rate is controlled by soil temperature; N release rate increases as the soils temperatures rise in the spring. This methodology aims to match crop need for N by "spoon-feeding" it with gradual N release from the ESN capsules. However, others tend to classify ESN simply as a slow-release product [27, 28] due to uncertainty in the level of control provided by the polymer coating. In fact, as water diffuses through a flexible, microthin polymer coating, the liquefied N is allowed to disperse out through the membrane and into the soil [29].

The common variations in soil characteristics and temperatures must be also considered. Soil and air temperatures, as well as crop development rate, do not increase linearly throughout the growing season. In most cropping systems, warmer/colder periods are likely; the speed of crop growth is known to vary considerably depending on the developmental growth stage. For instance, stem elongation (Zadoks 30–39) is the most rapid stage of wheat crop growth, resulting in much faster vegetative tissue growth and biomass production compared to other growth stages [30].

Studies conducted in a variety of crops have indicated that weather conditions, soil characteristics, application time, method, and crop rotations are the major factors influencing the effectiveness of the enhanced-efficiency [31–34]. Studies have shown that as all N has been dissolved within the ESN casing and the soil temperatures reach $20°C$, between 65 and 90% of N is released into the soil within a 30-day period [35]. Also, a comprehensive assessment of N release from the ESN granules revealed that the rate of N release is highly dependent on the soil type. A more rapid release of N was observed in clay soils compared with silt and sandy loam soils [35].

To meet a producers' GY and quality targets for specific crops and growing conditions, ESN can be used as a sole N source or as a blend with other fertilizers such as traditional urea. Typically, ESN blended with other soluble N products performs best as spring preplant applications in spring wheat. Research in potato also has shown that a one-time application of ESN can be as cost effective as multiple in-season applications of traditional N products with fertigation [36]. Better or similar potato GYs were observed with a single application of ESN (even when 25% less N was applied), compared to those obtained with the split urea application [37].

The ESN recommendations for a wide variety of field crops such as corn, canola, cotton, and wheat have been developed by the manufacturer. The manufacturer's guidelines for ESN use in spring wheat in the Great Plains area are application of N as 100% ESN in the fall prior to seeding.

TABLE 1: Initial soil chemical properties (0–30 cm), growing conditions, and summary of field activities at Conrad, Corvallis, and Kalispell, MT, in 2011–2013.

Field activity	2011	2012	2013
		Conrad	
Herbicide	GoldSky	GoldSky, Huskie	GoldSky, Huskie
Herbicide date	May 31	May 16	May 23
Sensing date	June 15	May 18	May 13
Topdress N date	June 15	May 21	May 14
Harvest date	August 30	August 16	September 10
Soil test N, kg ha^{-1}	22	56	51
Soil P, ppm (Olson)	18	12	32
Soil K, ppm	221	152	234
Organic matter, %	1.5	1.2	1.8
		Corvallis	
Herbicide	Bronate, Axial XL	Bronate, Axial XL	Supremacy, Axial XL
Herbicide date	June 13	May 28	May 28
Sensing date	June 15	June 8	June 5
Topdress N date	June 15	June 8	June 5
Harvest date	September 8	August 16	August 19
Soil test N, kg N ha^{-1}	37.5	44.3	51.6
Soil P, ppm	23	23	25
Soil K, ppm	423	423	398
Organic matter, %	2.9	2.9	2.2
		Kalispell	
Herbicide	Bronate, Axial XL	Bronate, Axial XL	—
Herbicide date	June 13	May 28	—
Sensing date	June 15	June 8	—
Topdress N date	June 15	June 8	—
Harvest date	September 8	August 16	—
Soil test N, kg N ha^{-1}	37.5	44.3	—
Soil P, ppm	23	23	—
Soil K, ppm	423	425	—
Organic matter, %	2.8	2.9	—

Alternatively, spring wheat can be fertilized with a blend (40–75% ESN + 25–60% urea) [25]. This paper summarizes the results from a three-year-long study that compared the effect of ESN and urea-ESN blend with urea, for spring wheat production in the Northern Great Plains. The specific objectives of the study were to evaluate urea, ESN, and ESN-urea blend along at-seeding and topdress N rates on spring wheat production, quality, and economic return in Montana.

2. Materials and Methods

2.1. Experimental Sites and Treatment Structure. Three field studies were initiated in the spring of 2011 and completed in the 2013 growing season (eight site-years). Field trials were conducted at one irrigated site, at Western Agricultural Research Center (Corvallis, 46°19′39.8532″, −114°05′08.6496″) and two dryland sites at the North Western Agricultural Research Center (Kalispell, 48°11′14.4312″,

−114°08′04.7436″) and at the Western Triangle Agricultural Research Center (Conrad, 48°19′11.6868″, −111°55′28.7256″).

The soils at Conrad, Corvallis, and Kalispell are Scobey clay loam (fine, smectitic, frigid Aridic Argiustolls), burnt fork loam (coarse-loamy, mixed, superactive, frigid Typic Haplustolls), and Creston silt loam (fine-silty, mixed, superactive Typic Haploborolls), respectively. Initial soil properties, growing conditions, and field activities are presented in Table 1.

Plots were arranged in a split-plot design with three N sources and at-seeding N rates in the main plot and topdress N in the subplot. An additional control (0 kg N ha^{-1}) treatment was included outside the factorial in the main plot. Each treatment was replicated 4 times at each site-year. The three N sources were urea, ESN, and a 50 : 50 blend of urea and ESN. The at-seeding N fertilizer rates were 56, 112, and 168 kg N ha^{-1} at Conrad and 112, 224, and 336 kg N ha^{-1} at Corvallis and Kalispell, representing low, medium, and high N rates, respectively, for each site. The at-seeding N rates were established based on the respective area's GY goal and current Montana State University's N fertilization guidelines for spring wheat. The topdress N was applied at 0 or 45 kg N ha^{-1} as granular broadcasted urea at Zadoks 30 (late tillering/beginning of stem elongation) growth stage.

Hard red spring wheat (cv. Choteau) was seeded into plots measuring 1.5 by 7.6 m at the seeding rate of 1,830,000 plants ha^{-1} for dryland sites and from 2,150,000 to 3,225,000 plants per ha^{-1} at the irrigated site, using the custom-made direct-seed drill with Conserva Pak™ openers manufactured by Swift Machining (Washougal, WA). At-seeding N fertilizer was sidebanded about 1.90 cm to the side and about 2.54 cm above the seed.

2.2. Measurements. At maturity, spring wheat was harvested with Hege 125 plot combine in 2011 and 2012 and Wintersteiger classic plot combine in 2013. The harvested grain was dried in the drying room for 14 days at the temperature of 35°C; then, the by-plot GY was determined. From each by-plot sample, 400 g subsample was drawn and analyzed by the Agvise Laboratories (Northwood, ND) for total grain N content using near infrared reflectance spectroscopy (NIR) with a Perten DA 7250 NIR analyzer (Perten Instruments, Inc., Springfield, IL).

From the GY and lab analysis results, grain N uptake (GNU) was determined using the following relationship:

$$\text{GNU} = \text{GY} * \text{N content of grain.} \qquad (1)$$

Furthermore, GP of spring wheat was calculated by multiplying the N content of grain by 6.5.

2.3. Statistical Analysis. All data were subjected to analysis of variance (ANOVA) using GLIMMIX procedure in SAS (SAS version 9.4, Cary, NC, USA). Before testing hypotheses, assumptions of normality and homogeneity of variance were checked for all measured variables using the UNIVARIATE procedure and Levene's Homogeneity of Variance Test, respectively. Very few outliers were identified and removed from the data.

After preliminary statistical analysis showed a significant site and year effects, data were analyzed for each site-year. In this three-factor study, the focus was on the two-way interaction effects (plus selected three-way interactions) of sources with at-seeding and topdress N rate treatment levels rather than the main effects. Whenever interactions were significant, tests of simple effects were generated and further dissected using the "slice" option under the least square (LS) means statement [38] in GLIMMIX for each factor level, at fixed level of the other factor. The test produces F value by extracting the appropriate rows from the coefficient matrix for each two-way interaction simple effect LS means [38, 39]. Then the "lines" statement under LS means statement was used to generate letter based pairwise LS means differences at $p < 0.05$ if not specified for each response variable.

Additionally, orthogonal polynomial contrasts were used to assess trends (linear and quadratic) in response variables against increasing rates of at-seeding N rate main effects and

their interaction effect with source or topdress N levels when significant at least at $p < 0.1$. Pearson product-moment correlation [40] analysis was used to assess the relationship of GY and GP. Pearson's correlation was performed for each site-year. If no significant interaction effects were observed, then main effects were discussed.

2.4. Partial Budget Economic Analysis. A partial budget economic analysis [41] was performed to assess the net benefits of the three sources. The marginal benefit and rate of return [42] of the ESN and blend relative to N from urea were also evaluated, using the optimum N rates for each significant site-year (168 kg N ha^{-1} at Conrad and 224 kg N ha^{-1} at Corvallis and Kalispell, Table 2). An additional partial budget analysis was performed to compare ESN at medium at-seeding N rate with urea at low at-seeding N rate along topdress N rate. For this analysis the medium at-seeding rates were 112 and 224 kg N ha^{-1} at Conrad and Kalispell, respectively. The low at-seeding rates were 56 and 112 kg N ha^{-1} at Conrad and Kalispell, respectively, along 45 kg ha^{-1} topdress N. The calculation assumed spring wheat grain GY price unadjusted and adjusted for GP premium and discount. Premium and discount values were obtained from West-Con Cooperative [43]. Price of wheat was set at $0.17 kg^{-1} [44]. Current price of N in urea and ESN was set at $1.31 and 1.71 kg^{-1}, respectively [45]. The cost of mixing urea and ESN blend was estimated at $3 ha^{-1}. The cost of topdress application of N was set to be $15.2 ha^{-1}. Partial budget equations were defined as follows [42]:

$$\text{NR} = \text{TR} - \text{TVC},$$
$$\Delta\text{NR} = \Delta\text{TR} - \Delta\text{TVC}, \qquad (2)$$
$$\text{MRR} = \frac{\Delta\text{TR}}{\Delta\text{TVC}},$$

where NR is net return, TR is total revenue, TVC is total variable cost, and ΔNR, ΔTR, and ΔTVC represent the change in net return, change in total revenue, and change in total variable cost due to ESN and ESN-urea blend, respectively, and RR is the rate of return that measures the dollar recovery to a dollar investment on ESN or blend.

3. Results

3.1. Grain Yield. In 2011, the lowest GY was observed at Conrad (2488 kg ha^{-1}), followed by Corvallis (2623 to 3161 kg ha^{-1}). Grain yield at Kalispell was the highest among the three experimental sites, ranging from 2898 to 4237 kg ha^{-1}. In 2012, GY was the highest among the three growing seasons for all experimental locations. Conrad was the highest producing site with GY ranging from 4035 to 5583 kg ha^{-1}. The GY for Kalispell and Corvallis was comparable but slightly higher at Corvallis. In 2013, GY was higher at Conrad (3699 to 4640 kg ha^{-1}) than Corvallis (1816 to 2623 kg ha^{-1}).

Nitrogen source affected GY significantly at Conrad in 2012 and Corvallis and Kalispell in 2011 (Table 2). At Conrad, ESN surpassed urea by 405 kg ha^{-1}. At Corvallis,

TABLE 2: Effect of N source and preplant and topdress N rate on spring wheat grain yield at Conrad, Corvallis, and Kalispell, MT, in 2011–2013 growing seasons.

Effect	Conrad			Corvallis			Kalispell	
	2011	2012	2013	2011	2012	2013	2011	2012
N source				kg ha^{-1}				
Urea	2488	5111$^{b£}$	4506	3161^a	4640	2623	3699^b	4506
ESN	2556	5515^a	4439	2690^b	4842	2488	3901^{ab}	4304
Blend	—	5447^{ab}	4708	2757^b	4640	2152	4035^a	4371
Prob > F	0.969	0.051	0.117	<0.001	0.699	0.452	0.003	0.517
At-seeding N rate, kg ha^{-1}								
0	2488	4035	3699	2623	3833	1816	2898	3833
Low†	2556	5178	4640	2757	4304	2556	4237	4371
Medium	2892	5447	4573	2959	5044	2556	4237	4371
High	3161	5582	4439	3026	4640	2152	4102	4371
Prob > F (trend)	0.006‡	0.003‡	0.055	0.048‡	0.050	0.051	0.114	0.945
Topdress N rate, kg ha^{-1}								
0	2389^{b§}	5380	4573	2690^b	4506^b	2354	3699^b	4371
45	2741^a	5447	4506	3026^a	4977^a	2488	4035^a	4439
Prob > F	<0.001	0.749	0.410	<0.001	<0.001	0.813	0.003	0.175

$^£$Down a column for each site-year source means followed by different letter are significant at $p < 0.05$.
†At-seeding N rates: low, medium, and high were 56, 112, and 168 kg ha^{-1} for Conrad and 112, 224, and 336 kg ha^{-1} for Corvallis and Kalispell.
‡ represents significant linear trend, for at-seeding N rate for each site-year.
§Down a column for topdress N rates, LS means followed by different letters are significant at $p < 0.05$ for each source.

urea resulted in 16% more GY than ESN and blend combined. In 2011, at Kalispell, the blend yielded 336 kg ha^{-1} more compared to urea alone (Table 2). Overall, urea-ESN blend resulted in a significantly higher GY at Kalispell.

At-seeding N rate significantly affected wheat GY at Conrad in 2011 and 2012 and at Corvallis in 2012 (Table 2). Wheat GY increased linearly from 2488 to 3161 kg ha^{-1} in 2011 and from 4035 kg ha^{-1} to 5582 kg ha^{-1} in 2012 at Conrad when at-seeding N rate increased from 0 to 168 kg ha^{-1}. Similarly, a linear increase from 2623 to 3026 kg ha^{-1} was observed in 2011 at Corvallis when at-seeding N rate increased from 0 to 336 kg ha^{-1} (Table 2). A quadratic trend was observed at Corvallis in 2012 and 2013 where GY reached peak at the medium N rate (224 kg ha^{-1}).

The topdress N treatment significantly affected GY in four of the eight site-years. At Conrad in 2011, Corvallis in 2011 and 2012, and Kalispell 2011, topdressing 45 kg N ha^{-1} resulted in 14.7, 12.5, 10.5, and 9.1% more GY than those plots that did not receive topdress N (Table 2).

The assessment of interaction effects of source by at-seeding N rate on GY showed significant ($p < 0.07$) effect at Conrad in 2013 and Kalispell in 2011. At Conrad, for 168 kg ha^{-1} at-seeding N rate, the blend source had 621 kg ha^{-1} more GY compared with urea and ESN combined. At this site-year, for urea, GY was linearly increased with increase in at-seeding N rate (Table 3(a)). At Kalispell, for 336 kg ha^{-1} at-seeding N rate, ESN had 563 kg ha^{-1} more GY than urea. Consistent with Conrad 2013, at Kalispell in 2011, GY decreased linearly with increase in at-seeding N rate from 112 kg ha^{-1} to 336 kg ha^{-1}. The source by topdress interaction showed significant differences at Conrad in 2011 ($p < 0.016$)

and 2013 ($p < 0.025$). In both years at this site, for each source, topdressing N resulted in more GY than not topdressing although statistically significant LS means difference was observed only for ESN source in 2011 and urea in 2013 (Table 3(a)). At Conrad in 2013, at the 45 kg ha^{-1} topdress N rate, the blend had 448 kg ha^{-1} more GY than urea.

Results of assessment of three-way interaction effects ($p < 0.15$) of source and at-seeding and topdress N rates on GY are presented in Table 3(b). At Conrad the maximum GY was obtained with the highest rate (168 kg ha^{-1}) at-seeding N rate combined with 45 kg ha^{-1} topdress N for the blend and ESN sources (Table 3(b)). For this site-year, the maximum yield for urea source was achieved with 120 kg ha^{-1} at-seeding N rate without topdress N. In 2013, at Conrad, the lowest at-seeding N rate combined with 45 kg ha^{-1} topdress N rate maximized yield for the ESN and urea. In contrast, GY was maximized with the application of 168 N ha^{-1} at-seeding N rate for the blend source although the values were not statistically different from other rates but the lowest at-seeding rate without topdress. At Kalispell in 2011, similar trend was observed for the ESN and urea where GY was maximized with the low (112 kg ha^{-1}) at-seeding rate plus 45 kg N ha^{-1} topdress rate. Furthermore, at Kalispell in 2011, GY was maximized with the application of 224 kg ha^{-1} at-seeding N rate and 45 kg ha^{-1} topdress N rate combination. But the value was not statistically different from the rest of the rates, except for the low rate without topdress N. Results suggest that low to medium at-seeding N rate along topdres N rate improved GY. In Montana, despite the short spring wheat growing season, yields did increase in this study as a result of topdressing. Splitting N improved yield but this may not translate to positive marginal

TABLE 3: (a) Simple effects of two-way interactions of source by at-seeding and topdress N rates for grain yield in selected site-years where interaction effect was significant at least at $p < 0.1$. (b) Simple effects of three-way interactions of source and at-seeding and topdress N rates on grain yield in selected site-years where interaction effect was significant at least at $p < 0.15$.

(a)

Site-year	At-seeding N rate, kg ha^{-1}	Blend	ESN	Urea
		kg ha^{-1}		
Conrad-2013	56	4536	4578	4855
	112	4695	4570	4452
	168	4838^{A†}	4200^{B}	4234^{B}
	Prob > F (trend)	0.156	0.754	0.019‡
Kalispell-2011	112	3656	4013	3785
	224	3878	3875	3603
	336	3749^{AB}	3905^{A}	3342^{B}
	Prob > F (trend)	0.745	0.689	0.05‡
	Topdress N rate, kg ha^{-1}			
Conrad-2011	0	2445	2255^{b§}	2470
	45	2668	2943^{a}	2613
Conrad-2013	0	4648	4379	4743^{a}
	45	4732^{A}	4519^{AB}	4284^{Bb}

†Across rows for each site-year and at-seeding N rate, LS means followed by different letters are significant at $p < 0.05$.
‡Significant linear trend for at-seeding N rate for each source.
§Down a column for topdress N rates, LS means followed by different letters are significant at $p < 0.05$ for each source.

(b)

At-seeding–topdress N rate, kg ha^{-1}	Blend	ESN	Urea	Blend	ESN	Urea
	Conrad-2011			Conrad-2013		
56–0	2523^{bc§}	1812^{c}	2027^{b}	4385^{b}	4301^{ab}	4755^{a}
56–45	2202^{c}	2268^{c}	2298^{b}	4688^{ab}	4855^{a}	4956^{a}
112–0	2008^{cB†}	2480^{bcB}	2932^{aA}	4620^{ab}	4486^{ab}	4805^{a}
112–45	2454^{bc}	2886^{b}	2885^{a}	4771^{abA}	4654^{abA}	4099^{bB}
168/0	2804^{ab}	2474^{bc}	2450^{ab}	4939^{aA}	4351^{abB}	4670^{abAB}
168–45	3349^{aAB}	3676^{aA}	2656^{abB}	4738^{abA}	4049^{bB}	3797^{bB}
	Kalispell-2011					
112–0	3478^{b}	3808^{b}	3681^{ab}			
112–45	3833^{ab}	4218^{a}	3889^{a}			
224–0	3842^{ab}	3755^{b}	3583^{ab}			
224–45	3915^{a}	3994^{ab}	3624^{ab}			
336–0	3625^{abAB}	3901^{abA}	3431^{bB}			
336–45	3872^{aA}	3910^{abA}	3253^{bB}			

§Down a column for each source, at-seeding–topdress N rate LS means followed by different lowercase letters are significant at $p < 0.05$.
†Across rows for each site-year, at-seeding–topdress N rate combination, source LS means followed by different uppercase letters are significant at $p < 0.05$.

return. We have found conflicting results from the different site-years regarding the effect of sources and N rates.

3.2. Grain Protein Content. In 2011, GP content values ranged from 13.9 to 15.6% at Corvallis, 11.2 to 13.8% at Kalispell, and 9.1 to 9.9% at Conrad. In 2012, excellent GP values were achieved at Kalispell (from 12.8 to 14.6%) and Corvallis (from 12.7 to 13.4%). At Conrad, the GP content was also high, ranging from 9.6 to 13.6%. Grain protein ranged from 14.7 to 17.3% at Corvallis and 11.0 to 14.4% at Conrad in 2013.

Grain protein content was significantly affected by N sources at Conrad in all the three years and at Kalispell in 2012 only (Table 4). At Conrad in 2011, the ESN and blend sources combined had 0.6% higher GP content than urea. Likewise, in 2013, the ESN source had 0.5% more GP content than the urea source. In contrast, in 2012 at Conrad, urea had 0.9 and 0.4% more GP content than ESN and blend, respectively. At Kalispell in 2012, the ESN and blend sources combined had 0.50% more GP content than urea. Wheat GP has increased with increase in at-seeding N rates at all site-years except at Corvallis in 2012. Accordingly, wheat GP values increased

TABLE 4: Effect of at N source and preplant and topdress N rate on spring wheat grain protein Conrad, Corvallis and Kalispell, 2011–2013 growing seasons.

Effect	Conrad			Corvallis			Kalispell	
	2011	2012	2013	2011	2012	2013	2011	2012
N source, kg ha^{-1}				%				
Urea	9.4$^{b\pounds}$	12.9^a	13.6^b	14.8	13.4	16.7	12.6	13.8^b
ESN	10.0^a	12.0^c	14.1^a	14.9	12.9	16.8	12.4	14.2^a
Blend	9.9^a	12.5^b	13.8^{ab}	14.9	12.7	16.8	12.4	14.4^a
Prob > F	0.0501	<0.001	0.011	0.732	0.123	0.887	0.656	<0.001
At-seeding N rate, kg ha^{-1}								
0	9.1	9.6	11.0	13.9	13.1	14.7	11.2	12.8
Low†	9.1	11.1	13.0	14.4	12.7	16.4	11.7	13.9
Medium	9.3	12.7	14.1	15.6	13.4	16.6	13.1	13.9
High	9.9	13.6	14.4	15.6	13.0	17.3	13.8	14.6
Prob > F (trend)	0.013‡	<0.001‡	<0.001‡	<0.001‡	0.078	0.003‡	<0.001‡	<0.001‡
Topdress N rate, kg ha^{-1}								
0	9.4	12.1^{b§}	13.6^b	14.4^b	13.0	16.9	12.1^b	14.4^b
45	9.5	12.9^a	14.1^a	15.3^a	13.0	16.6	12.9^a	13.9^a
Prob > F	0.918	0.001	0.001	0.001	0.968	0.251	0.001	0.001

$^\pounds$Down a column for each site-year source means followed by different letter are significant at $p < 0.05$.
†At-seeding N rates: low, medium, and high were 56, 112, and 168 kg ha^{-1} for Conrad and 112, 224, and 336 kg ha^{-1} for Corvallis and Kalispell.
‡Significant linear trend, for at-seeding N rate for each site-year.
§Down a column for topdress N rates, LS means followed by different letters are significant at $p < 0.05$ for each source.

linearly with increase in at-seeding N rate for all site-years where effects were significant (Table 4).

Source by at-seeding N rate interaction was significant at Conrad ($p < 0.02$) and Kalispell ($p < 0.001$) in 2012. At Conrad GP content increased consistently with increase in at-seeding N rate from 56 kg ha^{-1} to 168 kg ha^{-1} for all the three sources. However, a more steep increase was observed for urea than ESN or the blend (Table 5). At Conrad in 2012, at 1056 kg ha^{-1} at-seeding N rate, all sources exhibited low and insignificant GP content (11.1 to 11.2%). At the 112 kg ha^{-1} at-seeding N rate, GP content for blend and ESN was 1.2 and 0.7% lower than urea, respectively, while the difference was 1.4 and 0.6% for 168 kg ha^{-1} at-seeding N rate, in the same order.

At Kalispell in 2012, GP content linearly increased with at-seeding N rates for urea but decreased with increasing at-seeding N rate from 112 kg ha^{-1} t to 224 kg ha^{-1} and then peaked again with the ESN and blend sources (Table 5). For this site-year, at the 112 kg ha^{-1} at-seeding N rate, urea had the lowest GP content while ESN and the blend had similar GP content (Table 5).

Source by topdress N interaction was significant at Conrad in 2012 ($p < 0.05$), at Corvallis in all the three years ($p < 0.02$ to $p < 0.06$), and at Kalispell in 2012 ($p < 0.001$). At Conrad in 2012, without topdress N, ESN and urea combined resulted in 0.7% more GP content than the blend. With topdress N, urea resulted in 1.1 and 0.7% more GP content than the blend and ESN, respectively (Table 5). Additionally, at this site-year, topdress N resulted in 0.8, 0.6, and 1.1% more GP content for the blend, ESN, and urea, respectively, than plots that did not receive topdress N.

At Corvallis in 2011, with topdress N, the blend source had 0.5% more GP content than urea. Likewise, at this site-year, consistent across sources, topdressing N resulted in more GP content than no topdressing. Accordingly, topdress N resulted in 1.2, 1.2, and 0.6% more GP content for the blend, ESN, and urea sources, respectively, compared with no topdress.

At Corvallis in 2012, the effect of sources on the GP content at each topdress N was inconsistent. For this site-year, both blend and urea resulted in 0.6% more GP content than the ESN source with no topdress N while urea resulted in almost 1% more GP content than blend and ESN combined. For this site-year, for the blend source topdressed plots resulted in 0.7% less GP content than nontopdressed plots; however, when wheat was topdressed with N, 0.5% more GP content was obtained compared to nontopdressed plots (Table 5).

At Corvallis in 2013, without topdress N, urea had 0.8% less GP content than ESN and blend sources combined whereas, with topdress N, ESN and urea sources resulted in 0.7% more GP content than the blend (Table 5). For this site-year, topdress N resulted in 0.8% less and 0.6% more GP content for blend and urea sources, respectively.

At Kalispell in 2012, without topdress N, blend and ESN sources resulted in 1.2 and 1.0% more GP content than urea, respectively. Furthermore, for this site-year, when sources were blend and ESN, topdressed plots had 0.8 and 0.7% more GP content than plots that did not receive topdress N.

3.3. Nitrogen Uptake. Nitrogen source significantly affected GNU only at Corvallis and Kalispell in 2011 (Table 6). At

TABLE 5: Simple effects of two-way interactions of source by at-seeding and topdress N rates for wheat grin protein from selected site-years where interaction effect was significant at least at $p < 0.1$.

Site-year	At-seeding N rate, kg ha^{-1}	Source		
		Blend	ESN	Urea
		%		
Conrad-2012	56	11.1	11.1	11.2
	112	12.1^{C†}	12.6^B	13.3^B
	168	12.9^C	13.7^B	14.3^A
	Prob > F (trend)	0.012‡	0.021‡	0.011‡
Kalispell-2012	112	14.4^A	14.3^A	13.0^B
	224	13.7	14.2	13.9
	336	14.5	14.8	14.5
	Prob > F (trend)	0.023$^€$	0.047$^€$	0.009‡
	Topdress N, kg ha^{-1}			
Conrad-2012	0	11.6^{Bb§}	12.2^{Ab}	12.4^{Ab}
	45	12.4^{Ba}	12.8^{Ba}	13.5^{Aa}
Corvallis-2011	0	14.7^b	14.6^b	14.8^b
	45	15.9^{Aa}	15.8^{ABa}	15.4^{Ba}
Corvallis-2012	0	13.2^{Aa}	12.6^B	13.2^{Ab}
	45	12.5^{Bb}	12.8^B	13.7^{Aa}
Corvallis-2013	0	17.2^{Aa}	17.1^A	16.4^{Bb}
	45	16.4^{Bb}	17.1^A	17^{Aa}
Kalispell-2012	0	14.6^{Aa}	14.8^{Aa}	13.8^B
	45	13.8^b	14.1^b	13.8

†Across rows for each site-year, at-seeding N rate, or topdress N rate, LS means followed by different uppercase letters are significant at $p < 0.05$.

$‡$ & $€$ represent significant linear and quadratic trends, for at-seeding N rate for each source.

§Down a column for topdress N rates, LS means followed by different lowercase letters are significant at $p < 0.05$ for each source.

Corvallis, urea resulted in 16.3 kg ha^{-1} more GNU than ESN and blend combined. In contrast, ESN had 7.8 kg ha^{-1} more GNU than urea at Kalispell in 2011.

At-seeding N rate treatments significantly affected GNU at all site-years except at Corvallis 2013 (Table 6). Grain N uptake increased linearly with increase in at-seeding N rate at all site-years where the effect was significant except at Corvallis 2012. At Corvallis in 2012, a quadratic trend was observed where GNU peaked at 224 kg ha^{-1} at-seeding N rate. Likewise, plots supplemented with topdress N resulted in higher GNU than nontopdressed plots consistently across site-years where the effect was significant (Table 6).

Source by at-seeding N rate interaction was significant at Conrad in 2011 ($p < 0.05$) and 2013 (0.021) and at Kalispell ($p < 0.021$) in 2011. At Conrad GNU increased linearly with increase in at-seeding N rate for blend and ESN. In contrast, GNU was increased to 41.5 kg ha^{-1} at 112 kg ha^{-1} at-seeding N rate and decreased to 37.3 kg ha^{-1} with 168 kg N ha^{-1} at-seeding N rate for urea source (Table 7). Additionally, at the

112 and 168 kg ha^{-1} at-seeding N rates, sources showed significantly different response. Accordingly, at 112 kg ha^{-1} at-seeding N rate, blend had 8.5 and 8.8 less GNU than ESN and urea, respectively. In contrast, at 168 kg ha^{-1} at-seeding N rate, blend and ESN had 12.4 and 10.3 more GNU than urea, respectively (Table 7).

At Conrad in 2013, for the blend source, GNU increased linearly with increase in at-seeding N rate. At this site-year, at 168 kg ha^{-1} at-seeding N rate, the blend source had 10.6 kg ha^{-1} more GNU than ESN and urea combined. At Kalispell in 2011, GNU increased linearly with at-seeding N rate from 64.0 to 81.5 kg ha^{-1} when at-seeding N rate increased from 112 to 336 kg ha^{-1} for the blend source. At this site-year, at 112 kg ha^{-1} at-seeding N rate, ESN had 10.1 kg ha^{-1} more GNU than ESN. At 336 kg ha^{-1} at-seeding N rate, blend and ESN had 11.0 and 8.7 kg ha^{-1} more GNU than urea.

Significant source by topdress N rate interaction was observed only at Conrad in 2013. Accordingly, topdressing N increased GNU by 6.5 and 6.2 kg ha^{-1} compared with no topdressing. Alternatively, for urea GNU was decreased by 8.4 kg ha^{-1} in plots that received topdress N compared with that not receiving N.

3.4. Partial Budget Economic Analysis. Only three site-years, Conrad and Corvallis in 2012 and Kalispell in 2011, were considered for the economic analysis based on significant differences in GY (Table 8(a)). Net returns for GP-unadjusted revenue were higher for the blend source at Conrad in 2012 (Table 8(a)).

A slight revenue advantage was obtained with the blend at Kalispell. At Corvallis, urea outperformed both the ESN and blend. In fact, ESN performed poorly in all the three site-ears. At Conrad in 2012, ESN had lower marginal net return than blend ($19.02 ha^{-1} more). At Kalispell, a similar result was observed where the blend resulted in $8.29 ha^{-1} additional income compared to urea. At Conrad, MRR was $1.02 and 1.56 for ESN and blend, respectively, while at Kalispell the MRR was 0.38 and 1.17 for GP-unadjusted revenue. At Corvallis 2012, negative marginal rate of return was observed for both ESN and blend sources.

Similarly, protein-unadjusted and -adjusted partial budget analysis comparing medium at-seeding rate and low at-seeding rate along 45 kg ha^{-1} topdress N rates showed that ESN did not result in better marginal profit or marginal rate of return compared with urea at the specified rates (Table 8(b)).

4. Discussion and Conclusion

This study has enabled us to evaluate the N source effect on GY and GP in a wide range of growing environments. Average spring wheat GY has varied from 1816 to 5583 kg ha^{-1} (Table 2) and GP content ranged from 9.1 to 17.3% among the site-years (Table 4).

Spring wheat GY responded to N applied at the time of seeding at seven of eight site-years (Table 2); also higher GP content was achieved with at-seeding N application (Table 4). These results emphasize the importance of initial N fertilizer

TABLE 6: Effect of at N source, preplant and topdress N rate on spring wheat grain uptake Conrad, Corvallis and Kalispell, 2011–2013 growing seasons.

Effect	Conrad			Corvallis			Kalispell	
	2011	2012	2013	2011	2012	2013	2011	2012
N source				kg ha^{-1}				
Urea	51.6	117.7	106.5	96.4$^{a€}$	106.5	72.9	99.8^{b}	106.5
ESN	54.9	114.3	107.6	78.5^{b}	107.6	71.7	104.3^{ab}	105.4
Blend	54.9	116.6	109.9	81.8^{b}	102.0	67.3	107.6^{a}	107.6
Prob > F	0.620	0.328	0.401	<0.001	0.664	0.853	0.011	0.782
At-seeding N rate								
0	44.8	86.3	68.4	65.0	86.3	46.0	53.8	85.2
Low†	43.7	98.7	104.3	77.4	94.2	71.7	96.4	104.3
Medium	53.8	118.8	111.0	88.6	116.6	77.4	106.5	104.3
High	62.8	131.2	108.7	91.9	105.4	62.8	108.7	109.9
Prob > F	<0.001‡	<0.001‡	0.047‡	0.001‡	0.008$^{€}$	0.395	<0.001‡	0.051‡
Topdress N rate								
0	50.5^{b§}	112.1^{b}	106.5	79.6^{b}	99.8	70.62	98.7^{b}	107.6
45	57.2^{a}	120.0^{a}	108.7	91.9^{a}	111.0	70.6	109.9^{a}	105.4
Prob > F	0.049	<0.001	0.321	<0.001	<0.001	0.932	0.001	0.204

$^{€}$Down a column for each site-year source means followed by different letter are significant at $p < 0.05$.
†At-seeding N rates: low, medium, and high were 56, 112, and 168 kg ha^{-1} for Conrad and 112, 224, and 336 kg ha^{-1} for Corvallis and Kalispell.
‡ & € represent significant linear and quadratic trends, for at-seeding N rate for each site-year.
§Down a column for topdress N rates, LS means followed by different letters are significant at $p < 0.05$ for each source.

TABLE 7: Simple effects of two-way interactions of source by at-seeding and topdress N rates for N uptake from selected site-years where interaction effect was significant at least at $p < 0.1$.

Site-year	At-seeding N rate, kg ha^{-1}	Source		
		Blend	ESN	Urea
		kg ha^{-1}		
Conrad-2011	56	34.6	29.4	30.3
	112	32.7^{B†}	41.2^{A}	41.5
	168	49.7^{A}	47.6^{AB}	37.3^{B}
	Prob > F (trend)	0.05‡	0.008‡	0.045$^{€}$
Conrad-2013	56	89.5	94	95.2
	112	100.1^{A}	100.7^{B}	96.5^{B}
	168	104.8^{B}	93.9^{A}	94.5^{AB}
	Prob > F (trend)	0.022‡	0.165	0.754
Kalispell-2011	112	64^{A}	74.1^{AB}	68.7^{B}
	224	77.7	77.1	73.5
	336	81.5^{A}	80.2^{A}	71.5^{B}
	Prob > F (trend)	0.007‡	0.068‡	0.171
	Topdress N, kg ha$^{-1§}$			
Conrad-2013	0	94.9^{b}	93.1^{b}	99.1^{a}
	45	101.4^{Aa}	99.3^{Aa}	90.7^{Bb}

†Across rows for each site-year, at-seeding N rate, or topdress N rate, LS means followed by different uppercase letters are significant at $p < 0.05$.
‡ represents significant linear trend, for at-seeding N rate for each site-year.
§Down a column for topdress N rates, LS means followed by different lowercase letters are significant at $p < 0.05$ for each source.

application for wheat stand establishment and early-season crop development. The fact that only three out of eight site-years have benefited from the application of a high N rate at seeding suggests that the current university guidelines for spring wheat are too high for the typically observed GYs for the area. This is further reinforced by the finding that GY had not responded to at-seeding N application rate above the lowest rate at four out of eight site-years (Table 2). Furthermore, in four of eight site-years, GY was significantly increased with topdress N application (Table 2). Current work is underway at Montana State University to revise N fertilizer recommendations and adjust them to newly released varieties.

The 45 kg N ha^{-1} rate of topdress used in this study is a typically recommended rate for spring wheat grown in Montana for growers aiming to enhance GP values. The GP content was increased with topdress application of N in five of eight site-years (Table 4). This shows the significance of topdressing N to boost GP content that essentially improves wheat quality.

No consistent trend in GY associated with N source was observed in this study: out of eight site-years, urea resulted in higher GYs at three site-years, ESN produced higher GYs at three site-years, and the blend produced higher GYs at two site-years (with 1 site-year having no data available for the blend) (Table 2). These results are in agreement with previous findings [46, 47] who observed that ESN and blend have outperformed urea in some years but resulted in lower crop GY in other years.

No significant differences in GP content associated with N source were observed in most site-years; ESN and blend resulted in higher GP content compared to urea alone at three

TABLE 8: (a) Protein-unadjusted and -adjusted revenues and marginal returns for selected site-years where grain yield was significant. (b) Protein-unadjusted and -adjusted partial budget analysis comparing medium at-seeding rate (112 and 224 N kg ha^{-1} at Conrad and Kalispell, resp.) and low at-seeding rate (56 and 112 kg ha^{-1} at Conrad and Kalispell, resp.) along 45 kg ha^{-1} topdress N rates for selected site-years.

(a)

Site-year	Unadjusted for protein			Adjusted for protein		
	Urea	ESN	Blend	Urea	ESN	Blend
	Net return, $ ha^{-1}					
Conrad-2012	648.79	650.27	669.31	536.35	448.05	529.48
Corvallis-2012	243.93	74.26	126.42	209.16	15.08	55.68
Kalispell-2011	335.39	280.13	343.68	240.45	165.70	225.31
	Marginal profit, $ ha^{-1}					
Conrad-2012	—	**1.48**	**20.50**	—	−88.30	−6.85
Corvallis-2012	—	−169.67	−117.51	—	−194.08	−153.50
Kalispell-2011	—	−55.26	**8.29**	—	−74.75	−15.13
	Marginal rate of return, $ recovery $^{-1}$ investment					
Conrad-2012	—	**1.02**	1.56	—	−0.31	0.81
Corvallis-2012	—	−0.89	−1.41	—	−1.17	−2.14
Kalispell-2011	—	**0.38**	1.17	—	**0.17**	**0.69**

(b)

Site-year	Unadjusted for protein		Adjusted GY for protein	
	Urea	ESN	Urea	ESN
	Net return, $ ha^{-1}			
Conrad-2011	612.68	134.32	502.94	43.38
Conrad-2013	533.88	374.48	479.36	276.45
Kalispell-2011	352.49	255.31	252.67	145.16
	Marginal profit, $ ha^{-1}			
Conrad-2011	—	−426.36	—	−459.56
Conrad-2013	—	−85.00	—	−202.91
Kalispell-2011	—	−22.78	—	−107.51
	Marginal rate of return, $ recovery $^{-1}$ investment			
Conrad-2011	—	−8.20	—	−7.84
Conrad-2013	—	−1.14	—	−1.73
Kalispell-2011	—	−0.31	—	−0.45

of eight site-years evaluated (Table 4). While preplant or at-seeding applications of slow-release N products like ESN may be advantageous in winter wheat, it does not always result in improved GP in spring wheat. This may be mainly due to insufficient time for N release during shorter growing season.

Nitrogen uptake increased with increase in at-seeding N rates at seven of eight site-years; topdress N application increased N uptake in five of eight site-years (Table 6).

Nitrogen uptake was similar for all N sources, except for one site-year where slight increase in N uptake was observed with the blend application and for one site-year where the blend resulted in significantly greater N uptake (Table 6). It is possible that similar GY results achieved with urea, ESN, and blend were obtained because similar N losses (predominantly due to ammonia volatilization) have occurred from urea and ESN. As the soil temperatures increased, the rate of urea hydrolysis from the urea treatments increased [48]. At the same time, the dissolution of urea and its release from the

ESN capsules has also increased with higher soil temperatures.

This is where the controlled release versus delayed release discussion comes in, as an example; with average soil temperatures in Montana being approximately 68°F for June and July [29] it is expected that most of the urea has been released into the soil. In reality, we can expect comparable losses of N to occur from both urea and ESN treatments via ammonia volatilization. Thus, similar N amounts were available to wheat crop from all N sources applied.

The manufacturer's general recommendation for utilization of ESN for spring wheat in Great Plains region is application of N as 100% ESN in the fall prior to seeding wheat the next spring [49]. Another commonly recommended application scenario is a spring application as a blend (40–75% ESN + 25–60% urea). Our findings showed that ESN and urea-ESN blend performed well, but not better than urea alone. This is also supported by the partial budget economic analysis results. With GP-unadjusted revenue, by adopting ESN alone as N source, a farmer would lose money on investment but would recover costs by using urea-ESN blend. With GP-adjusted revenue, farmer would not recover investment costs from ESN or blend compared with urea. The losses associated with ESN fertilization strategy make it uneconomical for spring wheat in dryland or irrigated spring wheat production system given the assumptions of the partial budget analysis.

Farmers need to be cautious when adopting ESN for improving economic return although environmentally the ESN may be beneficial. An important aspect of fertilizer management strategy is to carefully assess growing conditions that might affect profitability of fertilization.

Competing Interests

The authors declare that they have no competing interests.

References

[1] J. Bruinsma, "The resource outlook to 2050: by how much do land, water and crop yields need to increase by 2050?" in *Proceedings of the Expert Meeting on How to Feed the World in 2050*, Rome, Italy, June 2009.

[2] FAO, *The State of Food Insecurity in the World (FAO)*, FAO, Rome, Italy, 2015.

[3] P. E. Fixen, "Potential biofuels influence on nutrient use and removal in the US," *Better Crops*, vol. 91, pp. 12–14, 2007.

[4] T. Roberts, "The role of fertilizer in growing the world's food," *Better Crops*, vol. 93, pp. 12–15, 2009.

[5] W. M. Stewart, *Fertilizer and Food Production*, International Plant Nutrition Institute (IPNI), Peachtree Corners, Georgia, 2009.

[6] W. Zhang and X. Zhang, "A forecast analysis on fertilizers consumption worldwide," *Environmental Monitoring and Assessment*, vol. 133, no. 1–3, pp. 427–434, 2007.

[7] M. Gupta and R. Khosla, "Precision nitrogen management and global nitrogen use efficiency," in *Proceedings of the 11th International Conference on Precision Agriculture*, Indianapolis, Ind, USA, July 2012.

[8] W. R. Raun and G. V. Johnson, "Improving nitrogen use efficiency for cereal production," *Agronomy Journal*, vol. 91, no. 3, pp. 357–363, 1999.

[9] O. S. Walsh, R. J. Christiaens, and A. Pandey, "Foliar-applied nitrogen fertilizers in spring wheat production," *Crops and Soils*, vol. 46, pp. 26–32, 2013.

[10] U. Singh, "Integrated nitrogen fertilization for intensive and sustainable agriculture," *Journal of Crop Improvement*, vol. 15, no. 2, pp. 259–288, 2006.

[11] USDA/NASS, *Crop Production 2013 Summary*, United State Department of Agriculture/National Agricultural Statistics Service, Washington, DC, USA, 2014.

[12] K. Girma and W. R. Raun, "Nutrient and water use efficiency," in *Handbook of Soil Science: Resource Management and Environmental Impacts*, P. M. Huang, Y. Li, and M. E. Sumner, Eds., pp. 405–421, CRS Press, Boca Raton, Fla, USA, 2nd edition, 2012.

[13] K. Olson-Rutz, C. Jones, and C. P. Dinkins, *Enhanced Efficiency Fertilizers*, Montana State University Extension, Bozeman, Mont, USA, 2011.

[14] A. Shaviv, "Advances in controlled-release fertilizers," *Advances in Agronomy*, vol. 71, pp. 1–49, 2001.

[15] W. Gordon, "Management of enhanced efficiency fertilizers," in *Proceedings of the North Central Extension-Industry Soil Fertility Conference*, Des Moines, Iowa, USA, November 2007.

[16] B. Gordon, "Nitrogen management for no-tillage corn and grain sorghum production," in *Proceedings of the Great Plains Soil Fertility Conference*, pp. 67–70, 2008.

[17] X. Gao, H. Asgedom, M. Tenuta, and D. N. Flaten, "Enhanced efficiency urea sources and placement effects on nitrous oxide emissions," *Agronomy Journal*, vol. 107, no. 1, pp. 265–277, 2015.

[18] J. L. Hatfield and R. T. Venterea, "Enhanced efficiency fertilizers: a multi-site comparison of the effects on nitrous oxide emissions and agronomic performance," *Agronomy Journal*, vol. 106, no. 2, pp. 679–680, 2014.

[19] S. A. Ebelhar, C. D. Hart, J. D. Hernandez, L. E. Paul, and J. J. Warren, "Evaluation of new nitrogen fertilizer technologies for corn," in *Proceedings of the Illinois Fertilizer Conference*, Peoria, Ill, USA, January 2007.

[20] Association of American Plant Food Control Officials (AAPFCO), *Stablized Fertilizers*, Association of American Plant Food Control Officials (AAPFCO), West Lafayette, Ind, USA, 2012.

[21] U. S. E. P. A. EPA, Inventory of U.S. Greenhouse Gas Emissions and Sinks: 1990–2013, Washington, DC, USA, EPA, 2015.

[22] A. D. Halvorson, C. S. Snyder, A. D. Blaylock, and S. J. Del Grosso, "Enhanced-efficiency nitrogen fertilizers: potential role in nitrous oxide emission mitigation," *Agronomy Journal*, vol. 106, no. 2, pp. 715–722, 2014.

[23] J. L. Hatfield and T. B. Parkin, "Enhanced efficiency fertilizers: effect on agronomic performance of corn in Iowa," *Agronomy Journal*, vol. 106, no. 2, pp. 771–780, 2014.

[24] M. L. Wilson, C. J. Rosen, and J. F. Moncrief, "A comparison of techniques for determining nitrogen release from polymer-coated urea in the field," *HortScience*, vol. 44, no. 2, pp. 492–494, 2009.

[25] Agrium, *Wheat Development and ESN Nitrogen Management*, Agrium Inc, Calgary, Canada, 2014.

[26] R. Keller, "Slow release nitrogen: sustainability could be reason for using slow-release N formulations and additives," Ag Professional, 2010, http://www.agprofessional.com/agprofessional-magazine/slow-release_nitrogen_120018994.html.

[27] D. Franzen, "Slow-release nitrogen fertilizers and nitrogen additives for field crops," in *Proceedings of the North Central Extension-Industry Soil Fertility*, Des Moines, Iowa, USA, November 2010.

[28] M. Ruark, "Understanding the value of slow-release fertilizers," in *Proceedings of the Wisconsin Crop Management Conference*, Madison, Wis, USA, January 2010.

[29] O. S. Walsh and R. J. Christiaens, "US West: urea, ESN, and urea-ESN blends performed equally well as nitrogen sources for spring wheat," *Crops and Soils*, vol. 47, pp. 26–31, 2014.

[30] M. Alley, P. Scharf, D. Brann, W. Baethgen, and J. Hammons, "Nitrogen management for winter wheat: principles and recommendations," in *Crop & Soil Environmental Sciences*, pp. 424–429, 1999.

[31] R. E. Blackshaw, X. Hao, R. N. Brandt et al., "Canola response to ESN and urea in a four-year no-till cropping system," *Agronomy Journal*, vol. 103, no. 1, pp. 92–99, 2011.

[32] R. E. Blackshaw, X. Hao, K. N. Harker, J. T. O'Donovan, E. N. Johnson, and C. L. Vera, "Barley productivity response to

polymer-coated urea in a no-till production system," *Agronomy Journal*, vol. 103, no. 4, pp. 1100–1105, 2011.

[33] A. D. Halvorson, S. J. Del Grosso, and C. P. Jantalia, "Nitrogen source effects on soil nitrous oxide emissions from strip-till corn," *Journal of Environmental Quality*, vol. 40, no. 6, pp. 1775–1786, 2011.

[34] R. H. McKenzie, A. B. Middleton, P. G. Pfiffner, and E. Bremer, "Evaluation of polymer-coated urea and urease inhibitor for winter wheat in southern Alberta," *Agronomy Journal*, vol. 102, no. 4, pp. 1210–1216, 2010.

[35] B. R. Golden, N. A. Slaton, R. J. Norman, C. E. Wilson Jr., and R. E. Delong, "Evaluation of polymer-coated urea for direct-seeded, delayed-flood rice production," *Soil Science Society of America Journal*, vol. 73, no. 2, pp. 375–383, 2009.

[36] M. L. Wilson, C. J. Rosen, and J. F. Moncrief, "Potato response to a polymer-coated urea on an irrigated, coarse-textured soil," *Agronomy Journal*, vol. 101, no. 4, pp. 897–905, 2009.

[37] C. M. Hutchinson, "Influence of a controlled release nitrogen fertilizer program on potato (*Solanum tuberosum* L.) tuber yield and quality," *Acta Horticulturae*, vol. 684, pp. 99–102, 2005.

[38] B. J. Winer, D. R. Brown, and K. M. Michels, *Statistical Principles in Experimental Design*, vol. 2, McGraw-Hill, New York, NY, USA, 1971.

[39] O. Schabenberger, T. G. Gregoire, and F. Kong, "Collections of simple effects and their relationship to main effects and interactions in factorials," *The American Statistician*, vol. 54, no. 3, pp. 210–214, 2000.

[40] G. Snedecor and W. Cochran, *Statistical Methods*, Oxford and IBH Publishing Co, Calcutta, India, 8th edition, 1994.

[41] T. Alimi and V. Manyong, *Partial Budget Analysis for on-farm Research*, vol. 65, IITA, 2000.

[42] M. E.-D. Soha, "The partial budget analysis for sorghum farm in Sinai Peninsula, Egypt," *Annals of Agricultural Sciences*, vol. 59, no. 1, pp. 77–81, 2014.

[43] West-Con-Cooperative, "Spring wheat proteins," https://s3.amazonaws.com/media.agricharts.com/sites/591/PDFs/SprProteins.pdf.

[44] USDA/AMS-WY, "Montana daily cash grain prices," http://www.ams.usda.gov/mnreports/bl_gr110.txt.

[45] Agrium, "ESN return on investment," http://www.smartnitrogen.com/roi-calculator.

[46] H. S. Weber and D. B. Mengel, "Use of nitrogen management products and practices to enhance yield and nitrogen use efficiency in no-till corn," in *Proceedings of the North Central Extension-Industry Soil Fertility Conference*, vol. 25, pp. 18–19, Des Moines, Iowa, USA, 2009.

[47] G. Randall and J. Vetsch, "Fall and spring-applied nitrogen sources for corn in Southern Minnesota," Tech. Rep., University of Minnesota Southern Research and Outreach Center, Waseca, Minn, USA, 2009.

[48] D. Yadav, V. Kumar, M. Singh, and P. Relan, "Effect of temperature and moisture on kinetics of urea hydrolysis and nitrification," *Australian Journal of Soil Research*, vol. 25, no. 2, pp. 185–191, 1987.

[49] O. Walsh, A. Pandey, and R. Christiaens, "Environmentally smart nitrogen performance in northern great plains spring wheat production systems," in *Proceedings of the Western Nutrient Management Conference*, vol. 11, Reno, Nev, USA, 2015.

22

Evaluating the Impact of Starter Fertilizer on Winter Canola Grown in Oklahoma

M. Joy M. Abit,[1] **Katlynn Weathers,**[2] **and D. Brian Arnall**[1]

[1]*Department of Plant and Soil Sciences, Oklahoma State University, Stillwater, OK 74078, USA*
[2]*P&K Equipment, Enid, OK 73701, USA*

Correspondence should be addressed to D. Brian Arnall; b.arnall@okstate.edu

Academic Editor: Mathias N. Andersen

Increased canola production costs and acres have driven Oklahoma (OK) farmers to ask more questions about their nutrient management recommendations in their production system. A study was conducted in 2011–2013 at Lahoma and Perkins, OK, to evaluate the effect of applying diammonium phosphate (DAP, 18-20-0:N-P-K) directly with seed on crop stand, grain yield, and grain quality of canola. In addition, the impact of proportion nitrogen (N) applied as a preplant and topdress was also evaluated. Diammonium phosphate was banded with the seed at planting at 0, 17, 34, 51, 67, and 84 kg DAP ha^{-1}. Remaining N was applied as urea (46-0-0) either as split (40% preplant and 60% topdress) application or as topdress only. Stand count reduction of up to 71% was observed with seed-placed DAP. However, loss of stand did not impair grain yield due to canola's ability to compensate for open areas via branching. Application of DAP of up to 84 kg ha^{-1} with seed may be possible; however, soil and climatic conditions should be considered when deciding how much DAP will be placed with seed. Moreover, when climatic conditions limit early season growth and favor late spring growth, applying all N at topdress (no preplant) tended to provide greater canola grain yield.

1. Introduction

Canola (*Brassica napus* L.) is an agronomic crop primarily grown for its seeds as a source of edible oil and animal meal qualities. In Oklahoma (OK), canola is grown in rotation with wheat to help disrupt wheat disease cycles and expand weed control options. It is typically seeded between September and October and harvested in May or June. Winter canola has grown considerably in Oklahoma where plantings for the crop years 2010 to 2015 doubled, from 24,000 to 56,000 hectares [1]. In 2015, Oklahoma was ranked as the second largest canola producing state in the US next to North Dakota.

Starter fertilizer is a small amount of fertilizer nutrients applied in close proximity to the seed at planting. In general, a seedling root system lacks the size and density to be able to intercept the necessary nutrients within the soil. Starter fertilizers enhance the development of emerging seedlings by placing a readily available supply of nutrients which the undeveloped root system of the seedling can easily access. Nitrogen (N) and phosphorus (P) are key nutrient components in a starter fertilizer. Phosphorus is important for promoting vigorous root growth. Phosphorus, however, is immobile in the soil. To be absorbed by the plant, the roots must be very close to the phosphate. Hence, P should be strategically placed close to the seed to obtain an early boost to growth. Nitrogen, on the other hand, is a mobile nutrient, so placement may not be as critical as P, but N in the starter fertilizer may help avoid early season N deficiency due to the slow release of N in organic matter particularly during cold conditions. Also, ammonium (NH_4, nitrogen in available form) from starter fertilizers can enhance P uptake from the starter and from the soil [2, 3].

It is a common practice for many Oklahoma winter wheat producers and crop producers with small acres of winter wheat to put down starter fertilizer in row with seed as they do not use additional starter attachments due to considerable equipment costs. A primary concern with in-furrow or seed-placed starter fertilizer in canola is the potential for salt injury to germinating seed, especially with N fertilizer. Rates of N that would normally cause little or no injury to wheat can cause severe injury and reduction in germination and emergence of canola when placed with the seed [4].

Nitrogen in nitrate (NO_3) form can damage canola seedlings by desiccation through salt effect. Ammonia toxicity from N-containing fertilizers also damages canola seedlings [5]. Phosphate, on the other hand, has no salt effect but common starter fertilizers are compound fertilizers containing both P and N.

In Oklahoma, the most commonly used starter fertilizers are diammonium phosphate (DAP, 18-20-0:N-P-K), monoammonium phosphate (MAP, 11-23-0:N-P-K), and ammonium polyphosphate (APP, 10-15-0:N-P-K). Among the three, DAP has become popular to farmers due to high N and P content, relatively lower prices, and greater availability, but the higher N component of DAP puts a limit on safe rates of seed-placed phosphate compared with MAP and APP.

An additional concern in the production of winter canola is the potential for excessive top growth and limited root system into winter dormancy due to N applied close to seeding and the impact this may have on winter hardiness and survivability. At this time, impacts of N applied on canola winter survivability are all speculative as there is little information on timing of N applications impact on winter canola.

The objectives of this research were to determine the yield response to fertilizer DAP applied with seed at planting; to identify the critical level at which salt injury negatively impacts stand and yield when DAP is applied with seed; and to evaluate the impact of N application method on canola yield.

2. Materials and Methods

2.1. Site Selection. Two sites were established in Oklahoma in 2011-2012 and 2012-2013. One was at the Cimarron Valley Research Station near Perkins, OK, USA (lat.: 35.99, long.: −97.033). The soil at Perkins location was Konawa (fine-loamy, mixed, active, Thermic Ultic Haplustalfs) and Teller (fine-loamy, mixed, active, Thermic Udic Argiustolls) loam soils. Teller and Konawa series soils are deep, well drained, and moderately permeable. Potential rooting depth is 2 to 3.5 m if there are no major restrictive layers. The average annual precipitation in this location is 94.13 cm with an average summer high temperature of 33°C and an average winter low temperature of −3°C. The other location was at the North Central Research Station near Lahoma, OK, USA (lat.: 36.38, long.: −98.10). The soil at the Lahoma location was Grant silt loam (fine-silty, mixed, super active, Thermic Udic Argiustolls). These soils are well drained, deep, and moderately permeable and rooting depth potential is similar to Teller and Konawa soil series. The average annual precipitation at this location is 82.02 cm with an average summer high temperature of 34.22°C and average winter temperature of 7.78° C. The Perkins location was a no-tillage cropping system, while the Lahoma location was a conventional tillage system. Both locations were established after wheat in both years.

2.2. Soil Sampling and Analysis. Soil samples were taken a month prior to planting to determine soil nutrient levels. In the second year, experimental plots were established adjacent to the previous year's test plots to prevent residual N and P

levels from affecting the test area. Top soil (top 0 to 15 cm) and subsoil (lower 15 to 30 cm) were collected for each soil sample that consisted of 15 to 20 cores. Soil samples were analyzed for soil pH, nitrate-N (NO_3-N), extractable P, potassium (K), sulfur (S), calcium (Ca), and magnesium (Mg). Soil test results are shown in Table 1. Soil samples were dried at 65°C overnight and ground to pass through a 2 mm sieve prior to extraction and analysis. The soil pH was measured using a combination electrode within a 1:1 ratio of soil to water suspension and Sikora buffer solution [6, 7]. Soil NO_3-N was extracted with a 2 M KCl solution and quantified by a Flow Injection Autoanalyzer [8]. Mehlich III solutions were used to extract plant available P, K, Ca, and Mg [9] and quantified using a Spectro CirOs ICP spectrometer [10]. Total N was determined using the LECO Truspec dry combustion carbon analyzer [11].

2.3. Planting and Treatment Establishment. A popular canola variety in Oklahoma, Dekalb brand "DKW 46-15," was planted with a John Deere 450 grain drill with double disk seed openers at 38.1 cm spacing at a target rate of 5.6 kg seeds ha^{-1}. Planting dates in 2011 were 26 and 27 September in Perkins and Lahoma locations, respectively. In 2012, canola was planted in 18 September at Perkins location and 2 October at Lahoma location. Plots were 6 m long by 2.5 m wide (6 rows) with a 6 m alley between replications. The experimental design was a randomized complete block with three replications. Table 2 lists the treatment structure used in this study. The starter fertilizer, DAP, was banded with the seed at planting at 0, 17, 34, 51, 67, and 84 kg DAP ha^{-1}. Remaining nitrogen was applied as urea (46-0-0) either as split (40% preplant and 60% topdress) application or as topdress only. All plots received a total of 140 kg N ha^{-1} except for the unfertilized check. For treatments that received split application, all topdress N were applied at 84 kg ha^{-1}; preplant N rates were computed based on the difference between the recommended N rate (140 kg ha^{-1}) and the topdress N (84 kg ha^{-1}) plus N from DAP. Preplant plus topdress and topdress only plots were also included for comparison. Preplant fertilization and topdress fertilization were applied as broadcast incorporated and broadcast, respectively. Topdressing was applied in the spring when canola was in the vegetative stage prior to bolting (stem elongation). Experimental plots were maintained weed-free using glyphosate. Insects were controlled using the Lambda-cyhalothrin insecticide as needed.

2.4. Data Collection. Canola stand counts were taken at each location two weeks after planting, by randomly placing a meter stick along the crop rows of each plot and counting the number of canola plants that emerged. Stand counts were collected at five locations per plot.

To determine the biomass production during the growing season, normalized difference vegetation index (NDVI) readings were obtained using GreenSeeker™ one week before and two weeks after topdress N application. In the second year, additional NDVI readings were collected four weeks after topdress N application. Normalized difference vegetation

TABLE 1: Preplant composite soil sample results in Perkins and Lahoma, OK, in 2011 and 2012. Samples were collected in the summer, a month prior to planting.

Location	Year	pH[†]	Buffer index	NO₃-N[‡] 0–15 cm	NO₃-N[‡] 15–30 cm	STP[§] 0–15 cm	STP[§] 15–30 cm	STK[¶] 0–15 cm	STK[¶] 15–30 cm	SO₄-S[††] 0–15 cm	Ca[‡‡] 0–15 cm	Mg[§§] 0–15 cm
								ppm				
Perkins	2011	4.5	6.6	27	—*	76	—	406	—	31	717	175
Perkins	2012	4.5	6.3	4.5	—	122	—	436	—	—	—	—
Lahoma	2011	6.1	7.0	11	7	32	—	474	—	—	—	—
Lahoma	2012	6.0	7.0	21	—	40	—	470	—	38	3326	1140

[†]pH: 1:1 soil:water.
[‡]NO₃-N: nitrate-nitrogen; 1 M KCl solution.
[§]STP: soil test phosphorus; Mehlich III.
[¶]STK: soil test potassium; Mehlich III.
[††]SO4-S: available sulfur; Mehlich III.
[‡‡]Ca: calcium; Mehlich III.
[§§]Mg: magnesium; Mehlich III.
*— (dash): no data available.

TABLE 2: Canola stand count as influenced by seed-placed diammonium phosphate (DAP) and nitrogen (N) application method two weeks after emergence at Lahoma and Perkins, OK, in 2011–2013.

Treatment	DAP with seed	N/P in DAP	N preplant	N topdress[†]	Stand count Lahoma	Stand count Perkins
		kg ha⁻¹			Plants m⁻²	
1	0	0/0	0	0	87	44
2	0	0/0	56	84	88	38
3	0	0/0	56N/14P	84	82	43
4	17	3/3.5	53	84	65	41
5	34	6/7	50	84	50	35
6	51	9/10.5	47	84	33	24
7	67	12/14	44	84	31	23
8	84	15/17.5	41	84	18	23
9	0	0/0	0	140	90	58
10	17	3/3.5	0	137	57	43
11	34	6/7	0	134	50	29
12	51	9/10.5	0	131	33	30
13	67	12/14	0	128	25	26
LSD					15	19

[†]All treatments received at total of 140 kg N ha⁻¹ except for the unfertilized check.

index values were collected twice from the middle of each plot approximately 70 to 100 cm directly above the crop canopy.

Seven to ten days prior to harvesting, four middle rows of each plot were swathed and pressed with a press wheel attached behind the swather to prevent seed shattering and for uniform seed drying. The swather was run low enough to get all the seed pods, leaving approximately 25 to 30 cm stubble height after swathing. A Massey Ferguson 8XP plot combine equipped with a Harvest Master yield monitor was used to harvest the grain and determine yields. Subsamples were collected and analyzed for protein and oil content using a Diode Array Near-Infrared instrument [12].

2.5. Statistical Analysis. Analysis of variance was performed on canola stand count, NDVI readings, grain yield, and oil and protein content using PROC MIXED (SAS Version 9.4, SAS Institute). Year, location, and treatments were considered

as fixed effects while replications were considered as random effect. Mean comparisons were separated using Fisher's Protected LSD at the $P \leq 0.05$ significance level. Homogeneity of variances and normality of distribution were tested and all data showed normal distribution and equal variances. Regression analyses were performed using Sigma Plot 13 procedures to evaluate the relationship between percent canola stand and seed-placed DAP rate and canola stand count and NDVI readings. Correlation coefficient analysis on stand count versus grain yield was done by using PROC CORR of SAS 9.4.

3. Results and Discussion

3.1. Climatic Conditions. The 2011-2012 sowing at Perkins and Lahoma occurred in what could be considered optimum conditions for canola germination and growth. Timely rainfall

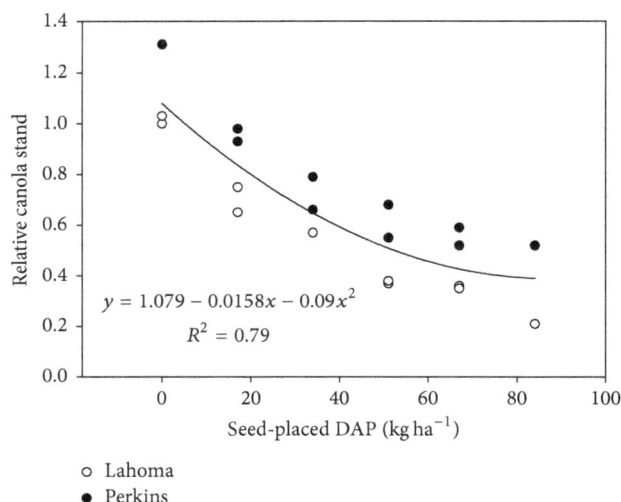

$$y = 1.079 - 0.0158x - 0.09x^2$$
$$R^2 = 0.79$$

o Lahoma
• Perkins

FIGURE 1: Canola stand relative to the untreated check as influenced by rate of seed-placed diammonium phosphate (DAP). Relative canola stand includes data from Lahoma and Perkins, OK, 2011–2013.

throughout October, November, and December, combined with the warmest winter on record, resulted in rapid canola growth. Temperatures during the 2011-2012 season were never cold. In mid-April, a wave of heat occurred in Oklahoma that quickly depleted soil water reserves, but by early May temperatures were near normal and moisture returned.

The 2012-2013 cropping season was generally a dry start for canola planting in Oklahoma. Drought conditions were observed in the fall of 2012. A few timely rains in September allowed good germination and rapid start for the crop but no substantial rain was received until early 2013. Rain in early 2013 was not much but enough to enable the crop to recover. Spring temperatures were generally cooler than normal which were beneficial for canola grain fill but delayed harvest by approximately two to three weeks compared to the previous cropping season.

3.2. Stand Count. Application of DAP with seed caused stand loss of canola plants in both locations (Table 2). Stand loss severity increased with increasing DAP rate (Figure 1). Loss of stand count from application of DAP at 51, 67, and 84 kg ha^{-1} was more pronounced than at 17 and 34 kg ha^{-1} DAP rate. Stand counts of plots applied with seed-placed DAP at 17 and 34 kg ha^{-1} were comparable with the unfertilized check at the Perkins location but caused 25 to 35% stand reduction at the Lahoma location. Seed-placed DAP at 51 and 67 kg ha^{-1} (contains 9 to 12 kg N ha^{-1}) caused 45 to 71% reduction in stand compared to the unfertilized check. Where no starter fertilizer was used, canola stand counts were approximately 38,000 to 58,000 and 82,000 to 90,000 plants per hectare in Perkins and Lahoma locations, respectively.

Between the two locations, a lesser stand count reduction was noted in Perkins (6–48%) than in Lahoma location (25–71%) although overall emergence was higher in Lahoma than in the Perkins location. Canola at Perkins location was

planted in a no-till system and the grain drill used was not ideally suited for no-till sowing. Moreover, emergence may have been influenced by low soil pH condition. It is also hypothesized that stand count difference between locations is due to temperature and soil moisture conditions. Temperatures at the Lahoma location had dropped below 10°C for four consecutive days, three days after seeding. Also, soil moisture was low (<0.12 mm) at seeding and in the first week after sowing, thereby increasing the chances of salt injury. Placing fertilizer in furrow increases salt concentration around the seed; if concentration is too high, the seed will be unable to germinate [13, 14]. Application of preplant broadcast N did not affect canola stand count, as number of seedlings in plots applied with preplant N was similar in plots with no preplant N applied.

3.3. NDVI Readings

3.3.1. 2011-2012. Normalized difference vegetation index readings recorded one week prior to topdress N and two weeks after topdress N had little variations within treatments (Figure 2). At both locations and at both sensing times, plants applied with N at preplant + topdress (treatment 2) and at topdress only (treatment 9) recorded NDVI values that were similar to the unfertilized check. Interestingly, plants supplied with preplant N and P + topdress N (treatment 3) yielded NDVI values that were significantly higher than the unfertilized check. The application of 17 kg ha^{-1} DAP + preplant N + topdress N provided the highest NDVI readings among the treatments regardless of sensing time. In addition, application of 34 kg ha^{-1} DAP + preplant N + topdress N gave higher NDVI values than the checks (unfertilized, preplant + topdress N, and topdress N only).

At Lahoma location, NDVI values decreased with increasing rate of DAP (Figure 2). This trend is similar to that of the crop stand count results. Values of NDVI were the lowest in plants applied with 84 kg ha^{-1} DAP compared with plants applied with 17 and 34 kg ha^{-1} but were comparable with the unfertilized check. Plants that received split application of N had higher NDVI values than plants with no preplant applications. NDVI values of plots receiving preplant N were either similar or higher compared to the unfertilized check. Similar results were reported in a wheat study using optical sensors and variable rate wherein split rate N provided higher NDVI values than N fertilizer applied at topdress only [15, 16]. Lower NDVI values were observed two weeks after topdress N at the Lahoma location most likely because sensing was delayed (due to unfavorable weather condition) and the crop had started to bolt.

At Perkins location, recorded NDVI values were generally low (0.19 to 0.31) due to low stand count and biomass. Soil analysis result in this location showed low pH of 4.5 (Table 1). Low soil pH reduces plant availability of several nutrients, increases levels of some elements to phytotoxic concentrations (i.e., Al^{3+} toxicity), and influences microbial activity or other soil properties [3, 17, 18]. These poor growth conditions can lead to reductions in root development which consequently causes slow vegetative growth and low total biomass per unit area. However, the low stand count due to

(a)

(b)

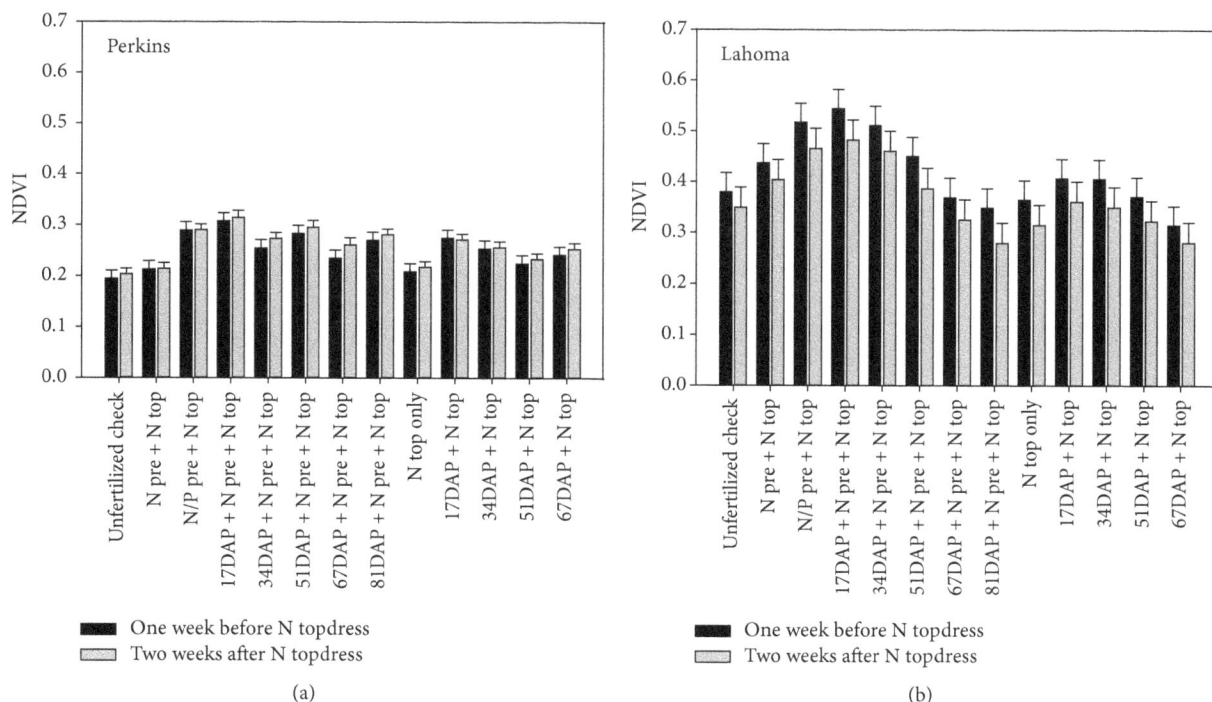

FIGURE 2: Normalized difference vegetation index (NDVI) values at one week before and two weeks after topdress nitrogen (N) as influenced by seed-placed diammonium phosphate (DAP) and N application method at Lahoma and Perkins, OK, in 2011-2012; pre: preplant; top: topdress.

seed-placed starter fertilizer (Table 2) did not affect NDVI values. Plants applied with seed-placed DAP had higher NDVI readings compared to plants with no seed-placed DAP except for the N/P preplant plus topdress treatment. The addition of DAP with the seed may have provided a readily available nutrient to the seedling causing enhanced root growth and consequently above-ground biomass.

3.3.2. 2012-2013. Normalized difference vegetation index values were similar in all treatments (data not shown). For both locations, little change was seen in NDVI values recorded two weeks after topdress compared to NDVI collected priorly (Figure 3). The low NDVI values were likely due to environmental factors like temperature and lack of precipitation. It was not until shortly after the second NDVI readings that temperatures started to increase and regular precipitation events occurred creating a suitable environment for rapid crop growth and nutrient uptake. This allowed plants to have an increase in biomass production which resulted in considerably higher NDVI values at four weeks after topdress N.

3.4. Grain Yield. Canola grain yields at Lahoma location ranged from 1048 to 1748 kg ha^{-1} and 568 to 2194 kg ha^{-1} in 2011-2012 and 2012-2013, respectively (Table 3). In 2011-2012, no significant difference in yield was observed in all treatments. The lack of yield variation between treatments may have been influenced by the uncommonly warm winter and timely rains in the spring, which is not typical in Oklahoma. Warm weather with adequate soil moisture during winter

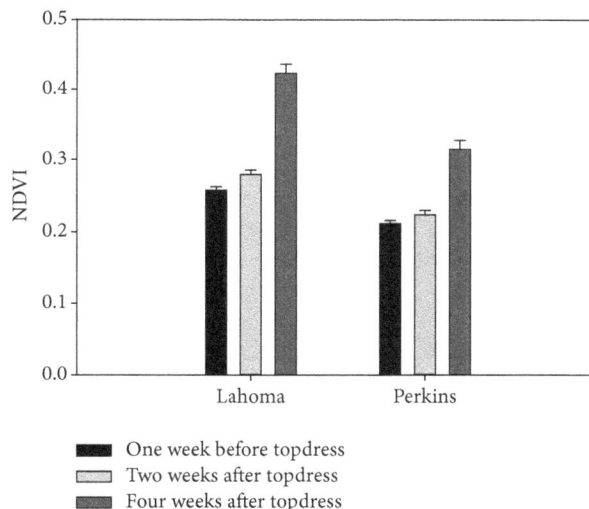

FIGURE 3: Normalized difference vegetation index (NDVI) values at one week before and two and four weeks after topdress nitrogen (N) at Lahoma and Perkins, OK, in 2012-2013. NDVI values are average across treatments.

likely increased the rate of mineralization, adding significant amounts of plant available N. Moreover, this condition allowed the plants to mature and produce additional branches and set additional seed pods due to the favorable weather pattern. In a study by Johnston et al. [19], canola was reported

TABLE 3: Canola grain yield as influenced by seed-placed diammonium phosphate (DAP) and nitrogen (N) application method at Lahoma and Perkins, OK, in 2011–2013.

Treatment	DAP with seed	N/P in DAP	Preplant N	Topdress N	Lahoma		Perkins	
					2011-2012	2012-2013	2011-2012	2012-2013
				kg ha^{-1}				
1	0	0/0	0	0	1558	568	467	259
2	0	0/0	56	84	1338	1753	543	364
3	0	0/0	56N/14P	84	1175	2194	395	1324
4	17	3/3.5	53	84	1048	1721	752	938
5	34	6/7	50	84	1290	1826	212	751
6	51	9/10.5	47	84	1718	1708	375	1120
7	67	12/14	44	84	1610	1764	272	1107
8	84	15/17.5	41	84	1399	1426	339	1064
9	0	0/0	0	140	1423	2106	387	309
10	17	3/3.5	0	137	1629	2017	400	997
11	34	6/7	0	134	1748	1976	336	1784
12	51	9/10.5	0	131	1257	2012	352	989
13	67	12/14	0	128	1649	1540	198	1163
LSD					NS	712	224	591

to produce more branches and compensate for lack of crop stand and allowed for maximum yield to be obtained when weather conditions were suitable for growth. In 2012-2013, the environment was less conducive to vegetative growth and propagation of branches; hence, differences in yields were observed among fertilizer treatments (Table 3). Regardless of DAP application, plants that were applied with topdress and no preplant N (treatments 10 to 13) tended to have greater yields than plants that received preplant (treatments 4 to 8). Lower grain yields in 2011-2012 compared to 2012-2013 may be attributed to delayed canola harvesting due to rain which may have resulted in shattering of pods even after swathing.

At Perkins location, grain yields were unusually low (198 to 752 kg ha^{-1}) during the first year (Table 3), which was likely enhanced by soil pH and by moisture deficit resulting from below-normal precipitation and above-normal temperatures during boot, bloom, and grain filling stages of the crop. These extreme environmental conditions during critical reproductive stages of canola can increase flower abortion and reduce translocation of assimilates to grain, collectively reducing grain yield [20]. In the second year, higher grain yields (259 to 1784 kg ha^{-1}) were observed in most treatments as there was an increase of available soil moisture during the period of bolting and seed fill. Plants that did not receive P fertilizer, however, had lower grain yields (259 to 364 kg ha^{-1}) than the rest of the fertilized plants (751 to 1163 kg ha^{-1}). This may be attributed to the low soil pH (4.5) condition at Perkins. At low pH, P is often tied by iron and aluminum rendering it unavailable for plant uptake [21]. Phosphorus is an important nutrient for winter survival of canola. At spring green up, plots that did not receive P had few to no plants that survived over the winter even though these plots had the highest recorded stand counts.

In general, canola grain yield was not significantly reduced despite the loss in stand count two weeks after planting. Stand loss was not well correlated with yield ($r^2 = 0.3$, $P \leq 0.00001$). This response indicates that canola could

compensate the level of stand loss observed without yield reductions, which was consistent with previous research [22].

3.5. Oil and Protein Content. Canola is grown primarily for oil; therefore, oil content is the most important parameter when assessing canola quality. No significant three- and two-way interactions were found; thus, oil content data are presented by year across locations. In 2011-2012, there were no significant differences among treatments; however, in 2012-2013, oil content differences were observed among treatments (Table 4). In general, plants that received P, whether through seed-placed DAP or preplant, had higher oil content (40.5 to 43.27%) than plants that were not applied with P (40.15 to 43.21%). Similar findings were also reported by Tomar et al. [23] and Gaydou and Arrivets [24] who observed that P application increased oil contents of soybean. Differences in oil content were also noted among plants applied with different rates of P. Plants with higher P applied produced more grain oil than with lesser P applied. These results suggest that P is an essential nutrient in increasing oil content in canola.

Protein content data was similar for both years. Seed protein content ranged from 20% in the unfertilized plot to 22% in plot which received the broadcast N and P fertilizer (Table 4). Significant differences were observed between fertilized and unfertilized plots but not among fertilized plots in both locations. Regardless of location, protein contents of all treatments were lower than the typical canola protein content of 23%, although protein content in whole canola seed is dependent on variety and growing conditions. According to Canola Council of Canada [25], P fertilization has little or no effect on canola quality. Studies conducted in western Canada showed that P fertilizer showed no specific effect on canola protein content except in very P-deficient locations.

4. Conclusions

This study demonstrated that application of DAP with the seed causes stand count reduction to canola. Placement

TABLE 4: Oil and protein content of canola as influenced by seed-placed diammonium phosphate (DAP) and nitrogen (N) application method in 2011–2013.

Treatment	DAP with seed	N/P in DAP	Preplant N	Topdress N	Oil content		Protein content	
					2011-2012	2012-2013	2011-2012	2012-2013
		kg ha^{-1}				%		
1	0	0/0	0	0	43.21	—*	19.89	20.32
2	0	0/0	56	84	42.43	40.15	20.93	21.12
3	0	0/0	56N/14P	84	43.14	41.14	20.82	21.50
4	17	3/3.5	53	84	42.97	40.62	21.01	21.97
5	34	6/7	50	84	43.21	40.94	21.00	21.62
6	51	9/10.5	47	84	43.27	41.38	20.94	21.28
7	67	12/14	44	84	43.16	41.28	21.98	21.69
8	84	15/17.5	41	84	43.06	41.39	21.20	21.42
9	0	0/0	0	140	42.86	—	21.05	21.33
10	17	3/3.5	0	137	42.87	40.50	21.32	21.95
11	34	6/7	0	134	42.89	41.22	21.12	21.90
12	51	9/10.5	0	131	43.21	41.25	20.95	21.56
13	67	12/14	0	128	43.03	41.28	21.04	21.48
LSD					NS	0.49	0.75	0.75

*— (dash): no data available.

of starter fertilizer with the seed even at the lowest rate of DAP (17 kg ha^{-1}) significantly reduced stand, but stand count reduction was not associated with yield. Since canola plants were able to compensate stand count losses through additional branching, application of DAP of up to 84 kg ha^{-1} (contains 15 kg N ha^{-1}) with seed may be possible. However, soil conditions that tend to increase potential damage to seed and environmental factors that induce stress to the crop should be considered when deciding how much DAP will be placed with seed. In addition, the targeted seeding rate of 5.6 kg ha^{-1} is well above the amount needed. Currently, some producers are using seeding rates as low as 3 kg ha^{-1} when using a grain drill and as low as 2.2 kg ha^{-1} when using a planter. While little to no yield reduction due to stand loss was observed in this study, it could be hypothesized that if seeding rate had been reduced a negative impact on yield would have been more likely. Thus, further research is needed to document impact of stand loss on reduced planting population in canola.

Broadcasting of N and P at preplant resulted in yield equal to or greater than the seed-placed treatments. When producers have some flexibility concerning application methods, broadcasting N and P at preplant may be the preferred method over seed-placed DAP because of lesser to no effect on canola stand.

Seasonal environment had a great impact on the effect of N application method. When climatic conditions limit early season growth and favor late spring growth, applying all N at topdress (no preplant) will tend to provide greater canola grain yield.

Phosphorus significantly influenced oil content but not protein content in canola. Plants with higher P applied produced more grain oil than with lesser P applied. This indicates that P is an essential nutrient in increasing oil content in

canola. Due to the lack of available research on fertility impact on canola seed oil production, further research needs to be conducted on these areas of canola production.

Competing Interests

The authors declare that there are no competing interests regarding the publication of this paper.

References

[1] [USDA ESMIS] USDA Economic, "Statistics and Market Information System, 'Crop production,'" 2016, http://usda.mannlib.cornell.edu/MannUsda/viewDocumentInfo.do? documentID= 1046.

[2] G. E. Welbaum, *Vegetable Production and Practices: Fertilization and Mineral Nutrition Requirements for Growing Vegetables*, CAB International, Boston, Mass, USA, 2015.

[3] N. C. Brady, *The Nature and Properties of Soil*, Macmillan, New York, NY, USA, 10th edition, 1990.

[4] Oilseeds Western Australia, *Growing Western Canola: An Overview of Canola Production in Western Australia*, Oil Industry Association of Western Australia, Belmont, Australia, 2006.

[5] Canola Council of Canada, "Canola encyclopedia: Seed and fertilizer placement," 2015, http://www.canolacouncil.org/canola-encyclopedia/crop-establishment/seed-and-fertilizer-placement/.

[6] J. T. Sims, "Lime requirement," in *Methods of Soil Analysis. Part 3: Chemical Methods*, D. L. Sparks, Ed., SSSA Book Series 5, SSSA and ASA, Madison, Wis, USA, 1996.

[7] F. J. Sikora, "A buffer that mimics the smp buffer for determining lime requirement of soil," *Soil Science Society of America Journal*, vol. 70, no. 2, pp. 474–486, 2006.

[8] Lachat Instruments, "Ammonia (phenolate) in 2 M KCl soil extracts (QuikChem Method 12-107-06-1-B)," in *QuikChem Automatic Ion Analyzer Methods Manual*, Lachat Instruments, Milwaukee Wis, USA, 1997.

[9] A. Mehlich, "Mehlich 3 soil test extractant: a modification of Mehlich 2 extractant," *Communications in Soil Science and Plant Analysis*, vol. 15, no. 12, pp. 1409–1416, 1984.

[10] P. N. Soltanpour, G. W. Johnson, S. M. Workman, J. B. Jones Jr., and R. O. Miller, "Inductively coupled plasma emission spectrometry and inductively coupled plasma-mass spectrometry," in *Methods of Soil Analysis. Part 3: Chemical Methods*, D. L. Sparks, Ed., SSSA Book Series 5, SSSA and ASA, Madison, Wis, USA, 1996.

[11] D. W. Nelson and L. E. Sommers, "Total carbon, organic carbon, and organic matter," in *Methods of Soil Analysis. Part 3: Chemical Methods*, D. L. Sparks, Ed., SSSA Book Series 5, SSSA and ASA, Madison, Wis, USA, 1996.

[12] J. Edwards, R. Kochenower, N. Dunford, R. Austin, B. Carver, and J. Ladd, "Current report: Protein content of winter wheat varieties in Oklahoma 2009," 2009, http://wheat.okstate.edu/variety-testing/wheat-protein/wheatprotein21352009web.pdf/view.

[13] D. B. Mengel, S. E. Hawkins, and P. Walker, "Phosphorus and potassium placement for no-till and spring plowed corn," *Journal of Fertilizer Issues*, vol. 5, pp. 31–36, 1988.

[14] B. J. Niehues, R. E. Lamond, C. B. Godsey, and C. J. Olsen, "Starter nitrogen fertilizer management for continuous no-till corn production," *Agronomy Journal*, vol. 96, no. 5, pp. 1412–1418, 2004.

[15] W. R. Raun, J. B. Solie, G. V. Johnson et al., "Improving nitrogen use efficiency in cereal grain production with optical sensing and variable rate application," *Agronomy Journal*, vol. 94, no. 4, pp. 815–820, 2002.

[16] R. L. Mahler, F. E. Koehler, and L. K. Lutcher, "Nitrogen source, timing of application, and placement: effects on winter wheat production," *Agronomy Journal*, vol. 86, no. 4, pp. 637–642, 1994.

[17] C. Meriño-Gergichevich, M. Alberdi, A. G. Ivanov, and M. Reyes-Díaz, "Al^{3+}-Ca^{2+} interaction in plants growing in acid soils: Al-phytotoxicity response to calcareous amendments," *Journal of Soil Science and Plant Nutrition*, vol. 10, no. 3, pp. 217–243, 2010.

[18] A. Pagani, J. E. Sawyer, and A. P. Mallarino, "Site-specific nutrient management for nutrient management planning to improve crop production, environmental quality, and economic return: chap. 8: soil pH and lime management," 2013, http://www.agronext.iastate.edu/soilfertility/nutrienttopics/4r/Site-SpecificNutrientManagementPlanning_ver2.pdf.

[19] A. M. Johnston, E. N. Johnson, K. J. Kirkland, and F. C. Stevenson, "Nitrogen fertilizer placement for fall and spring seeded *Brassica napus* canola," *Canadian Journal of Plant Science*, vol. 82, no. 1, pp. 15–20, 2002.

[20] L. Taiz and E. Zeiger, *Plant Physiology*, Sinauer Associates, Sunderland, Mass, USA, 5th edition, 2010.

[21] L. Busman, J. Lamb, G. Randall, G. Rehm, and M. Schmitt, "Nutrient management: the nature of phosphorus soils," 2009, http://www.extension.umn.edu/agriculture/nutrient-management/phosphorus/the-nature-of-phosphorus/.

[22] Canola Council of Canada, "Canol@Fact: Plant populations for profitability," 2005, http://www.canolacouncil.org/media/515841/plant_populations_for_profitability.pdf.

[23] S. S. Tomar, R. Singh, and P. S. Singh, "Response of phosphorus, sulphur and Rhizobium inoculation and growth, yield and quality of soybean," *Progressive Agriculture*, vol. 4, no. 1, pp. 72–73, 2004.

[24] E. M. Gaydou and J. Arrivets, "Effects of phosphorus, potassium, dolomite, and nitrogen fertilization on the quality of soybean. Yields, proteins, and lipids," *Journal of Agricultural and Food Chemistry*, vol. 31, no. 4, pp. 765–769, 1983.

[25] Canola Council of Canada, "Canola encyclopedia: Phosphorus fertilizer management," 2013, http://www.canolacouncil.org/canola-encyclopedia/fertilizer-management/phosphorus-fertilizer-management/.

PERMISSIONS

All chapters in this book were first published in IJA, by Hindawi Publishing Corporation; hereby published with permission under the Creative Commons Attribution License or equivalent. Every chapter published in this book has been scrutinized by our experts. Their significance has been extensively debated. The topics covered herein carry significant findings which will fuel the growth of the discipline. They may even be implemented as practical applications or may be referred to as a beginning point for another development.

The contributors of this book come from diverse backgrounds, making this book a truly international effort. This book will bring forth new frontiers with its revolutionizing research information and detailed analysis of the nascent developments around the world.

We would like to thank all the contributing authors for lending their expertise to make the book truly unique. They have played a crucial role in the development of this book. Without their invaluable contributions this book wouldn't have been possible. They have made vital efforts to compile up to date information on the varied aspects of this subject to make this book a valuable addition to the collection of many professionals and students.

This book was conceptualized with the vision of imparting up-to-date information and advanced data in this field. To ensure the same, a matchless editorial board was set up. Every individual on the board went through rigorous rounds of assessment to prove their worth. After which they invested a large part of their time researching and compiling the most relevant data for our readers.

The editorial board has been involved in producing this book since its inception. They have spent rigorous hours researching and exploring the diverse topics which have resulted in the successful publishing of this book. They have passed on their knowledge of decades through this book. To expedite this challenging task, the publisher supported the team at every step. A small team of assistant editors was also appointed to further simplify the editing procedure and attain best results for the readers.

Apart from the editorial board, the designing team has also invested a significant amount of their time in understanding the subject and creating the most relevant covers. They scrutinized every image to scout for the most suitable representation of the subject and create an appropriate cover for the book.

The publishing team has been an ardent support to the editorial, designing and production team. Their endless efforts to recruit the best for this project, has resulted in the accomplishment of this book. They are a veteran in the field of academics and their pool of knowledge is as vast as their experience in printing. Their expertise and guidance has proved useful at every step. Their uncompromising quality standards have made this book an exceptional effort. Their encouragement from time to time has been an inspiration for everyone.

The publisher and the editorial board hope that this book will prove to be a valuable piece of knowledge for researchers, students, practitioners and scholars across the globe.

LIST OF CONTRIBUTORS

Caiyun Lu, Chunjiang Zhao, Xiu Wang, Zhijun Meng, Jian Song, Guangwei Wu, Weiqing Fu, Jianjun Dong and Jiayang Yu
Beijing Research Center of Intelligent Equipment for Agriculture, Beijing 100097, China
Beijing Research Center for Information Technology in Agriculture, Beijing 100097, China

Milt McGiffen
Department of Botany and Plant Sciences, University of California, Riverside, CA 92521-0124, USA

Dan D. Fromme, Daniel Stephenson and Keith Shannon
LSU AgCenter, 8208 Tom Bowman Drive, Alexandria, LA 71302,USA

Trey Price
LSU AgCenter, 212AMacon Ridge Road, Winnsboro, LA 71295, USA

Josh Lofton
Oklahoma State University, Stillwater, OK 74078, USA

Tom Isakeit
Department of Plant Pathology and Microbiology, Texas A&M University, College Station, TX 77843, USA

Ronnie Schnell
Department of Soil and Crop Science, Texas A&M University, College Station, TX 77843, USA

Syam Dodla
LSU AgCenter, 262 Research Station Drive, Bossier City, LA 71112, USA

W. James Grichar
Texas A&M AgriLife Research and Extension Center, Corpus Christi, TX 78406, USA

Wang Xiukang
College of Life Science, Yan'an University, Yan'an, Shaanxi 716000, China

Xing Yingying
College of Life Science, Yan'an University, Yan'an, Shaanxi 716000, China
Key Laboratory of Agricultural Soil andWater Engineering in Arid and Semiarid Areas of Ministry of Education, Northwest A&F University, Yangling, Shaanxi 712100, China

Keenan C. McRoberts
Department of Animal Science, Cornell University, 149 Morrison Hall, Ithaca, NY 14853, USA

Quirine M. Ketterings
Department of Animal Science, Cornell Nutrient Management Spear Program, Cornell University, 323 Morrison Hall, Ithaca, NY 14853, USA

David Parsons
School of Land and Food, University of Tasmania, Private Bag 98,Hobart, TAS 7001, Australia

Tran Thanh Hai, Nguyen Hai Quan and Nguyen Xuan Ba
Hue University of Agriculture and Forestry, 102 Phung Hung Street, Hue, Vietnam

Charles F. Nicholson
Department of Supply Chain and Information Systems,The Pennsylvania State University, 467 Business Building, University Park, PA 16802, USA

Debbie J. R. Cherney
Department of Animal Science, Cornell University, 329 Morrison Hall, Ithaca, NY 14853, USA

Víctor García-Gaytán, Libia I. Trejo-Téllez and Gustavo Adolfo Baca-Castillo
Colegio de Postgraduados, Campus Montecillo, 56230 Texcoco, MEX, Mexico

Fernando Carlos Gómez-Merino and Soledad García-Morales
Colegio de Postgraduados, Campus C´ordoba, 94946 Amatl´an de los Reyes, VER, Mexico

Domingo J.Mata-Padrino, E. E. D. Felton,W. B. Bryan and D. P. Belesky
West Virginia University, Morgantown,WV 26506, USA

C. La Hovary
Department of Plant and Microbial Biology, North Carolina State University, Raleigh, NC 27695, USA

D. A. Danehower
Avoca, Inc., P.O. Box 129, 841 Avoca FarmRoad, MerryHill, NC 27957, USA

G. Ma, J. D. Williamson and J. D. Burton
Department of Horticultural Science, North Carolina State University, Raleigh, NC 27695, USA

C. Reberg-Horton
Department of Crop Science, North Carolina State University, Raleigh, NC 27695, USA

S. R. Baerson
USDA-ARS, Natural Products Utilization Research Unit, University, MS 38677, USA

Tami L. Stubbs
Department of Crop and Soil Sciences,Washington State University, 115 Johnson Hall, Pullman,WA 99164-6420, USA

Ann C. Kennedy
Northwest Sustainable Agroecosystems Research Unit, USDA-ARS, 215 Johnson Hall, Pullman,WA 99164-6421, USA

Olga S. Walsh
Department of Plant, Soil, and Entomological Sciences, Southwest Research and Extension Center, University of Idaho, 29603 U of I Lane, Parma, ID 83660, USA

Robin J. Christiaens
Private Enterprise, University of Idaho, 29603 U of I Lane, Parma, ID 83660, USA

Md. Quamruzzaman, Md. Jafar Ullah and Md. Fazlul Karim
Department of Agronomy, Sher-e-Bangla Agriculture University, Dhaka 1207, Bangladesh

Nazrul Islam, Md. Jahedur Rahman and Md. Dulal Sarkar
Department of Horticulture, Sher-e-Bangla Agriculture University, Dhaka 1207, Bangladesh

Mildred Osei-Kwarteng and Gustav Komla Mahunu
Department of Horticulture, Faculty of Agriculture, University for Development Studies, P.O. Box TL 1882, Tamale, Ghana

Joseph Patrick Gweyi-Onyango
Department of Agricultural Science and Technology, Kenyatta University, P.O. Box 4384400100, Nairobi, Kenya

A. B. Rosenani, R. Rovica, P. M. Cheah and C. T. Lim
Department of LandManagement, Faculty of Agriculture,Universiti Putra Malaysia (UPM), 43400 Serdang, Selangor, Malaysia

Yiyun Yan, Wengang Zuo, Weijie Xue and LijuanMei
College of Environmental Science and Engineering, Yangzhou University, Yangzhou 225009, China

Yanchao Bai
College of Environmental Science and Engineering, Yangzhou University, Yangzhou 225009, China
State Key Laboratory of Soil and Sustainable Agriculture, Institute of Soil Science, Chinese Academy of Sciences, Nanjing 210008, China
Institute of Biotechnology, Jiangsu Academy of Agricultural Sciences, Nanjing 210014, China

Chuanhui Gu
Department of Geology, Appalachian State University, Boone, NC 28608, USA

Yuhua Shan and Ke Feng
College of Environmental Science and Engineering, Yangzhou University, Yangzhou 225009, China
Jiangsu Collaborative Innovation Center for Solid OrganicWaste Resource Utilization, Nanjing 210095, China

Hédi Ben Ali
Agence de Promotion des Investissements Agricoles, 6000 Gabès, Tunisia

Moncef Hammami
Laboratory of Hydraulic, High School of Engineers of Rural Equipment, Medjez el Bab, Tunisia

Ahmed Saidi
National Research Institute of Rural Engineering, Water and Forests (INRGREF), Rue Hédi EL Karray El Menzah IV, BP 10, 2080 Ariana, Tunisia

Rachid Boukchina
Institut des R´egions Arides, 6000 Gabès, Tunisia

Mandeep K. Riar, Danesha S. Carley, Chenxi Zhang, Michelle S. Schroeder-Moreno, David L. Jordan and ThomasW. Rufty
Department of Crop Science, North Carolina State University, Raleigh, NC 27695, USA

Theodore M. Webster
Crop Protection and Management Research Unit, USDA-ARS, Tifton, GA 31793, USA

Dalila Trupiano, Claudia Cocozza, Carla Amendola, Giuseppe Lustrato, Francesca Fantasma, Roberto Tognetti and Gabriella Stefania Scippa
Dipartimento di Bioscienze e Territorio, Universit`a degli Studi del Molise, 86090 Pesche, Italy

Silvia Baronti, Sara Di Lonardo and Francesco Primo Vaccari
Istituto di Biometeorologia-Consiglio Nazionale delle Ricerche (IBIMET-CNR), 50145 Firenze, Italy

Abdulmajeed Hamza and Ezekiel Akinkunmi Akinrinde
Department of Agronomy, University of Ibadan, Ibadan, Nigeria

Yasuyuki Ishii and Sachiko Idota
Faculty of Agriculture, University of Miyazaki, Miyazaki 889-2192, Japan

Asuka Yamano
Graduate School of Agriculture, University of Miyazaki, Miyazaki 889-2192, Japan

Md. Sadek Hossain
Seed Distribution Division, Bangladesh Agricultural Development Corporation, Dhaka, Bangladesh

M. Mofazzal Hossain
Department of Horticulture, Bangabandhu Sheikh Mujibur Rahman Agricultural University, Gazipur 1703, Bangladesh

M. Moynul Haque
Department of Agronomy, Bangabandhu Sheikh Mujibur Rahman Agricultural University, Gazipur 1703, Bangladesh

Md. Mahabubul Haque
Farm Division, Bangladesh Agricultural Research Institute, Gazipur, Bangladesh

Md. Dulal Sarkar
Department of Horticulture, Sher-e-Bangla Agricultural University, Dhaka 1207, Bangladesh

Young-Mi Oh and Paul V. Nelson
Department of Horticultural Science, North Carolina State University, Raleigh, NC 27695-7609, USA

Dean L. Hesterberg
Department of Soil Science, North Carolina State University, Raleigh, NC 27695-7619, USA

Carl E. Niedziela Jr.
Department of Biology, Elon University, Elon, NC 27244, USA

Olga S. Walsh and Kefyalew Girma
Southwest Research & Extension Center, University of Idaho, 29603 U of I Lane, Parma, ID 83660-6699, USA

M. Joy M. Abit and D. Brian Arnall
Department of Plant and Soil Sciences, Oklahoma State University, Stillwater, OK 74078, USA

Katlynn Weathers
P&K Equipment, Enid, OK 73701, USA

Index

www.ingramcontent.com/pod-product-compliance
Lightning Source LLC
Chambersburg PA
CBHW080640200326
41458CB00013B/4686